REACTION ENGINEERING OF STEP GROWTH POLYMERIZATION

The Plenum Chemical Engineering Series

Series Editor
Dan Luss, *University of Houston, Houston, Texas*

COAL COMBUSTION AND GASIFICATION
L. Douglas Smoot and Philip J. Smith

ENGINEERING FLOW AND HEAT EXCHANGE
Octave Levenspiel

REACTION ENGINEERING OF STEP GROWTH POLYMERIZATION
Santosh K. Gupta and Anil Kumar

REACTION ENGINEERING OF STEP GROWTH POLYMERIZATION

Santosh K. Gupta
and
Anil Kumar
Indian Institute of Technology
Kanpur, India

PLENUM PRESS • NEW YORK AND LONDON

CHEMISTRY

7344-4418

Library of Congress Cataloging in Publication Data

Gupta, Santosh K.
  Reaction engineering of step growth polymerization.

  (The Plenum chemical engineering series)
  Includes bibliographies and index.
  1. Polymers and polymerization. I. Kumar, Anil. II. Title. III. Series.
TP156.P6G87   1987   668.9   86-30576
ISBN 0-306-42339-1

© 1987 Plenum Press, New York
A Division of Plenum Publishing Corporation
233 Spring Street, New York, N.Y. 10013

Printed in the United States of America

PREFACE

The literature in polymerization reaction engineering has bloomed sufficiently in the last several years to justify our attempt in putting together this book. Rather than offer a comprehensive treatment of the entire field, thereby duplicating earlier texts as well as some ongoing bookwriting efforts, we decided to narrow down our aim to step growth polymerization systems. This not only provides us the luxury of a more elaborate presentation within the constraints of production costs, but also enables us to remain on somewhat familiar terrain.

The style and format we have selected are those of a textbook. The first six chapters present the principles of step growth polymerization. These are quite general, and can easily be applied in such diverse and emerging fields as polymerization applications in photolithography and microelectronics. A detailed discussion of several important step growth polymerizations follows in the next five chapters. One could cover the first six chapters of this book in about six to eight weeks of a three-credit graduate course on polymerization reactors, with the other chapters assigned for reading. This could be followed by a discussion of chain-growth and other polymerizations, with which our material blends well. Alternately, the entire contents of this book could be covered in a course on step growth systems alone. Several problems have been included. Some help fortify the concepts developed, while some go farther into the unknown and may require the use of computers and modeling skills. It is assumed that for *optimal* use of this text, a student will have had *elementary* courses in calculus, transport phenomena, and reaction engineering. However, such knowledge is not essential. It is hoped that people with diverse backgrounds working in this area will be able to use this book without any disadvantage.

Any bookwriting effort of this nature must be, unfortunately, dated. Continuing advances are being made, particularly in such areas as polyurethanes. We have attempted to include a glimpse of some ongoing work in these areas to partially compensate for this drawback.

v

Several friends and colleagues have helped us in various ways in this activity and it is a pleasure to offer them our thanks. It is difficult to thank Professor Robert L. Laurence sufficiently for all the detailed comments and help he has provided us throughout this venture. Professor W. Harmon Ray has provided us with help in various ways and Professor K. S. Gandhi and Dr. H. K. Reimschuessel have provided us with useful comments on some chapters. One of us (S.K.G.) would like to thank Professor A. Varma for the generous use of all departmental facilities while he was a Visiting Professor at the University of Notre Dame and was, among other things, revising the manuscript. We also wish to acknowledge partial financial help received from the Curriculum Development Center of the Quality Improvement Program, IIT Kanpur, for manuscript preparation. Thanks are also due to U. S. Misra and Sherry DePoy for typing several drafts, to D. S. Panesar for the artwork, to John Foryt for help in assembling parts of the bibliography, and to Hari Ram, Debby Ciesiolka, and Jeanne Davids for other miscellaneous help associated with the project. All of these people helped us in our efforts, and always with a smile, thus making our work more pleasant.

To Shubhra and Renu we express our sincere appreciation for taking over all the duties of parenthood when Aatmeeyata, Akanksha, Chetna, and Pushkar yearned for their fathers' time. We renew our promises to them not to get involved again in bookwriting, a promise which has been broken more often than kept.

Santosh K. Gupta
Anil Kumar

Kanpur, India

CONTENTS

CHAPTER 3.   LINEAR STEP GROWTH POLYMERIZATION
             VIOLATING THE EQUAL REACTIVITY
             HYPOTHESIS

CHAPTER 4.   NONLINEAR STEP GROWTH
             POLYMERIZATION

CHAPTER 8. POLYESTER REACTORS

CHAPTER 9. URETHANE POLYMERS AND REACTION INJECTION MOLDING

CHAPTER 10. EPOXY POLYMERS

CHAPTER 11. POLYMERIZATION WITH FORMALDEHYDE

REACTION ENGINEERING OF STEP GROWTH POLYMERIZATION

INTRODUCTION

1.1. STEP AND CHAIN GROWTH POLYMERIZATIONS

Polymeric materials consist of long chain molecules formed by the chemical combination of small molecules called monomers. The reaction leading to the formation of polymer molecules is called polymerization. These have been broadly classified[1,2] as chain growth and step growth polymerizations. In chain growth polymerization, there are growth centers in the reaction mass, to which monomer molecules add on *successively*. Usually, a high molecular weight product is produced right from the very beginning of the reaction with the quantity of unreacted monomer in the reaction mass decreasing slowly with time. Depending upon the nature of the growth centers, chain growth polymerizations can be further classified into radical, cationic, anionic, and stereoregular polymerizations.[1-8] In step growth polymerization, on the other hand, the growth of molecules usually occurs through the reaction of functional groups, e.g., $-COOH$, $-NH_2$, $-OH$, etc., located on the molecules. Polymer formation can occur through this mechanism only when the monomer has at least two functional groups. When there are more than two functional groups in the monomer molecules, the resultant polymer is either branched or cross-linked in structure, while for bifunctional monomers, linear polymer molecules are formed. In contrast to chain growth polymerization, the growth of the polymer molecules in this case occurs by the reaction between *any two* molecules (whether polymeric or monomeric) in the reaction mass, and there is a slow increase in the average molecular weight of the polymer. Since the growth process is random in nature, the reaction mass, in general, consists of oligomers having different chain lengths at any instant of time. The molecular weights of the polymer formed are usually not as high as for chain growth polymerizations, and relatively small amounts of unreacted monomer are present after the beginning of the reaction.

The classification of polymerization reactions into chain and step growth types parallels, to some extent, the earlier classification of Carothers[9] into addition and condensation polymerizations. The present-day classification overcomes some serious shortcomings inherent in the earlier one, as discussed by Flory.[1] By this more general definition, the polymerizations of phthalic anhydride with diols (polyesters), of diisocyanates with diols (polyurethanes), etc. all fall within the category of step growth polymerization, even though they are not condensation polymerizations. More recently, these polymerizations have been classified into random and sequential types,[10,11] since these words are statistically more meaningful. In this book, however, the terms "step growth" and "chain growth polymerizations" have been used.

In the early years (1940s and 1950s), the major research effort in the area of polymerizations was concentrated on understanding the kinetics and mechanisms of the major reactions involved. A considerable amount of experimental rate and equilibrium data was reported, and several generalizations were made based on these. One of the more important of these was the hypothesis that the reactivity of the growth centers is independent of its chain length (sometimes called the equal reactivity hypothesis).[1] In recent years, the effects of other physical aspects which are important in industrial reactors have been investigated. Some of these include the following:

1. The study of the effect of side reactions, e.g., cyclic oligomer formation in nylon 6 polymerization, diethylene glycol end group (DEG) formation in PET, etc. The presence of some of the products of side reactions, even though in low concentrations, leads to important changes in the physical properties and so affects the end use of the polymer. For example, DEG end groups in PET decrease the tensile strength of the polymer but increase its dyeability. Some side reactions may also lead to a "quenching" of the functional groups, thus giving an effect which is similar to that produced by the termination step of chain growth polymerization. Kinetic data on the side reactions are usually proprietary and it is only recently that some information on these has appeared in the open literature.

2. The study of the diffusion of low molecular weight by-products through the usually viscous reaction mass. The equilibrium constants for PET, nylon 66 polymerization, etc., are unfavorable and a high vacuum has to be applied to the reaction mass in order to get polymers of reasonably high average molecular weights. Mass transfer resistances in bulk polymerization are usually high after some stage, and special equipment is necessary to obtain high molecular weight products. The lack of proper accounting of the vaporization

of the condensation byproducts in the early stages of polymerization has, at times, led to apparently erroneous rate constants determined experimentally.

3. The study of the temperature gradients in the high-viscosity, low-conductivity polymerizing medium. The study of heat transfer effects is important at times for the proper design and analysis of step growth polymerization reactors, even though these effects are not as significant as in the case of some chain growth polymerizations.

4. The study of the effect of high viscosity on the rate constants at advanced stages of polymerization.

In view of these developments, it is now possible to simulate, optimize, and control a polymer reactor with reasonable precision.

The explosion of information in this area, particularly in the last several years, has led to the emergence of an entirely new area of specialization in chemical engineering, called polymer reaction engineering. Several books and review articles emphasizing different aspects of polymerization have appeared in recent years, of which some are listed.[12-24] In most of these, the emphasis has been on chain growth polymerization. In this text, attention has been focused mainly on the reaction engineering of step growth polymerization in view of the recent explosion of information in this area. An effort has been made to streamline the current thinking on the reactor design of polymers reacting this way.

1.2. STEP GROWTH POLYMERIZATION

In step growth polymerization wherein reaction between functional groups are involved, the starting monomer can either be bifunctional [for example, amino-caproic acid, $H_2N(-CH_2-)_5COOH$, or hexamethylene adipamide, $HOOC(-CH_2-)_4CONH(-CH_2-)_6-NH_2$, represented as ARB, with A and B the reactive functional groups] or multifunctional [e.g., pentaerythritol, $C(-CH_2OH)_4$, with two $-CH_2OH$ groups self condensing]. In the case of the former, the polymer formed is linear in structure, whereas in the case of the latter it is either branched or has a network structure. In either case, the polymerization can be represented schematically as

$$P_n + P_m \rightleftarrows P_{n+m} + W, \qquad n, m = 1, 2, 3, \ldots \qquad (1.2.1)$$

where P_n is a polymer molecule having n repeating monomeric units and W is a low molecular weight condensation by-product (at times, there is no *W* formed). Equation (1.2.1) is strictly valid for systems where the two reactive functional groups are situated on the same monomer molecule. However, there are cases when the two functional groups are located on different monomers, as in the polymerization of terephthalic acid

(HOOC—⟨◯⟩—COOH) with ethylene glycol (HOC_2H_4OH), represented as ARA + BR′B, with A and B the reactive functional groups, or phthalic anhydride

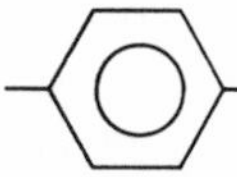

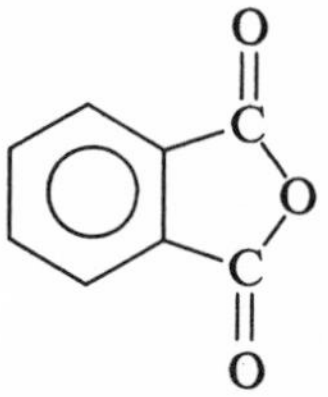

with glycerol

$$(OH-CH_2-CH-CH_2-OH)$$
$$\qquad\qquad\quad |$$
$$\qquad\qquad\quad OH$$

In such cases, the representation of Eq. (1.2.1) is not exact and the mathematical analysis of such systems is more complex.

As is obvious, the product from the reactor consists of several homologs in varying concentrations irrespective of whether it is the polymerization of bifunctional or multifunctional monomers. The presence of an *infinite* array of products having the same "chemical" composition is a characteristic of polymerization reactions, one which is not present in reactors for low molecular weight products, and leads to interesting (and at times confounding) ramifications in reactor design. Because of the presence of so many species, the product is described in terms of a *distribution*, called the molecular weight distribution (MWD) or the chain length distribution (CLD). The physical properties of a given polymeric product depend not only upon the chemical nature of the repeating units, but also upon its MWD. In fact, this property is commercially exploited to obtain various grades of the same "chemical" polymer to suit specific applications.

1.3. MOLECULAR WEIGHT DISTRIBUTION

In sharp contrast to nonpolymeric compounds, any given polymer sample consists of several homologs differing only in the chain length, *n*.

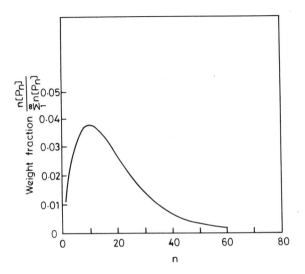

Figure 1.1. A typical weight fraction distribution plot. Note that even though the graph is continuous, only discrete values of n are permitted.

These products are normally described by a plot (or an analytical function) of the mole or mass fraction of the oligomer P_n versus the chain length n. The former is sometimes called the mole-fraction distribution and the latter the weight-fraction distribution. Both these are equivalent representations of the MWD. Strictly speaking, this is a discrete function of n, but is often represented by a continuous function. Figure 1.1 shows a typical plot of the MWD.

Any molecular weight distribution function can be equivalently represented by its moments. The kth *unnormalized* moment, λ_k, of the distribution is defined by

$$\lambda_k = \sum_{n=1}^{\infty} n^k[P_n], \qquad k = 0, 1, 2, \ldots \qquad (1.3.1)$$

for a discrete distribution, where the square brackets on P_n refer to the molar concentration of species P_n. If the distribution can be approximated by a continuous function, these moments can be written as

$$\lambda_k = \int_{n=0}^{\infty} n^k[P_n]\, dn, \qquad k = 0, 1, 2, \ldots \qquad (1.3.2)$$

The infinite series of moments describe the MWD completely. However, the moments are preferred over the MWDs because only the first three or four of them are usually sufficient to characterize any distribution well enough. In addition, the zeroth, first, and second moments have physical significance as discussed below and their ratios can be measured experimentally.

From Eq. (1.3.1), the first three unnormalized moments, λ_0, λ_1, and λ_2 are given by

$$\lambda_0 = \sum_{n=1}^{\infty} [P_n] \tag{1.3.3a}$$

$$\lambda_1 = \sum_{n=1}^{\infty} n[P_n] \tag{1.3.3b}$$

$$\lambda_2 = \sum_{n=1}^{\infty} n^2[P_n] \tag{1.3.3c}$$

Thus, at any given time in the reaction mass, λ_0 gives a measure of the total moles of polymer per unit volume whereas λ_1 is proportional to the total mass per unit volume. One can write expressions for the number-average and weight-average chain lengths,[†] μ_n and μ_w as well as for the polydispersity index, Q (which is a measure of the breadth of the distribution, and is related to its variance), as

$$\mu_n \equiv \frac{\lambda_1}{\lambda_0} = \frac{\sum_{n=1}^{\infty} n[P_n]}{\sum_{n=1}^{\infty} [P_n]} \tag{1.3.4a}$$

$$\mu_w \equiv \frac{\lambda_2}{\lambda_1} = \frac{\sum_{n=1}^{\infty} n^2[P_n]}{\sum_{n=1}^{\infty} n[P_n]} \tag{1.3.4b}$$

$$Q = \frac{\mu_w}{\mu_n} \tag{1.3.4c}$$

In these definitions of μ_n, μ_w, and Q, the concentration of the monomer ($[P_1]$) has been included in the summations. These averages are, therefore, useful when the output of the reactor is directly utilized and are quite commonly used in step growth polymerizations. On multiplication with the molecular weight of the repeat unit, μ_n and μ_w give the number-average molecular weight, $\bar{M}_n$, and the weight-average molecular weight, $\bar{M}_w$, respectively. Sometimes, as for example when the monomers are carcinogenic, the reaction product is subjected to high vacuum to volatilize

---

[†] The notation DP_n (degree of polymerization) is also used in the literature instead of μ_n, but the latter will be used in this text.

the monomer. Under these circumstances the number and weight average chain lengths, μ_{n1} and μ_{w1}, with the concentration of monomer omitted from the summations, are of significance in determining the product properties. These averages and the corresponding polydispersity index, Q_1, are given by[18]

$$\mu_{n1} \equiv \frac{\sum_{n=2}^{\infty} n[P_n]}{\sum_{n=2}^{\infty} [P_n]} \tag{1.3.5a}$$

$$\mu_{w1} \equiv \frac{\sum_{n=2}^{\infty} n^2[P_n]}{\sum_{n=2}^{\infty} n[P_n]} \tag{1.3.5b}$$

$$Q_1 \equiv \mu_{w1}/\mu_{n1} \tag{1.3.5c}$$

(Alternatively, one may redefine the mer index suitably, e.g., with P_2 being redefined as P_1, and use appropriate equations for μ_{n1}, etc.)

The study of the MWDs or their various moments in the analysis of polymerization reactors is necessary since the physical properties of the product depend significantly on them. For example, the viscosity of molten polymers is proportional to the weight-average molecular weight[†] (at the same temperature and concentration), and for $\bar{M}_w$ above a certain critical value, the viscosity usually varies[3] as $\bar{M}_w^{3.4}$. This is important in estimating the power requirements for pumping the reaction mass, say, through a tubular reactor (where $\bar{M}_w$ increases as the reaction progresses), a spinnerette to obtain fibers, or extruders and molds. The viscosity also determines the heat transfer coefficient, and in view of this the study of $\bar{M}_w$ is necessary to design the cooling or heating system for the reactor. In the solid state too, the molecular weight averages determine various mechanical properties of the polymer. For example,[2] the aliphatic polyester formed from ω-hydroxydecanoic acid has little strength or spinnability when μ_n is about 25, gives long but extremely weak fibers which can be cold drawn when μ_n is about 55, but can be spun easily and cold drawn to strong fibers (tensile strength about 131 MN/m^2) when μ_n is about 100. Such behavior of increasing mechanical strength (to an almost asymptotic value) with increasing μ_n is a characteristic of most polymers.[13] Unfortunately, the current understanding on the exact *quantitative* relationship between the MWD (or μ_n and μ_w) and various important properties of a polymer, e.g., tear strength, resilience, adhesive tack, friction, flex fatigue, impact strength, hardness, etc., is not highly developed and only a few important rules of thumb are available.[25] Because of this shortcoming there is still an element of "art"

[†] Rigorously speaking, viscosity depends on the viscosity average molecular weight, which is slightly lower ($\sim$10%–20%) than $\bar{M}_w$.

present in the design of polymerization reactors. In polymerization reaction engineering, at present, it is, therefore, assumed that one is interested in producing a specified MWD product, without focusing attention on how this MWD is related to the various physical properties.[12] Future research in the area of structure–property relations for polymers would help remove this shortcoming. The other design aspects important in nonpolymeric reactors also apply for polymer reactors, namely, maximizing the yield and purity of products (with respect to side products), minimizing reaction time, etc.

The effect of the MWD and its averages on the physical properties of the product is even more sharply manifested in the case of nonlinear polymers. In the case of step growth polymerizations with monomers having functionalities of three or more, highly branched molecular structures are obtained. In fact, at some stage before completion of the reaction, it is observed that $\bar{M}_w$ (and so the viscosity) rises sharply to infinity. This state, called the gel point, is characterized by the formation of at least one infinitely large cross-linked molecule (the gel) extending over the entire reaction mass, interspersed within which are several smaller finite molecules (called sol). Further progress of the reaction leads to the gradual linking up of the molecules in the sol fraction with the gel so that the total weight of the sol decreases with time. The cross-linked gel behaves like a rubber[3] and has an elastic modulus which depends on the average molecular weight between consecutive cross-links. Also, the dynamic loss modulus of these network structures is related to the concentration and average chain length of the "dangling" or pendant chains, i.e., chains which are linked to the gel only at one end. The study of the polymerization of such nonlinear systems must obviously focus attention upon this detailed molecular information.

1.4. MEASUREMENT OF MWD AND ITS MOMENTS[1,3,26–30]

The various techniques of obtaining the MWD and its moments are briefly described in this section so as to give the reader some idea of these. With the advent of size exclusion or gel permeation chromatography,[30] the entire MWD of a polymer sample can be easily determined experimentally. This helps in monitoring the reactor and is of immense use in determining experimental values of various rate constants associated with polymerization reactions. This method consists of depositing a polymeric sample at one end of a series of two to four columns packed with porous beads (usually of different porosities in different columns), and then eluting it with a solvent (see Fig. 1.2). The smaller molecules diffuse into the bead at the feed-end, while the solvent carries forward the larger molecules which cannot penetrate the pores. As the flow of the solvent continues, a reverse

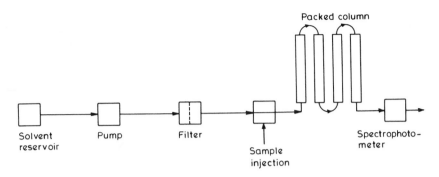

Figure 1.2. Schematic of a gel permeation chromatograph.

concentration gradient is established and the smaller polymer molecules diffuse out of the pores, get carried downstream, and rediffuse into the pores further downstream. Thus, there is a continual diffusion into and out of the pores, the extent of this diffusion being determined by the hydrodynamic size (related to the molecular weight) of the molecules. The largest molecules get eluted first, while the smallest ones are eluted last. Under the conditions of the experiment, the volume of the solvent that must be eluted before a particular sample appears at the downstream end of the column depends on its hydrodynamic volume and so on its molecular weight. Concentration sensitive detectors (e.g., conductivity, refractive index, or IR absorbance measurements) are used along with calibration functions relating the molecular weight to the elution volume to get the MWD. More recently, low-angle laser light scattering detectors have been used[30] in series with concentration sensitive detectors to obtain the absolute molecular weights and so make the technique give more reliable results, even for branched polymers. A typical system[31] for nylon 6, for example, uses a series of two GMIXH 4 (Toyo Soda, Japan) columns with the solvent as a mixture of 15% (by weight) m-cresol, 84.75% chloroform and 0.25% benzoic acid at 35°C, and a flow rate of about 1 ml/min. The sample concentration can be 1.4 g/liter and refractive index detectors can be used. Similarly for the determination of the MWD for epoxy resins,[32] three Styrogel columns (nominal porosities of 250, 500 and 2000 Å) followed by a fourth column packed with Sephadex LH-20 gel are used. The solvent is tetrahydrofuran at ambient temperature and refractive index detectors can be used.

The number-average molecular weight can be obtained experimentally using osmometry (Fig. 1.3). The osmotic pressure, π, of a dilute (~1%) polymer solution is obtained at several concentrations using commercially available automatic osmometers (in which servo devices are used to adjust the pressure on the solvent side so that no flow occurs across the semiper-

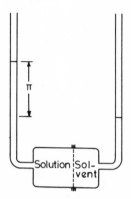

Figure 1.3. Osmotic pressure measurements.

meable membrane). $\bar{M}_w$ is obtained using the extrapolated value (at polymer concentration, $c \rightarrow 0$) and the relation

$$\lim_{c \to 0} \frac{\pi}{c} = \lim_{c \to 0} \frac{\mu_1^0 - \mu_1}{V_1 c} = \frac{RT}{\bar{M}_n} \qquad (1.4.1)$$

where μ_1 is the chemical potential of the solvent in the solution of concentration c, μ_1^0 is the corresponding value for the pure solvent at the same temperature, V_1 is the molar volume of the solvent, R is the universal gas constant, and T is the temperature. This method is useful for $\bar{M}_n$ values from about 10,000 to 500,000.

Vapor-pressure osmometers may be used for measuring values of $\bar{M}_n$ below about 20,000. In this method, a drop of polymer solution and a drop of solvent are placed on adjacent thermisters in an insulated chamber. A differential amount of solvent distills over to the solution, thereby leading to a change in the temperatures (the extrapolated value of which as $c \rightarrow 0$ is related to $\bar{M}_n$). Equations for this colligative property in terms of $\mu_1 - \mu_1^0$ are available in standard texts.[1,3]

Another useful and inexpensive technique to measure the average molecular weight is dilute solution viscometry. The viscosities of dilute solutions of the polymer are determined using the Ubbelohde viscometer (based on the time of flow of a fixed volume through a capillary tube; see Fig. 1.4). The intrinsic viscosity, $[\eta]$, is obtained by extrapolation and is related to the average molecular weight[1,3] as

$$[\eta] \equiv \lim_{c \to 0} \frac{\eta - \eta_0}{\eta_0 c} = \phi(\langle L^2 \rangle_0 / M)^{3/2} M^{1/2} \alpha^3 \simeq K \bar{M}^a \qquad (1.4.2)$$

In Eq. (1.4.2), η_0 is the viscosity of the pure solvent, and K and a are the

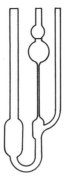

Figure 1.4. Sketch of an Ubbelohde viscometer.

Mark–Houwink constants for a given polymer–solvent–temperature combination. $\langle L^2 \rangle_0$ is the mean square end-to-end length of the molecule, ϕ is a universal constant, and α is the excluded volume factor representing the expansion of the molecule over its value under theta conditions because of solvent–polymer and polymer–polymer interactions.[1,3] Extensive tabulations of K and a are available.[33] This method gives the viscosity-average molecular weight, which, as mentioned earlier, is slightly lower than $\bar{M}_w$.

Another method of measuring the average molecular weight is light scattering. In this technique, the intensity of monochromatic light scattered by dilute polymer solutions, at an angle θ (with respect to the direction of propagation) is measured by a photometer. The relevant equations are[1,3]

$$\lim_{\theta \to 0} \frac{\mathscr{K}c}{\mathscr{R}_\theta} = \frac{1}{\bar{M}_w} + 2A_2 c + \cdots \tag{1.4.3a}$$

$$\lim_{c \to 0} \frac{\mathscr{K}c}{\mathscr{R}_\theta} = \frac{1}{\bar{M}_w} \left[1 + \frac{16\pi^2 \sin^2(\theta/2)}{3\lambda^2} \langle S^2 \rangle + \cdots \right] \tag{1.4.3b}$$

where

$$\mathscr{K} = \frac{2\pi^2 \tilde{n}_0^2 (d\tilde{n}/dc)^2}{N_{av} \lambda^4}$$

$$\mathscr{R}_\theta = \frac{r^2}{1 + \cos^2 \theta} \frac{i_s(\theta)}{I_0} \tag{1.4.4}$$

In Eqs. (1.4.3) and (1.4.4), A_2 is the second virial coefficient, $\langle S^2 \rangle$ is the mean square radius of gyration of the polymer molecule in solution, $\tilde{n}_0$ is the refractive index of the pure solvent, $\tilde{n}$, the refractive index of the solution,

N_{av} the Avogadro number, λ and I_0, the wavelength and intensity of the incident radiation, r is the distance at which the scattered radiation is measured, and $i_s(\theta)$ is the intensity of the scattered radiation from the solution *minus* that scattered by the pure solvent, per unit volume of the solution. $d\tilde{n}/dc$ is obtained using a differential refractometer. A double extrapolation technique[1,3] ($c \to 0$ and $\theta \to 0$) gives $\bar{M}_w$.

An indirect but relatively easier experimental technique of determining the number average degree of polymerization is to find the concentration of functional groups, [A], by titration. If the initial concentration of the functional groups in the reaction mass is $[A]_0$, and the concentration at time t is [A], then, for ARB polymerization, μ_n is related to these as follows:

$$\mu_n = \frac{[A]_0}{[A]} \tag{1.4.5}$$

This technique is commonly used to follow step growth polymerizations experimentally. For example,[34] for nylon 6, the carboxyl end group concentrations can be obtained by titrating a ($\sim 2\%$) polymer solution in benzyl alcohol (at $\sim 180°C$), with a (~ 0.05 mol/liter) KOH benzyl alcohol solution. Similarly, amino end groups can be obtained by titrating a polymer solution in m-cresol at room temperature, with an (~ 0.05 mol/liter) aqueous solution of p-toluene sulfonic acid.

Besides these methods, there are other analytical techniques like mass spectroscopy, liquid chromatography, etc., which can be used to determine the concentrations of polymer and various side-products formed, e.g., cyclic oligomers.

1.5. TYPES OF REACTORS[22]

Several types of reactors are used commonly for carrying out polymerizations, each having their own advantages and characteristics. The batch reactor is the most versatile of these and has been used extensively, particularly for speciality polymers with low production capacities. It consists of a stirred autoclave, with provision for heating and/or cooling and for withdrawing or feeding samples and the contents, and usually is connected to a reflux condensor so that no material is removed by evaporation. Some examples of step growth polymerizations carried out in such reactors include nylon 6, phenol-formaldehyde, urea-formaldehyde, melamine-formaldehyde, etc. In the case of PET polymerization, methanol or ethylene glycol evaporates during polymerization, and because of this the volume of the reaction mass decreases continuously with time. Such reactors are called semibatch reactors because the condensation product is allowed to leave the reaction mass. The only disadvantage of both these reactors is that their operation is batchwise, because of which they are being rapidly replaced by continuous reactors in newer, high-capacity plants. One example of a continuous flow reactor is a tubular reactor. In these, the Reynolds number is usually high (i.e., flow is turbulent) and the velocity is almost uniform across the cross section of the tube. Where this is not so (which is usually the case when the reaction mass is highly viscous), internal grids may be provided to ensure that the flow profile becomes uniform. These reactors can, therefore, be modeled as plug-flow reactors (PFRs), shown in Fig. 1.5a. They can also be jacketed or may be provided with internal coils to remove or provide heat to the reaction mass. Examples of the use of these reactors can be found in the production of nylon 6 and nylon 66 polymers.

A characteristic of PFRs is the lack of agitation, which is a mixed blessing. Because of the absence of mixing, step-growth polymers with lower polydispersity indices can be produced. However, lack of mixing may lead to problems when the reaction mass (of low thermal conductivity) becomes viscous. The continuous-flow, stirred tank reactor (CSTR) overcomes this problem, and is shown in Fig. 1.5b. These reactors can have good heat transfer characteristics and deposit formation on the walls can be avoided

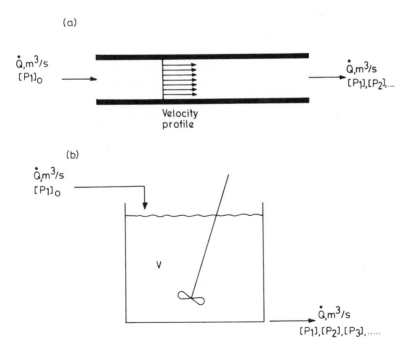

Figure 1.5. Schematic representation of (a) a plug flow reactor (PFR) and (b) a continuous flow, stirred tank reactor (CSTR) with pure monomer feed.

if required, by appropriate design of stirrers, e.g., by use of anchor or ribbon agitators. The residence time distribution of such reactors is extremely wide and so these reactors cannot be used to attain high conversions. However, a series of CSTRs followed by a PFR exploits the advantages of both these reactors. Such cascades are now very commonly used for polymerizations. In fact, some industrial reactors with more complex flow patterns can also be modeled as cascades of several reactors. The VK reactor for nylon 6 polymerization can be modeled as a CSTR (with the mixing provided by the evaporating water) followed by a plug-flow reactor. It may be emphasized that heat transfer effects in step-growth polymerizations are not nearly as critical as in chain polymerizations.

Though several polymerizations can be carried out in these common reactors, some step growth polymerizations have characteristics which dictate the use of special reactors. In some cases, the equilibrium constant of the step growth polymerization is low, and to drive the reaction in the forward direction, the by-product formed [see Eq. (1.2.1)] must be removed continuously by applying high vacuum. Usually, this must be accomplished in a reaction mass which is highly viscous and this calls for special reactors in which the surface area exposed to vacuum is increased by the use of special mechanical devices. Several methods have been used to achieve this. Bubbles of inert gas may be passed through the reaction mass (e.g., in nylon 6 formation), or blades can be used to produce *thin* layers of the reaction mass on surfaces wherein vacuum is applied. Details of such reactors are given in later chapters. In the polymerization of urethanes, the reaction rates are very high and reaction takes place even when the monomers are being mixed and pumped into molds. Special equipment is again required in which the reaction is almost completed by the time the material fills the mold. This operation is termed reaction injection molding (RIM), and it is evident that mixing and flow equations must be solved simultaneously with those for chemical reaction to model these complex situations. Yet another example of special equipment required for some polymerizations is the finishing stage of nylon 6, nylon 66, and PET polymerizations. The final product from conventional reactors consists of an equilibrium mixture of polymer, unreacted monomer, and the condensation by-product. Higher molecular weight polymer is obtained by heating chips or flakes of the material *below its melting point* in a stream of hot gases in a fluidized bed or in a drier operating under vacuum. The monomer and the condensation by-product diffuse out, and further reaction takes place inside the solid. This leads to high molecular weight product useful, for example, for tire cords and for molding purposes in the case of nylon 6. The polymerization here takes place in the solid state and severe diffusional limitations characterize these reactions.[35,36] The progress of these reactions may be as much controlled by the diffusion of the condensation by-product as by the morphology of the solid. It is thus seen that one has to devise reactors depending on the characteristics of the system to be polymerized. Several of these industrially important polymerization reactors will be modeled in this text.

1.6 CONCLUSIONS

In this chapter, an attempt has been made to highlight the special problems present in polymerization reactors. A review of common reactors and how they overcome these special problems is also presented. The stage is now set to go into the detailed modeling of some systems. Simple ARB polymerizations in various reactors are considered first in order to establish the important principles, and then the effects of unequal reactivity, diffusion, etc., are gradually incorporated. Thereafter, some industrial reactors like those producing PET, nylon 6, epoxies, urethanes, etc., are discussed. In several cases, the models developed have been tested either against data on industrial reactors or on pilot plants, and have thus been validated.

REFERENCES

1. P. J. Flory, *Principles of Polymer Chemistry*, 1st ed., Cornell University Press, Ithaca, New York (1953).
2. R. W. Lenz, *Organic Chemistry of Synthetic High Polymers*, 1st ed., Wiley, New York (1967).
3. A. Kumar and S. K. Gupta, *Fundamentals of Polymer Science and Engineering*, 1st ed., Tata McGraw-Hill, New Delhi, India (1978).
4. P. E. M. Allen and C. R. Patrick, *Kinetics and Mechanisms of Polymerization Reactions*, 1st ed., Ellis Horwood, Chichester (1974).
5. J. Furukawa and O. Vogl, *Ionic Polymerization, Unsolved Problems*, 1st ed., Marcel Dekker, New York (1976).
6. T. Keii, *Kinetics of Ziegler-Natta Polymerization*, 1st ed., Kodansha, Tokyo (1972).
7. J. Boor, *Ziegler-Natta Catalysts and Polymerizations*, 1st ed., Academic, New York (1979).
8. J. C. W. Chien, *Coordination Polymerization*, 1st ed., Academic, New York (1975).
9. W. H. Carothers, Studies on polymerization and ring formation. I. An introduction to the general theory of condensation polymers, *J. Am. Chem. Soc.* **51**, 2548-2559 (1929).
10. J. L. Stanford and R. F. T. Stepto, Rate theory of irreversible linear random polymerization, *J. Chem. Soc., Faraday Trans. I.* **71**, 1292-1307 (1975).
11. R. F. T. Stepto, in *Developments in Polymerization* (R. N. Haward, Ed.), 1st ed., Vol. 3, pp. 81-141, Applied Science Pub., Barking, UK (1982).
12. M. V. Tirrell and R. L. Laurence, *Polymer Reaction Engineering*, Academic, New York, in preparation.
13. G. Odian, *Principles of Polymerization*, 2nd ed., Wiley, New York (1981).
14. J. L. Throne, *Plastics Process Engineering*, 1st ed., Marcel Dekker, New York (1979).
15. J. A. Biesenberger and D. H. Sebastian, *Principles of Polymerization Engineering*, 1st ed., Wiley, New York (1983).
16. S. K. Gupta and A. Kumar, Simulation of step-growth polymerizations, *Chem. Eng. Commun.* **20**(1-2), 1-52 (1983).

17. W. H. Ray and R. L. Laurence, in *Chemical Reactor Theory* (L. Lapidus and N. R. Amundson, Eds.), 1st ed., pp. 532–582, Prentice Hall, Englewood Cliffs, New Jersey (1977).

18. W. H. Ray, On the mathematical modeling of polymerization reactors, *J. Macromol. Sci., Rev. Macromol. Chem.* **C8**(1), 1–56 (1972).

19. S. L. Liu and N. R. Amundson, Analysis of polymerization kinetics and the use of a digital computer, *Rubber Chem. Tech.* **34**, 995–1133 (1961).

20. W. H. Ray, in *Chemical Reaction Engineering—Plenary Lectures* (J. Wei and C. Georgakis, Eds.), 1st ed., pp. 101–133, ACS Symp. Ser. No. 226, American Chemical Society, Washington, DC (1983).

21. T. C. Bouton and D. C. Chappelear, Eds., *Continuous Polymerization Reactors*, AIChE Symp. Ser., 160 (1976).

22. H. Gerrens, On selection of polymerization reactors, *Germ. Chem. Eng.* **4**, 1–13 (1981).

23. M. V. Tirrell, R. Galvan, and R. L. Laurence, in *Chemical Reaction and Reactor Engineering* (J. J. Carberry and A. Varma, Eds.) 1st ed., pp. 735–738, Marcel Dekker, New York, (1986).

24. B. W. Brooks, Polymerization reactors, *Rev. Chem. Eng.* **1**, 403–430 (1983).

25. F. K. Lautenschlaeger and M. Myhre, Classification of properties of elastomers using the optimum property concept. I. Introduction, *J. Appl. Polym. Sci.* **24**, 605–634 (1979).

26. F. Rodriguez, *Principles of Polymer Systems*, 2nd ed., McGraw-Hill, New York (1982).

27. F. W. Billmeyer, *Textbook of Polymer Science*, 2nd ed., Wiley, New York (1971).

28. E. L. McCaffery, *Laboratory Preparation for Macromolecular Chemistry*, 1st ed., McGraw-Hill, New York (1970).

29. D. Braun, H. Cherdron and W. Kern, *Techniques of Polymer Synthesis and Characterization*, 1st ed., Wiley, New York (1972).

30. T. Provder, ed., *Size Exclusion Chromatography*, ACS Symp. Ser., Vol. 138 (1980).

31. K. Tai, Y. Arai, H. Teranishi, and T. Tagawa, The kinetics of hydrolytic polymerization of ε-caprolactam. IV. Theoretical aspect of the molecular weight distribution, *J. Appl. Polym. Sci.* **25**, 1789–1792 (1980).

32. L. Antal, L. Füzes, G. Samay, and L. Csillag, Kinetics of epoxy resin synthesis on the basis of GPC measurements, *J. Appl. Polym. Sci.* **26**, 2783–2786 (1981).

33. J. Brandrup and E. H. Immergut, Eds., *Polymer Handbook*, 2nd ed., Wiley, New York (1975).

34. K. Tai, H. Teranishi, Y. Arai, and T. Tagawa, The kinetics of hydrolytic polymerization of ε-caprolactam, *J. Appl. Polym. Sci.* **24**, 211–224 (1979).

35. L. B. Sokolov, *Solid Phase Polymerization, Synthesis by Polycondensation*, 1st ed., Israel Program for Scientific Translations, Jerusalem (1968).

36. F. C. Chen, R. G. Griskey, and G. H. Beyer, Thermally induced solid state polycondensation of nylon 66, nylon 6-10 and polyethylene terephthalate, *AIChE J.* **15**, 680–685 (1969).

EXERCISES

1. Derive or deduce the following:

a. $\sum\limits_{n=2}^{\infty} n^k \sum\limits_{m=1}^{n-1} a_m b_{n-m} [P_m][P_{n-m}] = \sum\limits_{m=1}^{\infty} \sum\limits_{n=1}^{\infty} a_m b_n (m+n)^k [P_m][P_n], \qquad k = 0, 1, 2$

b. $\sum\limits_{n=2}^{\infty} \sum\limits_{m=1}^{\infty} a_m a_{n-m} [P_m][P_{n-m}] = \left(\sum\limits_{n=1}^{\infty} a_n [P_n] \right)^2$ (special case of a)

c. $\sum\limits_{n=2}^{\infty} n^k \sum\limits_{m=1}^{n-1} [P_m][P_{n-m}] = \sum\limits_{n=1}^{\infty} [P_n] \sum\limits_{m=1}^{\infty} (m+n)^k [P_m]$ (special case of a)

d. $\sum\limits_{n=1}^{\infty} n^k \sum\limits_{m=n+1}^{\infty} [P_m] = \sum\limits_{n=2}^{\infty} \left(\sum\limits_{m=1}^{n-1} m^k \right)[P_n]$

e. $\sum\limits_{n=2}^{\infty} s^n \sum\limits_{m=1}^{n-1} [P_m][P_{n-m}] = G^2(s, t) \equiv \left(\sum\limits_{n=1}^{\infty} s^n [P_n] \right)^2$

f. $\sum\limits_{n=1}^{\infty} n^2 [P_{n\pm1}] = \lambda_2 \mp 2\lambda_1 + \lambda_0$; similarly obtain an expression for $\sum\limits_{n=1}^{\infty} n^2 [P_{n\pm2}]$

g. $\sum\limits_{n=1}^{\infty} \sum\limits_{m=0}^{n-1} [Q_m][P_{n-m-1}] = \left(\sum\limits_{n=0}^{\infty} [Q_n] \right) \left(\sum\limits_{m=0}^{\infty} [P_m] \right)$

where n, m, k are integers, s is the transform variable of the generating function (or transform) $G(s, t) \equiv \sum_{n=1}^{\infty} s^n [P_n]$ and a_n, b_n are functions of n. Q_m and P_m are two different families of polymers. Express the answers for $k = 2$ in terms of the moments λ_0, λ_1, λ_2, etc., where possible. These summations will prove useful in solving the exercises given in later chapters.

2. Compute $\bar{M}_n$ and $\bar{M}_w$ for polymer samples having the following compositions:

Molecular weight	Sample A mole fraction	Sample B mass fraction
10^3	0.5	0.5
10^5	0.5	0.5

3. Osmotic pressure and η/η_0 data at 30°C on dilute polymethylmethacrylate (atactic) solutions in acetone have been obtained at different concentrations and are given below:[26]

c (g/100 ml)	π (cm solvent)	η/η_0
0.275	0.457	1.170
0.338	0.592	
0.344	0.609	1.215
0.486	0.867	
0.896	1.756	1.629
1.006	2.098	
1.199	2.710	1.892
1.536	3.725	
1.604	3.978	2.330
2.108	5.919	2.995
2.878	9.713	

The density of acetone is 0.780 g/cm^3. Use a plot of $(\pi/c)^{1/2}$ vs. c to obtain the limiting value of π/c in Eq. (1.4.1) and thus determine $\bar{M}_n$. Similarly, plot

$$\frac{1}{c}\left(\frac{\eta}{\eta_0} - 1 \right) \quad \text{vs.} \quad c$$

to obtain $[\eta]$. Using[33] $K = 7.7 \times 10^{-5}$ deciliter/g and $a = 0.70$ for this system, obtain an estimate of the molecular weight from viscosity data.

LINEAR STEP GROWTH POLYMERIZATION FOLLOWING THE EQUAL REACTIVITY HYPOTHESIS

2.1. INTRODUCTION[1,2]

In order to form linear polymer chains in step growth polymerization, the starting monomer must be bifunctional. The reacting functional groups A and B can be on the same molecule as in ARB, where R is an alkyl or aryl group. Some examples of this kind of monomers are hexamethylene adipamide $[COOH-(CH_2)_4-CONH-(CH_2)_6-NH_2]$, which is sometimes called nylon 66 salt, and aminocaproic acid $[COOH-(CH_2)_5-NH_2]$, consisting of functional groups $-COOH$ and $-NH_2$, which can react to give a $-CONH-$ linkage. It is easy to see that at any time, for this case of polymerization, the concentrations of functional groups A and B are equal. If P_m and P_n denote polymer molecules having m and n monomeric units, respectively, the polymerization can be represented schematically by

$$P_m + P_n \rightleftharpoons P_{m+n} + W, \qquad m, n = 1, 2, \ldots \qquad (2.1.1)$$

where W is the condensation product and P_{m+n} is a polymer molecule of length $(m + n)$ formed as a result of the reaction between the functional groups of P_m and P_n.

Linear step growth polymerization can also be carried out when bifunctional monomers of the type ARA and BR'B react, as for example, the reaction between terephthalic acid COOH—⟨O⟩—COOH and ethylene glycol $(OH-CH_2CH_2OH)$ in polyester formation. In this case, the various oligomers present in the reaction mass are distinguished by their end groups and can be classified into three types: A$\sim\sim$A, B$\sim\sim$B and A$\sim\sim$B, where $\sim\sim$ denotes the polymer chain. If these species are denoted by P_{AA_n}, P_{BB_n}, and P_{AB_n}, respectively, where n denotes the number of repeat units in the chain, by analogy with ARB polymerization, the chain growth steps can be represented by

$$P_{AB_m} + P_{AB_n} \rightleftharpoons P_{AB_{(m+n)}} + W$$

$$P_{AA_m} + P_{BB_n} \rightleftharpoons P_{AB_{(m+n)}} + W$$

$$(2.1.2)$$

$$P_{AB_m} + P_{BB_n} \rightleftharpoons P_{BB_{(m+n)}} + W$$

$$P_{AA_m} + P_{AB_n} \rightleftharpoons P_{AA_{(m+n)}} + W \qquad m, n = 1, 2, \ldots$$

If monomers ARA and BR'B are present in equimolar ratio, it can be shown that the set of reactions in Eq. (2.1.2) collapse into a single one and the overall polymerization can once again be represented on an average by Eq. (2.1.1).[1,2]

As pointed out in Chapter 1, polymerization can either be carried out in batch or continuous reactors. As higher and higher throughputs are desired, the economy of large reactors demand that continuous reactors be used, and usually either tubular reactors or tanks with agitators are employed. These can be idealized as plug flow (PFRs) or homogeneous continuous-flow stirred tank reactors (HCSTRs). It can be easily proven that the performances of PFRs and batch reactors are identical,[3] and in view of this, the discussion in this book has been mainly focused on the modeling of batch reactors and HCSTRs for polymeric systems.

The difficulty of modeling these ideal reactors for step growth polymerization systems arises because the growth steps [as in Eqs. (2.1.1) and (2.1.2)] consist of infinite elementary reactions. In general, the rate constants for these elementary reactions can take on any value depending upon the chemical nature of the reacting functional groups and the chain lengths m and n of the interacting molecules P_m and P_n. The chain length dependence of these rate constants must be modeled first to reduce the complexity of mathematical modeling. In the literature there exist several models, and in this chapter the discussion is restricted to the systems following the equal reactivity hypothesis. A systematic approach has been developed to obtain the MWDs of the polymer formed in various types of commercial reactors since these are related to the physical properties.

Several strategies can be adopted to obtain the desired molecular weight distribution of the polymer formed in a given reactor. One can add monofunctional compounds in the feed, which has the effect of reducing the average molecular weight of the polymer formed. If it is desired to have very high molecular weight of the polymer, it can be done either by having larger reactor residence times or higher temperatures of polymerization or both. Thus, these serve as very important design variables which affect the MWD of the polymer formed. Alternatively, a part of the product stream may be mixed with the feed. This is called recycling and is known to have important ramifications on the MWD. In this chapter, the effects of all these variables have been analyzed with the purpose of establishing a rational design strategy for reactors manufacturing polymers formed through step growth polymerization.

2.2. RATE CONSTANTS FOR ELEMENTARY STEP GROWTH POLYMERIZATION REACTIONS

The reaction between *small* molecules A and B in a homogeneous condensed phase

$$A + B \xrightarrow{k} C \qquad (2.2.1)$$

proceeds mechanistically in several steps.[3-5] Molecules of A and B first diffuse from the bulk of the reaction mass to each other's proximity and then interact to form a chemical bond. This process can be written schematically as

$$A + B \underset{\text{diffusion}}{\overset{\text{Bulk}}{\rightleftharpoons}} [A \cdots B] \xrightarrow{\text{Chemical}} AB \qquad (2.2.2)$$

The reaction of functional groups in step growth polymerization differs slightly from the mechanism given in Eq. (2.2.2) because of the long chain nature of the polymer molecules. Polymer molecules in the reaction mass are present in a highly coiled state, and the functional groups, usually located at the ends of the molecules, are buried within these coils. Thus, chemical reaction between functional groups may not occur even when two molecules collide. An additional step of the coordinated movement of nearby segments must simultaneously occur within the colliding molecules, to bring the reacting functional groups to within close proximity for chemical reaction. This coordinated motion of segments is called *segmental diffusion*. Thus, if ⋙A and ⋙B are two polymer molecules with A and B representing the reacting functional groups, the overall process of polymerization can be written as

$$(2.2.3)$$

Since the bulk as well as the segmental diffusions are dependent on the chain lengths of the two polymer molecules in the reaction, the overall rate constants for these may be expected to depend on the chain lengths. Therefore,

$$P_m + P_m \underset{k'_{p,m+n}}{\overset{k_{p,mn}}{\rightleftharpoons}} P_{m+n} + W, \qquad m, n = 1, 2, \ldots, \qquad (m+n) = 2, 3, \ldots$$

$$(2.2.4)$$

where the forward rate constant, $k_{p,mn}$ is expected to be some complex function of the chain lengths m and n of the molecules involved. It may be noted that the reverse reaction involves a large and a small molecule, and the corresponding rate constant, $k'_{p,m+n}$, is a function of the chain length of P_{m+n} only.

2.3. EQUAL REACTIVITY HYPOTHESIS

It is very difficult to determine $k_{p,mn}$ and $k'_{p,m+n}$ from experimental measurements of the rate of polymerization alone. However, model experiments[1,6] of the following type reveal the nature of the chain length dependence of these rate constants. The kinetics of the esterification of mono and diacid in *excess* of ethanol

$$H(CH_2)_n COOH + C_2H_5OH \xrightarrow{HCl} H(CH_2)_n COOC_2H_5 + H_2O$$

$$(2.3.1a)$$

$$COOH(CH_2)_n COOH + 2C_2H_5OH \xrightarrow{HCl} C_2H_5OCC(CH_2)_n COOC_2H_5$$
$$+ 2H_2O \qquad (2.3.1b)$$

have been studied and the rate of esterification, $\mathscr{R}_e$, measured using compounds having different values of n. $\mathcal{R}_e$ is correlated through the following expression:

$$\mathcal{R}_e = -\frac{d[\text{COOH}]}{dt} = k[\text{COOH}] \qquad (2.3.2)$$

where [COOH] represents the total molar concentration of acid end groups in the reaction mass and can be easily measured by titration. The rate constant, k, has been evaluated for various values of n in Eq. (2.3.1), and the experimental results are given in Table 2.1. From this, it can be seen that for monobasic acids, k reaches an asymptotic value indicating the absence of chain length dependence beyond $n \geq 8$. A similar conclusion is reached for dibasic acids. This has been explained physically by Flory[1] as follows. Even though the bulk diffusion of the molecules becomes slower with increase in chain length, they stay close to each other for a longer time, during which several more collisions of A and B *groups* occur because of segmental motion. These two effects apparently cancel each other. The model experiments of Table 2.1 simulate the polymerization reactions of Eq. (2.2.4) very well and reveal the functional dependence of rate constants $k_{p,mn}$ and $k'_{p,m+n}$ on the chain length of reacting species.

Based on these model experiments, the rate constants, $k_{p,mn}$ and $k'_{p,m+n}$ in step growth polymerization, as a first approximation, can approximately be assumed to be independent of the chain lengths of the polymer molecules involved. This is sometimes known as the equal reactivity hypothesis and represents a simple model for the forward and reverse rate constants. Evidently, this approximation may be a poor representation for low conversions, when the average chain length in the reaction mass has not reached

TABLE 2.1.[a] Rate Constants for the Esterification of Mono[H(CH$_2$)$_n$COOH] and Dibasic Acids [HOOC(CH$_2$)$_n$COOH] in Excess of Ethanol, 25°C

n	Monobasic acid (10^4 k, liters/mole sec)	Dibasic acid (10^4 k, liters/mole sec)
1	22.1	–
2	15.3	6.0
3	7.5	8.7
4	7.4	8.4
5	7.4	7.8
6	—	7.3
7	7.5	—
8	7.4	—
Higher	7.6	—

[a] References 1, 6.

some critical value (about 8 in the case of Table 2.1). However, for larger conversions, the equal reactivity hypothesis is expected to model the step growth polymerization much better.

2.4. KINETIC MODEL OF IRREVERSIBLE STEP GROWTH POLYMERIZATION UNDER THE EQUAL REACTIVITY HYPOTHESIS

As discussed earlier, a chemical reaction can occur only when the reacting molecules collide. The rate of reaction, $\mathcal{R}$, can thus be written in terms of the product of the collision frequency, $\omega_{m,n}$, between P_m and P_n and the probability of their reaction, $Z_{m,n}$. Therefore,

$$\mathcal{R} = \alpha \omega_{m,n} Z_{m,n} \tag{2.4.1}$$

where α is a constant of proportionality. From the equal reactivity hypothesis, the probability of reaction, $Z_{m,n}$, is independent of m and n. Thus if Z is the probability of reaction between the *reacting functional groups* and P_m and P_n can react in s distinct ways, $Z_{m,n}$ would be equal to (sZ). Additionally, the collision frequency $\omega_{m,n}$ between two dissimilar molecules is proportional to $[P_m][P_n]$ whereas that between P_m and P_m is proportional to $[P_m]^2/2$ (square brackets representing molar concentrations). This additional factor of 2 arises because in writing $\omega_{m,n}$ proportional to $[P_m]^2$, the total number of collisions have been counted twice. Thus, if k_p is the rate constant *associated with the reaction of functional groups*, the kinetic model under the equal reactivity hypothesis can be written as

$$k_{p,mn} = \begin{cases} \dfrac{\mathcal{R}}{[P_m][P_n]} = sk_p; & m \neq n; \qquad m, n = 1, 2, \ldots \quad (2.4.2a) \\[4mm] \dfrac{\mathcal{R}}{[P_m]^2} = sk_p/2; & m = n; \qquad n = 1, 2, 3, \ldots \quad (2.4.2b) \end{cases}$$

For linear chains with functional groups A and B located at the chain ends, there are two distinct ways in which the polymer chains can react, as shown in Fig. 2.1. Therefore,

$$k_{p,mn} = \begin{cases} 2k_p; & m \neq n, \qquad m, n = 1, 2, 3, \ldots \quad (2.4.3a) \\[3mm] k_p; & m = n, \qquad n = 1, 2, 3, \ldots \quad (2.4.3b) \end{cases}$$

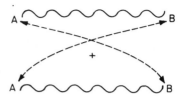

Figure 2.1. Different ways in which linear polymer chains can react.

This simplification makes it quite easy to write mole balance equations for individual *molecular species* in isothermal batch reactors in terms of k_p, the *functional group* reactivity.

As will be discussed later, step growth polymerizations are often carried out under conditions of high vacuum so that the overall reaction is driven in the forward direction to give polymer of high molecular weight. Under these circumstances, the elementary reactions of Eq. (2.1.1) can be assumed to be irreversible and can be written as

$$P_m + P_n \xrightarrow{2k_p} P_{m+n} + W; \quad m \neq n; \quad m, n = 1, 2, \ldots \quad (2.4.2a)$$

$$P_m + P_n \xrightarrow{k_p} P_{2m} + W; \quad m = n; \quad n = 1, 2, \ldots \quad (2.4.4b)$$

The monomer P_1 is consumed whenever it reacts with any molecular species of the reaction mass and its mole balance equation is given by

$$\frac{d[P_1]}{dt} = -k_p\{2[P_1][P_1] + 2[P_1][P_2] + 2[P_1][P_3] \ldots\}$$
$$= -2k_p[P_1][P] \quad (2.4.5)$$

where

$$[P] = \sum_{i=1}^{\infty} [P_i] \quad (2.4.6)$$

Since each of the molecules $P_1, P_2, \ldots$ contain one A and one B functional group, $[P]$ represents the total concentration of functional groups in the reaction mass. The polymer molecule P_n ($n \geq 2$) is formed whenever a P_r ($r < n$, $r = 1, 2, \ldots$) reacts with P_{n-r} whereas it is consumed whenever it reacts with any molecule in the reaction mass. The corresponding balance equation is thus

$$\frac{d[P_n]}{dt} = -2k_p[P_n][P] + \frac{1}{2}2k_p \sum_{r=1}^{n-1} [P_r][P_{n-r}], \quad n \geq 2 \quad (2.4.7)$$

Equations (2.4.5) and (2.4.7) can be summed to obtain the time variation of [P] as (see Exercise 2.2)

$$\frac{d[P]}{dt} = -k_p[P]^2 \tag{2.4.8}$$

Since there is one unreacted A and one unreacted B functional group per polymer molecule, it is possible to rewrite Eq. (2.4.8) as

$$\frac{d[A]}{dt} = \frac{d[B]}{dt} = -k_p[A]^2 = -k_p[B]^2 = -k_p[A][B] \tag{2.4.9}$$

This can be integrated for isothermal batch reactors as

$$\frac{1}{[A]} - \frac{1}{[A]_0} = k_p t \tag{2.4.10}$$

where $[A]_0$ is the molar concentration of A at $t = 0$ and t is the time of polymerization. The conversion, p_A, of functional groups A is then given by

$$p_A \equiv \frac{[A]_0 - [A]}{[A]_0} = \frac{k_p[A]_0 t}{1 + k_p[A]_0 t} \tag{2.4.11}$$

It may be noted that in obtaining Eq. (2.4.11), the reactor temperature has been assumed to be constant, but one need not assume that. For a nonisothermal reactor, it is possible to define a variable τ such that

$$\tau = \int_0^t k_p(T)\, dt \tag{2.4.12}$$

and Eq. (2.4.9) can once again be suitably integrated to obtain a similar relation as in Eq. (2.4.11). Equation (2.4.9) also shows that the set of infinite elementary reactions written in terms of *molecular species* $P_1, P_2, \ldots$ collapses into a single one and Eq. (2.4.4) can be equivalently written in terms of the reaction between *functional groups* as

$$-A + -B \xrightarrow{k_p} -AB- + W \tag{2.4.13}$$

The representation of the step growth polymerization reactions in terms of the reactions between functional groups and the reactivity, k_p, of functional groups is more general than for the restricted case for which it has been derived above. In fact, it leads to a considerable simplification of the analysis of polymerization reactions. It can be shown mathematically that a similar representation as in Eq. (2.4.14) is possible for the polymerization of nonstoichiometric amounts of (ARA + BRB) monomers where [A] need not be equal to [B] (see Exercise 2.3).

2.5. EXPERIMENTAL VERIFICATION OF THE KINETIC MODEL

Historically, the kinetic model discussed above was first verified by Flory[1,7] on the catalyzed and noncatalyzed polyesterifications of adipic acid with ethylene glycol. In uncatalyzed polymerization, the acid end groups of adipic acid were found to act as a catalyst. As a result, the overall rate of polymerization, $\mathscr{R}_p$, was written by Flory as second order in [COOH] and first order in [OH], i.e.,

$$\mathscr{R}_p \equiv -\frac{d[COOH]}{dt} = k_1[COOH]^2[OH] \tag{2.5.1}$$

If equal moles of adipic acid and ethylene glycol are taken in the feed, the above expression reduces to a third-order rate expression,

$$-\frac{d[COOH]}{dt} = k_1[COOH]^3 \tag{2.5.2}$$

On integration of this equation, the functional group conversion, p_A, can be derived to be

$$\frac{1}{(1-p_A)^2} = 2[COOH]_0^2 k_1 t + \text{const} \tag{2.5.3}$$

where $[COOH]_0$ is the initial concentration of COOH groups in the reaction mass.

On the other hand, if the polyesterification is carried out in the presence of a suitable catalyst (for example, p-toluene sulfonic acid), the rate of polymerization is given by

$$\mathscr{R}_p \equiv -\frac{d[COOH]}{dt} = k_1'[COOH][OH] \tag{2.5.4}$$

which, on integration, for an equimolar ratio of COOH and OH groups, yields

$$1/(1-p_A) = [COOH]_0 k_1' t + \text{const} \tag{2.5.5}$$

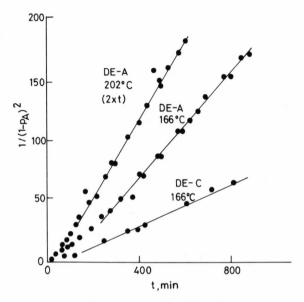

Figure 2.2. Uncatalyzed polymerization of ethylene glycol-adipic acid (DE-A) and ethylene glycol-caproic acid (DE-C). $2 \times t$ indicates that the time axis is to be stretched by a factor of 2.[7,8]

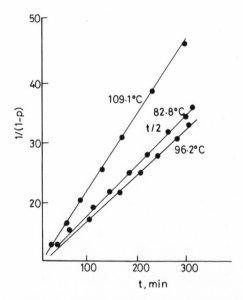

Figure 2.3. Catalyzed polymerization of decamethylene glycol with adipic acid with 0.1 equiv % of toluene sulfate catalyst. $t/2$ indicates that the time scale is to be compressed by a factor of 2.

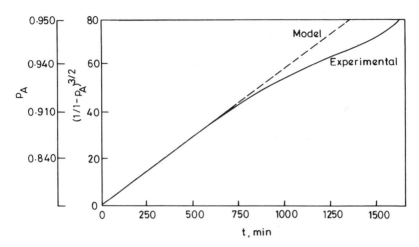

Figure 2.4. The $2\frac{1}{2}$ order plot[13] of the uncatalyzed polymerization of adipic acid with ethylene glycol at 166°C with experimental results of Flory.

Experimental results of Flory[1,7] for uncatalyzed and catalyzed polyesterifications have been plotted in Figs. 2.2 and 2.3 in accordance with Eqs. (2.5.3) and (2.5.4). Figure 2.2 shows that the fit of experimental data by Eq. (2.5.3) is good only beyond the values of $1/(1 - p_A)^2$ of about 25, which means that the kinetic model is a good representation only after approximately 80% conversion! Similar conclusions are reached from Fig. 2.3.

In more recent studies[12] on the polymerization of adipic acid with ethylene glycol, a $2\frac{1}{2}$-order kinetic expression has been proposed which has been claimed to fit the entire range of polymerization. Figure 2.4 shows this particular plot and it is found that the experimental data actually spread as an S-shaped curve around the theoretical line. This perhaps indicates that the $2\frac{1}{2}$-order kinetic model is also not appropriate.[13]

Lin et al.,[14-16] contrary to these studies, observe that the general kinetic expression for the polyesterification of adipic acid and ethylene glycol should be written as

$$-\frac{d[COOH]}{dt} = k[COOH][OH][H^+] \qquad (2.5.6)$$

where $[H^+]$ is the concentration of hydrogen ions in the medium of the reaction mass due to the ionization of the acid. Since only small quantities of adipic acid (0.390 moles per mole of ethylene glycol at 180°C) are dissociated, $[H^+]$ does not depend upon the total adipic acid added but on the concentration of ethylene glycol. In other words

$$[H^+] = k_h[OH] \qquad (2.5.7)$$

where k_h is a proportionality constant. If ethylene glycol and adipic acid are fed to the batch reactor in the molar ratio of r ($\equiv[OH]_0/[COOH]_0$), Eq. (2.5.6) reduces to

$$-\frac{d[COOH]}{dt} = k_a[COOH](a + [COOH])^2 \qquad (2.5.8)$$

where

$$k_a = kk_h \qquad (2.5.9a)$$

$$a = (r-1)[COOH]_0 \qquad (2.5.9b)$$

Equation (2.5.8) can now be integrated to give

$$\ln\frac{r - p_A}{1 - p_A} - \frac{r-1}{(1 - p_A)} = a^2 k_a t + \left(\ln r - \frac{r-1}{r}\right) \qquad (2.5.10)$$

where p_A is the conversion of COOH groups. Equation (2.5.10) has been

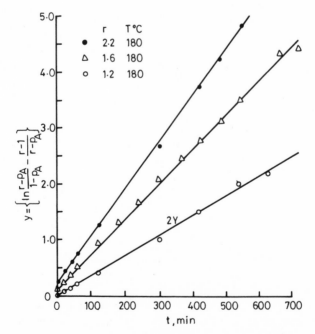

Figure 2.5. Uncatalyzed polymerization of nonequimolar quantities of adipic acid and ethylene glycol.[14] (Reprinted from Ref. 14 with permission of John Wiley & Sons, New York.)

plotted in Fig. 2.5 along with the experimental data on the polymerization of adipic acid and ethylene glycol.[14] A good fit is obtained in the conversion range of 0%–99% even though different lines are obtained for different r.

For acid catalyzed polymerization of adipic acid for any r, however, the kinetic expression found to fit the experimental data was

$$-\frac{d[COOH]}{dt} = k''[COOH]^2 \qquad (2.5.11)$$

which on integration gives

$$\frac{1}{1 - p_A} = 1 + [COOH]_0 k'' t \qquad (2.5.12)$$

In Fig. 2.6, experimental data[14] have been plotted as $1/(1 - p_A)$ versus t. The plot gives a decent straight line except for the shortcoming that k''

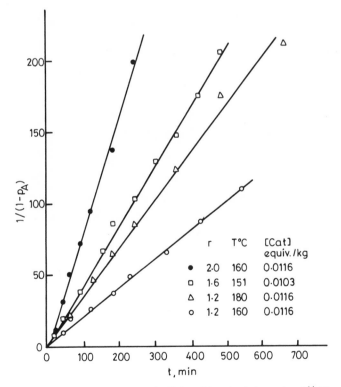

Figure 2.6. Catalyzed polymerization of adipic acid and ethylene glycol.[14] (Reprinted from Ref. 14 with permission of John Wiley & Sons.)

depends not only on temperature but also on r. Subsequently, Lin *et al.*[16] have developed a semiempirical approach to evaluate k_a and k'' as a function of r and have given this functional relation for the polymerizations of adipic and succinic acids with ethylene glycol. These can be used for design purposes till more fundamental models become available.

2.6. IRREVERSIBLE ARB POLYMERIZATIONS WITH MONOFUNCTIONAL COMPOUNDS IN BATCH REACTORS[17-20]

In several situations, as in nylon 6 and polyester reactors, monofunctional compounds, sometimes called modifiers, are used to control the molecular weight distribution of the polymer formed. In this section, mathematical techniques are developed to account for the presence of such compounds.

The feed to the batch reactor is assumed to consist of monomer AR_1B and modifier AR_2X, where A and B are the reacting functional groups and X is an inert group which does not participate in the polymerization. The immediate effect of the presence of AR_2X is to make the molar concentrations of A and B functional groups nonequimolar in the reaction mass. The MWD of the polymer formed is now obtained.

On polymerization of AR_1B monomer with modifier AR_2X for some time, the reaction mass consists of two kinds of polymer molecules: those which have ends of A and B functional groups and those with A and X end groups. These are distinguished and denoted as

$$P_n: A(BA)_{\overline{n-1}}B \qquad (2.6.1a)$$

$$P_{nx}: A(\underline{\quad}BA)_{\overline{n-1}}X \qquad (2.6.1b)$$

For convenience, R_1 and R_2 have not been written above, and n in P_n and P_{nx} denotes the chain length of these molecules. This distinction between the molecules is necessary because these two classes of molecules react differently.

P_m can react with P_n as well as with P_{nx}, whereas P_{mx} cannot react with P_{nx}. The various elementary reactions taking place in the reaction mass can be written as

$$P_m + P_n \xrightarrow{2k_p} P_{m+n}; \qquad m \neq n; \qquad m, n = 1, 2, \ldots \quad (2.6.2a)$$

$$P_m + P_m \xrightarrow{k_p} P_{2m}; \qquad m = 1, 2, \ldots \quad (2.6.2b)$$

$$P_m + P_{nx} \xrightarrow{k_p} P_{(m+n)x}; \qquad m, n = 1, 2, \ldots \quad (2.6.2c)$$

The rate constant for the reaction (2.6.2c) has been written as k_p in view of the fact that the functional groups involved in the reaction between P_m and P_{nx} are exactly the same as those in the reaction between P_m and P_n. Additionally, the number of distinct ways s [Eq. (2.4.2a)] for the reaction between P_m and P_{nx} is 1. It is further observed that in Eq. (2.6.2c), $m = n$ does not lead to the condition of collision of identical molecules, and, therefore, there is no reaction equivalent to Eq. (2.6.2b) for monofunctional compounds.

With the kinetic scheme given in Eq. (2.6.2), mole balance equations for each homolog can be written for batch reactors

$$\frac{d[P_1]}{dt} = -2k_p[P_1] \sum_{i=1}^{\infty} [P_i] - k_p[P_1] \sum_{i=1}^{\infty} [P_{ix}] \qquad (2.6.3a)$$

$$\frac{d[P_{1x}]}{dt} = -k_p[P_{1x}] \sum_{i=1}^{\infty} [P_i] \qquad (2.6.3b)$$

$$\frac{d[P_n]}{dt} = k_p \sum_{r=1}^{n-1} [P_r][P_{n-r}] - k_p[P_n]\left\{2 \sum_{i=1}^{\infty} [P_i] + \sum_{i=1}^{\infty} [P_{ix}]\right\} \qquad (2.6.3c)$$

$$\frac{d[P_{nx}]}{dt} = k_p \sum_{r=1}^{n-1} [P_{rx}][P_{n-r}] - k_p[P_{nx}] \sum_{i=1}^{\infty} [P_i], \qquad n \geqslant 2 \qquad (2.6.3d)$$

The infinite set of mole balance equations written in Eqs. (2.6.3) must be solved simultaneously to obtain the molecular weight distribution. Several mathematical techniques can be used to do so, of which one is discussed herein. An analytical solution can be obtained using generating functions $G(s, t)$ and $G_x(s, t)$ defined as

$$G(s, t) \equiv \sum_{n=1}^{\infty} s^n[P_n] \qquad (2.6.4a)$$

$$G_x(s, t) \equiv \sum_{n=1}^{\infty} s^n[P_{nx}] \qquad (2.6.4b)$$

where s is a dummy variable less than one. Using these, the infinite set of equations in Eqs. (2.6.3) can be collapsed into a single equation which can, in turn, be easily solved for $G(s, t)$ and $G_x(s, t)$. As shown in Appendix 2.1, where some of the important properties of generating functions are discussed, it is possible to determine either the entire MWD or the various moments of the distributions from these. Before this is done, Eq. (2.6.3) is rewritten in terms of time τ defined by Eq. (2.4.12). Equations (2.6.3a) and

(2.6.3b) are multiplied by s and Eqs. (2.6.3c) and (2.6.3d) by s^n and appropriately summed to give

$$\frac{\partial G(s, \tau)}{\partial \tau} = -(2[P] + [P_x])G(s, \tau) + G^2(s, \tau) \qquad (2.6.5a)$$

$$\frac{\partial G_x(s, \tau)}{\partial \tau} = -[P]G_x(s, \tau) + G(s, \tau)G_x(s, \tau) \qquad (2.6.5b)$$

It may be noted that in the limit as $s \rightarrow 1$, $G(s, \tau)$ and $G_x(s, \tau)$ approach [P] and [P_x], respectively. Therefore Eq. (2.6.5) gives

$$\frac{d[P]}{d\tau} = \frac{\partial G(1, \tau)}{\partial \tau} = -[P]([P] + [P_x]) \qquad (2.6.6a)$$

$$\frac{d[P_x]}{d\tau} = \frac{\partial G_x(1, \tau)}{\partial \tau} = -[P][P_x] + [P][P_x] = 0 \qquad (2.6.6b)$$

Equation (2.6.6b) states that [P_x] is constant with τ. This is because in the reaction mechanism of Eq. (2.6.2), P_{nx} and P_{mx} do not react. As a result, the number of moles of monofunctional oligomers (P_{nx}) remains unchanged, even though its number average chain length increases with the time of polymerization.

To solve Eq. (2.6.5), one defined y and y_x as

$$y(s, \tau) = G(s, \tau)/[P] \qquad (2.6.7a)$$

$$y_x(s, \tau) = G_x(s, \tau)/[P_x] \qquad (2.6.7b)$$

Their time dependence can be derived to be

$$\frac{\partial y}{\partial \tau} = [P]y(y - 1) \qquad (2.6.8a)$$

$$\frac{\partial y_x}{\partial \tau} = [P]y_x(y - 1) \qquad (2.6.8b)$$

These can be easily solved to give

$$y = \frac{y_0(1 - p_A)}{1 - p_A y_0} \qquad (2.6.9a)$$

$$y_x = \frac{y_{x0}(1 - p_A)}{1 - p_A y_0} \qquad (2.6.9b)$$

where y_0 and y_{x0} are the values of y and y_x for the feed and p_A is the conversion of A-functional groups, which is given by

$$p_A = \frac{[P]_0 - [P]}{[P]_0 + [P_x]_0} \qquad (2.6.10)$$

The special case when the feed does not consist of any monofunctional compound, i.e., $[P_x]_0 = 0$, is now considered. Then, the concentrations of functional groups A and B would be equal. If the feed consists of monomer only, y_0 in Eq. (2.6.9), is nothing but s. This gives y at any time τ of polymerization to be

$$y = \frac{(1 - p_A)s}{1 - p_A s}$$

$$= (1 - p_A) \sum_{n=1}^{\infty} p_A^{n-1} s^n \qquad (2.6.11)$$

Evidently, the definition of y in Eq. (2.6.7a) gives the coefficient of s^n to be $[P_n]/[P]$. On further observing that $P = (1 - p_A)[P]_0$, Eq. (2.6.11) gives

$$\frac{[P_n]}{[P]_0} = (1 - p_A)^2 p_A^{n-1} \qquad (2.6.12)$$

This is the MWD of the polymer formed after time τ of polymerization with only monomer and no monofunctional compound in the feed. It is interesting to find that the MWD is *completely* characterized by the conversion, p_A, of functional group A in the reaction mass. This result holds whether the reactor is operated isothermally or nonisothermally, with the conversion related to the time of polymerization by Eq. (2.4.11). In terms of the reactor operation, one cannot control the conversion and the MWD of the polymer independently.

Historically, Flory[1,21,22] derived Eq. (2.6.12) using statistical arguments in which the conversion of the functional groups in the reaction mass was taken to be the same as the probability of finding a reacted functional group on a polymer chain. If a polymer molecule, P_n, of chain length n is examined, it is found that it has $(n - 1)$ reacted A groups and one unreacted A group. Thus, the probability of finding a P_n molecule in the reaction mass (which is the same as $[P_n]/[P]$) is given by (probability of finding an unreacted A group) $\times$ (probability of finding a reacted A group),$^{n-1}$ or $(1 - p_A)p_A^{n-1}$. If $[P]$ is eliminated using Eq. (2.6.26), the MWD is found to be the same as in Eq. (2.6.30). This is commonly called Flory's most probable distribution. Equation (2.6.30) gives the mole fraction distribution whereas the weight fraction distribution, w_n, can be derived to be

$$w_n = \frac{n[P_n]}{\sum_1^\infty n[P_n]} = \frac{n[P_n]}{[P]_0} = n(1 - p_A)^2 p_A^{n-1} \tag{2.6.13}$$

These have been plotted[21,22] in Figs. 2.7a and 2.7b for various values of p_A.

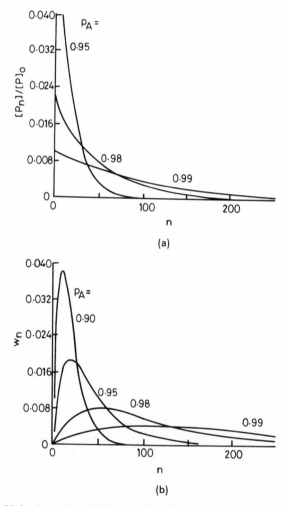

(a)

(b)

Figure 2.7. Molecular weight distribution for ARB polymerization with pure monomer feed at various conversions,[21] p_A. (a) Mole fraction distribution. (b) Weight fraction distribution. (Reprinted from Ref. 21 with permission of the American Chemical Society, Washington, DC.)

The mole fraction distribution becomes broader as p_A increases, and for a given p_A, it is found to decrease monotonically with n. As opposed to this, the weight fraction distribution undergoes a maximum and the curves broaden as p_A increases.

If the feed to the batch reactor consists of higher oligomers in addition to the monomer, the kinetic approach presented in this section can be easily extended. An example to illustrate the effect of the composition of the feed on the MWD is now presented. It is assumed to consist of monomer and dimer in the molar ratio of $1 : r_1$. For such a feed, y_0 in Eq. (2.6.9) is given by

$$y_0 = \frac{s(1 + r_1 s)}{(1 + r_1)} \tag{2.6.14}$$

which yields

$$y = (1 - p_A) \frac{\dfrac{s(1 + r_1 s)}{(1 + r_1)}}{1 - s p_A \dfrac{1 + r_1 s}{1 + r_1}}$$

$$= (1 - p_A) \sum_{m=0}^{\infty} \sum_{n'=0}^{m+1} p_A^m r_1^{n'} \binom{m+1}{n'} (1 + r_1)^{-(m+1)} s^{(m+n'+1)} \tag{2.6.15}$$

This expression can be expanded term by term for various values of m and n', and the coefficients of s^n are collected to give

$$\frac{[P_n]}{[P]_0(1 - p_A)^2} = \begin{cases} \displaystyle\sum_{p=0}^{n/2}{}^{n-q}C_q\, p_A^{n-q-1} \dfrac{r_1^q}{(1 + r_1)^{n-q}} & \text{for even } n \quad (2.6.16a) \\[4mm] \displaystyle\sum_{q=0}^{(n-1)/2}{}^{n-q}C_q\, p_A^{n-q-1} \dfrac{r_1^q}{(1 + r_1)^{n-q}} & \text{for odd } n \quad (2.6.16b) \end{cases}$$

This has been plotted in Fig. 2.8 to show the effect of feed composition on the MWD. On the same diagram, Flory's distribution has also been given and the latter is found to be sharper for the same p_A than predicted by Eq. (2.6.16). It is thus found that by controlling feed, one can exercise some degree of control on MWD of the polymer.

The various moments of the MWD can now be examined using the results of Appendix 2.1. The number and weight average chain lengths, μ_n and μ_w, of the polymer formed can be found by evaluating the combined

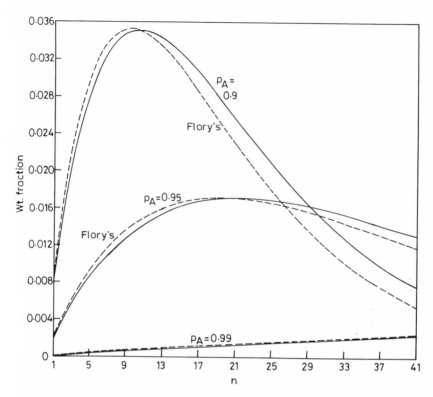

Figure 2.8. Weight fraction distribution for ARB polymerization with the feed containing 10% mole percent dimer in monomer.

kth moments, λ_k^c, for the generalized feed. This is nothing but the sum of λ_k and λ_{kx}, where these are the kth moments of the distributions of polymer, P_n, and monofunctional oligomers, P_{nx}.

The analysis of this section has shown the effect of the feed composition on the product distribution as the conversion is increased. Since the amount of monofunctional compound $[P_{nx}]_0$ in the feed determines the maximum conversion (and hence the average chain length of the polymer formed) it is necessary to control the amount of monofunctional compound in the feed carefully. Also, the MWD of the polymer formed can be given by Flory's distribution in Eq. (2.6.13) only when the feed is pure monomer.

If the feed to the batch reactor consists of monomer only (without monofunctional compounds), μ_n and μ_w can be derived as

$$\mu_n = \frac{1}{(1 - p_A)} \qquad (2.6.17a)$$

$$\mu_w = 1 + \frac{2p_A}{(1 - p_A)} \qquad (2.6.17b)$$

The polydispersity index, Q, is then given by

$$Q = (1 + p_A) \qquad (2.6.18)$$

which approaches a value of 2 for 100% conversion.

The effect of monofunctional compounds on the number average chain length has been derived by Flory through statistical arguments as follows. It assumes that the number of functional groups A and B at $t = 0$ are N_{A0} and N_{B0} and the purpose of the monofunctional compound is only to enhance the ratio $r(= N_{A0}/N_{B0})$ to a value larger than one. This can be approximated by the nonequimolar polymerization of (AA + BB) monomers, for which case the number average chain length can be derived as

$$\mu_n = \frac{\text{total number of molecules at } t = 0}{\text{total number of molecules at } t} = \frac{1 + r}{2r(1 - p_A) + (1 - r)} \qquad (2.6.19)$$

Even for $p_A = 1$, μ_n attains a limiting finite value of $(1 + r)/(1 - r)$ in contrast to infinity, as predicted by Eq. (2.6.17a) for polymerization without monofunctional compounds.

Irreversible step growth polymerization in batch reactors has been analyzed by several workers using various other mathematical techniques as well. These are now briefly described. In one mathematical technique a Z transform function $Z(z, t)$ is defined as[17,18,23]

$$Z(z, t) = \sum_{n=1}^{\infty} z^{-n} P_n \qquad (2.6.20)$$

where z is an arbitrary variable whose absolute value is larger than or equal to one. On comparison of this equation with Eq. (2.6.5), it is found that the moment generating method and the Z-transform technique are identical if one recognizes the mapping of $s = z^{-1}$. All theorems valid for moment generating function hold for Z-transform function. However, the latter is becoming more popular because more extensive tables for transforming various functions and obtaining the corresponding inverses are available.[19]

In the continuous variable technique,[24-26] an attempt is made to express the infinite set of ordinary differential equations in Eq. (2.6.3) by a single partial differential equation. This is done by defining a continuous variable, $P(n, t)$, of chain length n as well as time t, whose variation with respect to n and t are governed by the mechanism of polymerization. Zeman and

Amundson[24-26] have demonstrated this technique for the following simplified reaction mechanism for chain-growth polymerization:

$$P_n + P_1 \rightarrow P_{n+1} \qquad (2.6.21)$$

It may be recalled that step growth polymerization cannot be described by the successive addition of monomer as postulated above, and this method is virtually useless for this mechanism of polymerization.

Finally, the MWD for irreversible step growth polymerization can also be obtained analytically by direct integration. If the polymerization is carried out without monofunctional compound in the feed (i.e., $[P_x]_0 = 0$), the mole balance equations for various oligomers in a batch reactor can be easily written and the total moles of polymer, $[P](=\sum_{n=1}^{\infty}[P_n])$, can be solved to obtain Eq. (2.4.11). With this, mole balance equations for P_1, P_2, etc. can be sequentially integrated to obtain the same MWD as given in Eq. (2.6.12). It would be shown later that for reversible polymerizations, the MWD can be obtained only by solving the entire set of equations for P_1, P_2 ... simultaneously, and thus require numerical procedures.

The analysis presented in this section shows that the molecular weight distribution in batch reactors is completely characterized by the conversion, p_A, of the functional group in the reaction mass. Thus, the only degree of freedom that exists in reactor design is to choose the reactor residence time, τ. For a given τ, p_A can be uniquely obtained and the MWD is completely fixed by the conversion. To enhance the controllability of the reactor, it is important that the number of degrees of freedom be increased, and this can be done by having a recycle loop as discussed in the next section.

2.7. POLYMERIZATION IN PLUG FLOW REACTORS (PFR) WITH RECYCLE[22,23]

A plug flow reactor is shown in Fig. 2.9a and mole balance equations for the various oligomers on the differential volume in the reactor can be written as

$$\mathring{Q}\Delta[P_n] = (\mathscr{R}_{P_n})A\Delta x; \qquad n = 1, 2, \ldots \qquad (2.7.1)$$

where $\mathscr{R}_{P_n}$ is the rate of production of P_n under the reaction conditions existing in the differential element. In Eq. (2.7.1), $\mathring{Q}$ is the volumetric flow rate and is assumed to be constant. On substituting $t = (Ax/\mathring{Q})$ in Eq. (2.7.1), the mole balance equations reduce to the mole balance equation for P_n in batch reactors. Thus, the performance of batch reactors becomes identical to that for PFRs under this transformation of variables.

(a) Schematic representation of tubular reactor

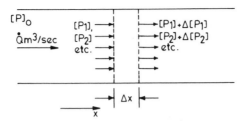

(b) Tubular reactor with recycle

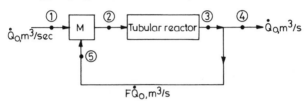

Figure 2.9. (a) Schematic representation of a tubular reactor; (b) tubular reactor with recycle.

Since the mole balance equations for a tubular reactor are the same as those for a batch reactor, there is only one degree of freedom (namely, residence time) when the feed to the reactor is pure monomer. In order to increase the number of degrees of freedom, higher oligomers could be added in the feed, and the one convenient way of doing that is to recycle a part of the product stream and mix it with the feed. In this section, step growth polymerization of $(AR_1B + AR_2X)$ in a tubular reactor with recycle has been analyzed.

In Fig. 2.9b, it is assumed that a fraction F of the product stream is recycled. The fresh feed, which need not necessarily be pure monomer, is assumed to flow at a rate of $\mathring{Q}$ m³/sec to the mixer M. In this figure, positions 1–5 where mole balances must be made have been marked. These positions differ in t as defined in Eq. (2.7.2). Generating functions $G(s, t_i)$ and $G_x(s, t_i)$ for species P_n and P_{nx} can be defined for different positions as in Section 2.6 as

$$G(s, t_i) = \sum_{n=1}^{\infty} s^n [P_n]_i, \qquad i = 1, 2, \ldots, 5 \qquad (2.7.2a)$$

$$G_x(s, t_i) = \sum_{n=1}^{\infty} s^n [P_{nx}]_i, \qquad i = 1, 2, \ldots, 5 \qquad (2.7.2b)$$

where index i refers to the various locations shown in Fig. 2.9b. Corresponding normalized generating functions $y_i(z, t_i)$ and $y_{xi}(z, t_i)$ defined as

$$y_i(s, t_i) = \frac{G(s, t_i)}{[P]_i}, \qquad i = 1, 2, \ldots, 5 \qquad (2.7.3a)$$

$$y_{xi}(s, t_i) = \frac{G_x(s, t_i)}{[P_x]_i}, \qquad i = 1, 2, \ldots, 5 \qquad (2.7.3b)$$

where

$$[P]_i = \sum_{n=1}^{\infty} [P_n]_i, \qquad i = 1, 2, \ldots, 5 \qquad (2.7.4a)$$

$$[P_x]_i = \sum_{n=1}^{\infty} [P_{nx}]i, \qquad i = 1, 2, \ldots, 5 \qquad (2.7.4b)$$

The overall conversion, p_B of the functional group B in the reactor can be written in terms of its concentrations at points 4 and 1 as

$$p_B \equiv (1 - f) = 1 - \frac{[P]_4}{[P]_1} \qquad (2.7.5)$$

As earlier, the feed (i.e., point 1) is assumed to consist of the monomer AR_1B and the monofunctional compound AR_2X in the molar ratio of $1 : r$:

$$r = \frac{[P_x]_1}{[P]_1} \qquad (2.7.6)$$

In terms of these quantities, the molar flow rates of functional group B and monofunctional compounds (MF) at various locations can be easily written. These have been summarized in Table 2.2 and the reader can confirm these easily.

After carrying out the overall mole balances as above, mole balance equations for various oligomers can be written around the mixer. From these one can easily derive

$$y_1(s, t_1) + Ffy_5(s, t_5) = (1 + F)\frac{1 + fF}{(1 + F)} y_2(s, t_2) \qquad (2.7.7a)$$

$$y_{x1}(s, t_1) + Fy_{x5}(s, t_5) = (1 + F)y_{x2}(s, t_2) \qquad (2.7.7b)$$

Since the batch and plug flow reactors are governed by the same differential equations, the performance of the latter would be given by

TABLE 2.2. Total Moles and Concentrations of Mono- and Bifunctional Oligomers at Various Locations in Figure 2.7b

Location	Flow rate (m^3/sec)	Moles of B $\overline{\text{sec}}$	Moles of MF $\overline{\text{sec}}$	$G(s = 1, t_i)$ $=[P]_i$ $(\text{moles}/\text{m}^3)$	$G_x(s = 1, t_i)$ $=[P_x]_i$ $(\text{moles}/\text{m}^3)$
1	$\mathring{Q}_0$	$\mathring{Q}_0[P]_1$	$\mathring{Q}_0[P]_1 r$	$[P]_1$	r
2	$\mathring{Q}_0(1 + F)$	$[P]_1 \mathring{Q}_0\{1 + (1 - p_B)F\}$	$\mathring{Q}_0(1 + F)[P]_1 r$	$[P]_1 \dfrac{1 + F(1 - p_B)}{(1 + F)}$	r
3	$\mathring{Q}_0(1 + F)$	$\mathring{Q}_0[P]_1(1 + F)$ $(1 - p_B)$	$\mathring{Q}_0[P]_1(1 + F)r$	$(1 - p_B)[P]_1$	r
4	$\mathring{Q}_0$	$\mathring{Q}_0[P]_1(1 - p_B)$	$\mathring{Q}_0[P]_1 r$	$(1 - p_B)[P]_1$	r
5	$\mathring{Q}_0 F$	$\mathring{Q}_0 F(1 - p_B)[P]_1$	$\mathring{Q}_0 F[P]_1 r$	$(1 - p_B)[P]_1$	r

Eq. (2.6.9) as

$$y_3(s, t_3) = (1 - p_s)\frac{y_2(s, t_2)}{1 - p_s y_2(s, t_2)} \qquad (2.7.8a)$$

and

$$y_{x3}(s, t_3) = (1 - p_s)\frac{y_{x2}(s, t_2)}{1 - p_s y_2(s, t_2)} \qquad (2.7.8b)$$

where p_s is the conversion of A groups across the reactor *alone* (i.e., across points 2 and 3). p_s can be calculated with the help of Table 2.2 as

$$p_s = \frac{1 - f}{1 + fF + r(1 + F)} \qquad (2.7.9)$$

$y_2(s, t_2)$ and $y_{x2}(s, t_2)$ can now be eliminated between Eqs. (2.7.7) and (2.7.8) to find

$$(1 + fF)y_3(s, t_3) - p_s fF y_3^2(s, t_3) - p_s y_1(s, t_1)y_3(s, t_3)$$

$$= \{y_1(s, t_1) + fF y_3(s, t_3)\}(1 - p_s) \qquad (2.7.10a)$$

$$(1 + F)y_{x3}(s, t_3)\{(1 + fF) - p_s[y_1(s, t_1) + fF y_3(s, t)]\}$$

$$= (1 + fF)(1 - p_s)\{y_{x1}(s, t_1) + F y_{x3}(s, t_3)\} \qquad (2.7.10b)$$

These equations can now be used to find the various moments of the MWD. Equation (2.7.10a) is differentiated with respect to $(\ln s)$ and in the

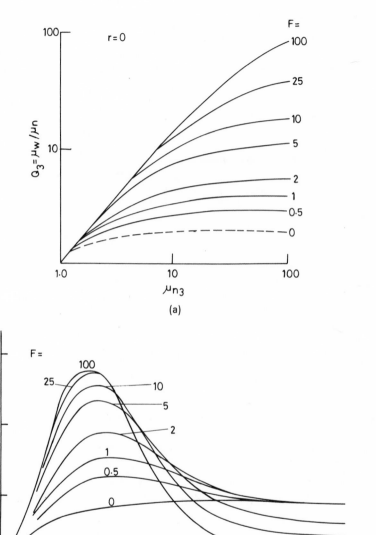

Figure 2.10. Polydispersity index, Q_3, of the product for various recycle ratios, F.[22] (a) Q_3 versus μ_{n_3} with no monofunctional compound in feed. (b) Q_3 of the polymer fraction (i.e., A⌇B type molecules). (c) Q_3 of the overall polymer. For (b) and (c), 5% monofunctional compounds used in the feed. (Reprinted from Ref. 17 with permission of the American Chemical Society, Washington, DC.)

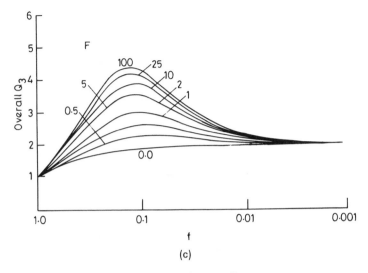

Figure 2.10 (*continued*)

limit as $s \to 1$ gives

$$(1 + fF)\mu_{n,3} - 2p_s fF\mu_{n,3} - p_s\mu_{n,1} - p_s\mu_{n,3} = (1 - p_s)(\mu_{n,1} + fF\mu_{n,3}) \quad (2.7.11)$$

where $\mu_{n,i}$ $(i = 1, 2, \ldots, 5)$ are the number average chain lengths of the P_n distribution at point i. This equation gives the first moment at position 3 in terms of the feed conditions. Redifferentiating Eq. (2.7.10), and taking the limit as $s \to 1$, gives an equation for the second moment. Similar equations may be derived for the moments of the P_{nx} distribution. Thus, if the feed composition is specified, $\mu_{n,3}$, $\mu_{w,3}$, and Q_3 at location 3 can be calculated. In these computations, the contribution due to the monofunctional oligomers could be included or excluded and results have been plotted in Fig. 2.10.

In Fig. 2.10a, the polydispersity index, Q_3, of the polymer *without* monofunctional oligomers (i.e., $[P_x]_0 = 0$) at point 3 has been plotted as a function of the number average chain length. In this diagram, the recycle ratio F has been treated as a parameter and $F = 0$ corresponds to Flory's distribution for which Q_3 attains the asymptotic value of 2. For recycle ratios greater than zero, Q_3 first quickly rises, ultimately attaining an asymptotic value which is always greater than 2. As F is increased, higher and higher asymptotic values of Q_3 are reached. It is observed that one can now independently choose the conversion and the Q of the final polymer (by choosing the residence time and recycle ratio), something which was not possible without the use of recycle.

In generating Fig. 2.10b the feed has been assumed to consist of 5% monofunctional compound, and the effect of conversion, f, on Q_3 has been examined with the recycle ratio F as the parameter. In Fig. 2.10b, Q_3 of P_n and in Fig. 2.10c the overall Q_3 have been plotted as a function of f. For a given recycle ratio F, on decreasing f, the Q_3 of the P_n distribution is found to increase considerably beyond the value of 2 initially, and then falls after undergoing a maximum. The overall polydispersity index, Q_3, also undergoes a maximum similarly, but all the curves for different recycle ratios F merge into a single one for small values of f as seen in Fig. 2.10c.

In this section, a simple kinetic scheme of polymerization has been taken for which it was possible to derive an analytical expression for Q. In most of the polymerization systems like polyesters or nylon-6, polymer formation involves several side reactions and $\mu_{n,3}$ and Q_3 can be found only numerically. Results of this section show that the recycle ratio and the conversion in the plug flow reactor (PFR) have considerable effect on the polymer formed, and both these can now be manipulated to obtain the desired physical properties.

2.8. ARB POLYMERIZATION IN HOMOGENEOUS CONTINUOUS-FLOW STIRRED TANK REACTORS (HCSTRs)[17,27–30]

It has already been pointed out that for high rates of production, continuous reactors are preferred over batch reactors. Among these, plug flow reactors have the same mole balance equations as those for batch reactors and have already been discussed. In this section, attention is focused on the analysis of HCSTRs.

In homogeneous continuous-flow stirred tank reactors, the feed, usually consisting of monomer, is continuously fed in and the product consisting of various oligomers is withdrawn at the same flow rate. The reactor is assumed to be well mixed (micromixed) and the composition of the product stream is assumed to be identical to that within the vessel.

In the analysis of the HCSTR presented in this section, the reactor is assumed to be unsteady and isothermal. Steady state equations can easily be deduced from these simply by equating the terms accounting for time variation to be zero. The mole balance equations of various oligomers in the reactor[31,32] carrying out *irreversible* step growth polymerization can be easily derived as[29]

$$V\frac{d[P_1]}{dt} = \overset{\circ}{Q}[P_1]_0 - \overset{\circ}{Q}[P_1] - V(2k_p[P_1][P]) \qquad (2.8.1a)$$

$$V\frac{d[P_n]}{dt} = -\mathring{Q}[P_n] + V\left\{ k_p \sum_{r=1}^{n-1} [P_r][P_{n-r}] - 2k_p[P_n][P] \right\}, \qquad n \geq 2$$

$$(2.8.1b)$$

where V is the volume of the reactor and $[P]$ represents the total moles of polymer molecules as defined in Eq. (2.4.6). These equations can be written in terms of dimensionless variables

$$\frac{dC_1}{d\tau} = 1 - C_1 - C_1 C\tau^* \qquad (2.8.2a)$$

$$\frac{dC_n}{d\tau} = -C_n + \tfrac{1}{2}\tau^* \sum_{r=1}^{n-1} C_r C_{n-r} - C_n C_n C\tau^*, \qquad n \geq 2 \quad (2.8.2b)$$

where

$$C_n = [P_n]/[P_1]_0, \qquad n = 1, 2, \ldots \qquad (2.8.3a)$$

$$\tau = t/(V/\mathring{Q}) \equiv t/\bar{\theta} \qquad (2.8.3b)$$

$$\tau^* = 2k_p[P_1]_0(V/Q) = 2k_p[P_1]_0\bar{\theta} \qquad (2.8.3c)$$

$$C = \sum_{i=1}^{\infty} C_i \qquad (2.8.3d)$$

In the study of HCSTRs under unsteady state operation, there are two common initial conditions. In the first case, the reactor is assumed to be filled with an inert solvent before the monomer feed is started. This initial conditions, IC1, would then be

IC1:   At $t = 0$

$$C_1 = C_2 = C_3 = \cdots = 0 \qquad (2.8.4)$$

In the second case, the reactor vessel is assumed to be filled with monomer at $t = 0$, in which case the initial conditions IC2 would be

IC2:   At $t = 0$

$$C_1 = 1 \qquad (2.8.5a)$$

$$C_2 = C_3 = \cdots = 0 \qquad (2.8.5b)$$

To find the MWD as a function of dimensional time τ, C in Eq. (2.8.3d) is evaluated first. To be able to do so, Eq. (2.8.2) is summed for all values of n

$$\frac{dC}{d\tau} = 1 - C - \tfrac{1}{2}\tau^* C^2 \tag{2.8.6}$$

The two initial conditions IC1 and IC2 are

$$\text{IC1:} \quad C = 0 \quad \text{at } \tau = 0 \tag{2.8.7a}$$

$$\text{IC2:} \quad C = 1 \quad \text{at } \tau = 0 \tag{2.8.7b}$$

Equation (2.8.8), governing C, can now be easily solved as

$$C(\tau) = \frac{\theta' - 1}{\tau^*} \frac{1 - e^{-\theta'\tau}}{1 - \delta e^{-\theta'\tau}} \quad \text{for IC1} \tag{2.8.8a}$$

$$C(\tau) = \frac{\theta' - 1}{\tau^*} \frac{1 - (\beta/\delta) e^{-\theta'\tau}}{1 - \beta e^{-\theta'\tau}} \quad \text{for IC2} \tag{2.8.8b}$$

where

$$\theta' = (1 + 2\tau^*)^{1/2} \tag{2.8.9a}$$

$$\delta = \frac{1 - \theta'}{(1 + \theta')} \tag{2.8.9b}$$

$$\beta = \frac{1 - \theta' + \tau^*}{1 + \theta' + \tau^*} \tag{2.8.9c}$$

$C(\tau)$ can now be substituted in Eq. (2.8.4a) to solve for C_1, C_2 etc. This can in principle be done, but beyond C_2, the integration becomes cumbersome.

Numerical integration of Eq. (2.8.2) for various values of n can be easily carried out.[29] Since for IC1, concentrations of all species at $t = 0$ are zero, the curves for C_1, C_2 etc. all start from the origin. On the other hand, for IC2, the curve for C_1 would start from 1. The mass fraction of w_1 ($=[P_1]/\sum_{n=1}^{\infty} n[P_n]$) has been plotted in Fig. 2.11 with residence time τ^* as a parameter. As expected, for a given τ^*, w_1 falls (or rises) quickly, but after a short transient period of a few residence times, it reaches the asymptotic, steady state value.

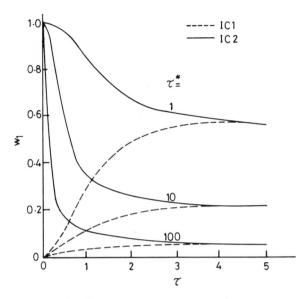

Figure 2.11. Mass fraction of monomer versus τ for various τ^* for IC1 and IC2 initial conditions in the product stream of HCSTRs.[29]

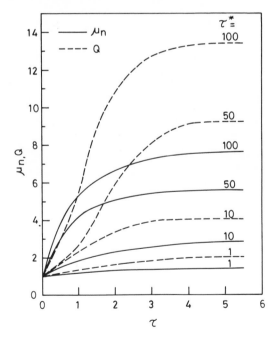

Figure 2.12. μ_n and Q of the polymer as a function of τ for various τ^* and IC2 initial condition in HCSTRs.[29]

From the molecular weight distribution so calculated, the number average chain length μ_n and polydispersity index Q can be calculated, and for IC2, these have been plotted[29] in Fig. 2.12. As seen from this figure, the transient period is small and the curves attain the steady state asymptotic value which is dependent only upon the dimensionless residence time, τ^*. Calculations show that the steady state results are independent of initial conditions IC1 and IC2, which is intuitively expected to be so.

The results for steady state HCSTRs[17] can be easily derived by setting $dC_n/d\tau = 0$ in Eq. (2.8.2). One obtains

$$C_s = \frac{\theta' - 1}{\tau^*} \tag{2.8.10a}$$

$$C_{1s} = \frac{1}{\theta'} \tag{2.8.10b}$$

$$C_{ns} = \frac{\tau^*}{2(1 + \tau^* C_s)} \sum_{r=1}^{n-1} C_{rs} C_{(n-r)s}, \qquad n = 2, 3, 4, \ldots \tag{2.8.10c}$$

Through successive elimination from Eq. (2.8.10), one can solve for C_{ns} as

$$C_{ns} = g_n \frac{(\tau^*/2)^{n-1}}{(\theta')^{2n-1}}, \qquad n = 1, 2, \ldots \tag{2.8.11}$$

where

$$g_1 = 1 \tag{2.8.12a}$$

$$g_n = \sum_{r=1}^{n-1} g_r g_{n-r}, \qquad n = 2, 3, 4, \ldots \tag{2.8.12b}$$

Thus, the values of g_n can be successively calculated and are given in Table 2.3.

TABLE 2.3.[a] Values of g_m in Eq. (2.8.11)

m	g_m	m	g_m	m	g_m
1	1	6	42	11	16,796
2	1	7	132	20	$17,672,628 \times 10^2$
3	2	8	429	30	$10,022,411 \times 10^8$
4	5	9	1430		
5	14	10	4862		

[a] Reference 20.

The number and weight average chain lengths, μ_{ns} and μ_{ws}, and the polydispersity index, Q_s, at steady state can now be derived and are given by

$$\mu_{ns} = \frac{\tau^*}{\theta' - 1} \tag{2.8.13a}$$

and

$$\mu_{ws} = 1 + \tau^* \tag{2.8.13b}$$

$$Q_s = \frac{(\tau^* + 1)(\theta' - 1)}{\tau^*} \tag{2.8.13c}$$

The polydispersity index Q_s in HCSTRs can take any numerical value depending upon τ^* and is not limited to the value of 2 as found for batch reactors. The MWDs of the polymer obtained from batch reactors and HCSTRs have been compared in Fig. 2.13 and are found to be substantially different in nature. In the latter, there is a considerably higher amount of unreacted monomer and the weight fraction distribution does not show a maximum. Practically speaking, this excess of unreacted monomer means that an HCSTR cannot be used as a "finishing" reactor. The irreversible polymerization of ARB monomer in an HCSTR has also been solved by Abraham[30] using Z transforms and it leads to similar results.

One of the major assumptions made in the analysis of HCSTRs is that the fluid elements in the reactor are completely *micromixed* and so have the same concentrations. Tracer experiments[31,32] have shown that in real stirred tank reactors, various fluid elements stay in it for different times, i.e., there is distribution of residence times. In an HCSTR, the distribution is exponential in nature, but since the fluid elements are mixed on a *microscopic* level, the output concentration and the concentration of every fluid element within the reactor are identical. The concept of micromixing is an idealization which is rarely achieved in practice. Another model for stirred tank reactors is the *segregated* CSTR (SCSTR), in which the fluid elements are completely segregated on a microscopic level and thus behave like individual batch reactors. An SCSTR is usually modeled as a set of an infinite number of small batch reactors in parallel, with their residence times distributed as in HCSTRs, as

$$f(t) = \frac{1}{\bar{\theta}} \exp(-t/\bar{\theta}) \tag{2.8.14}$$

where $f(t)$ is the fraction of the batch reactors with residence time lying

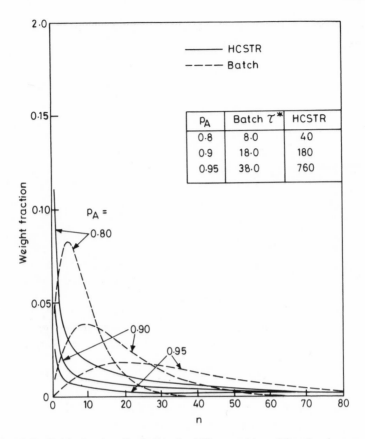

Figure 2.13. Weight fraction distribution for 90% conversion of functional groups in batch reactors and HCSTRs.

between t and $(t + dt)$. The average residence time $\bar{\theta}$ is defined in Eq. (2.8.3b). The average value of concentration of the i-mer, $\bar{C}_i$, in SCSTRs is given by

$$\bar{C}_i = \frac{1}{\bar{\theta}} \int_0^\infty C_{ib}(t) \, e^{-t/\bar{\theta}} \, dt \qquad (2.8.15)$$

where $C_{ib}(t)$ is the concentration of the i-mer from a *batch* reactor of residence time t. The kth moment for SCSTRs could be calculated as

$$\bar{\lambda}_k = \sum_{n=1}^\infty n^k \bar{C}_n = \frac{1}{\bar{\theta}} \int_0^\infty \lambda_{kb}(t) \, e^{-(t/\theta)} \, dt \qquad (2.8.16)$$

Thus $\bar{\mu}_n$, $\bar{\mu}_w$, and $\bar{Q}$ for SCSTRs can be calculated for ARB polymerization

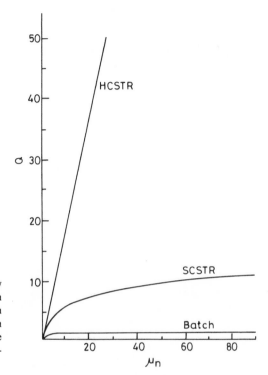

Figure 2.14. Mean polydispersity index versus number average chain length of the polymer formed in various reactors.[33] (Reprinted from Ref. 33 with permission of the American Chemical Society, Washington, DC.)

as done by Tadmore and Biesenberger.[33] In Fig. 2.14, the polydispersity indexes for batch reactors, HCSTRs, and SCSTRs have been plotted as a function of $\bar{\mu}_n$. For batch reactors, Q is limited to a value of 2, whereas for HCSTRs, it continues to increase to a very large value. The values of $\bar{Q}$ for SCSTRs lie between the two limits and approach an asymptotic value larger than 2 for large $\bar{\mu}_n$. As stated, real stirred tank reactors have a degree of micromixing lying between those of an HCSTR and an SCSTR and may also have a different residence time distribution. Several models have been suggested to account for these but have not been applied to step growth polymerization. It may be pointed out that in *all* these cases, once $\bar{\mu}_n$ is fixed, $\bar{Q}$ cannot be independently controlled.

2.9. REVERSIBLE ARB POLYMERIZATION

The analysis presented in Sections 2.6–2.8 has been limited to irreversible polymerizations mainly because it was then possible to derive an analytical solution. Most polymerizations, however, are reversible and should in fact be written as

$$P_m + P_n \xrightleftharpoons[k_p']{k_p} P_{m+n} + W, \qquad m, n = 1, 2, \ldots \qquad (2.9.1)$$

where W is the condensation product. The reverse reaction is the reaction between a reacted —AB— bond and W. It may be noted that P_1 is consumed when it reacts with any molecule in the reaction mass and is formed when a P_n ($n \geq 2$) reacts with W. P_3, P_4, etc. have two —AB— bonds (at their two ends) which, on reaction with W, give P_1. P_2 however has only one bond, but it splits to give two molecules of P_1. Thus the mole balance equation for P_1 for a batch reactor is given by

$$\frac{d[P_1]}{dt} = -2k_p[P_1] \sum_{i=1}^{\infty} [P_i] + 2k_p'[W] \sum_{n=2}^{\infty} [P_n]$$

$$= -2k_p[P_1][P] + 2k_p'[W]([P] - [P_1]) \qquad (2.9.2)$$

where k_p and k_p' are the rate constants for the reaction involving *functional groups* as shown below

$$\sim\!\!\sim\!A + B\!\sim\!\!\sim \xrightleftharpoons[k_p']{k_p} \sim\!\!\sim\!AB\!\sim\!\!\sim + W \qquad (2.9.3)$$

A molecule of P_n ($n \geq 2$) of chain length n is formed when (a) an oligomer P_r ($r < n$) reacts with P_{n-r} through the forward reaction, and (b) an oligomer $P_m\{m \geq (n + 1)\}$ reacts with W through the reverse reaction. On the other hand, P_n is consumed when (c) it reacts with any molecule through the forward reaction and (d) any of the reacted —AB— bonds of P_n interacts with W through the reverse reaction. The mole balance equation for P_n in batch reactors is therefore given by

$$\frac{d[P_n]}{dt} = -2k_p[P_n] \sum_{m=1}^{\infty} [P_m] + k_p \sum_{r=1}^{n-1} [P_r][P_{n-r}]$$

$$+ 2k_p'[W] \sum_{m=n+1}^{\infty} [P_m] - k_p'(n - 1)[P_n][W], \qquad n \geq 2 \qquad (2.9.4)$$

Appropriate weighting factors have been incorporated in Eqs. (2.9.2) and (2.9.4) to account for various reactions as well as for avoiding the duplication of molecular collisions as was done for irreversible polymerization. These equations can be summed to find the time variation of [P] as

$$\frac{d\lambda_0}{dt} \equiv \frac{d[P]}{dt} = -k_p[P]^2 + k_p'[W] \sum_{n=2}^{\infty} (n - 1)[P_n] \qquad (2.9.5)$$

The term $\sum_{n=2}^{\infty} (n - 1)[P_n]$ in Eq. (2.9.5) represents the total concentration of the reacted $-AB-$ bonds. Thus, one has

$$\frac{d[P]}{dt} = \frac{d[A]}{dt} = \frac{d[B]}{dt} = -k_p[A][B] + k_p'[W][-AB-] \qquad (2.9.6)$$

once again justifying the representation of the reaction between molecular species $P_1, P_2, \ldots,$ [Eq. (2.9.1)] by the reaction of functional groups alone as shown in Eq. (2.9.3), in terms of the functional group reactivities.

For reversible polymerization, Flory[34] and Tobolsky[35] have shown through statistical arguments that the molecular weight distribution given by Eq. (2.6.12) would still hold. Subsequent numerical computations[36,37] have shown that Flory's distribution holds as long as the equal reactivity hypothesis holds and k_p and k_p' are constant. If pure monomer is the feed to the batch reactor (initial concentration $[P_1]_0$) and p_A is the conversion of functional groups, then

$$\frac{1}{[P_1]_0} \frac{dp_A}{dt} = k_p(1 - p_A)^2 - k_p' p_A^2 \qquad (2.9.7)$$

where the condensation product liberated is assumed to remain within the batch reactor. This can be easily integrated to give p_A as[32]

$$p_A = \frac{K}{K - 1} - \frac{K^{1/2}}{K - 1} \frac{(K - 1) + (K^{1/2} - 1)\, e^{-2\tau/\sqrt{K}}}{(K - 1) - (K^{1/2} - 1)\, e^{-2\tau/\sqrt{K}}} \qquad (2.9.8)$$

where

$$K = k_p / k_p' \qquad (2.9.9a)$$

$$\tau = k_p[P_1]_0 t \qquad (2.9.9b)$$

Computer calculations of the MWD for ARB polymerization have been performed by several workers and are found to require considerable computation time even on fast computers. In view of this, sometimes, Eqs. (2.9.2) and (2.9.4) are reduced to moment equations as follows. It is recognized, that the zeroth moment, λ_0, is the same as [P] and is, therefore, given by Eq. (2.9.5). The first moment, λ_1, gives the total number of repeat units at any time in the reaction mass and is time invariant. If the feed to the reactor is pure monomer, then $\lambda_1 = [P_1]_0$ independent of time. Thus, the zeroth and first moments are quite easily determined.

The second moment, λ_2, can be found by appropriate summation of Eqs. (2.9.2) and (2.9.4):

$$\frac{d\lambda_2}{dt} = -2k_p\lambda_0 \sum_{n=1}^{\infty} n^2[P_n] + k_p \sum_{n=2}^{\infty} n^2 \sum_{r=1}^{n-1} [P_r][P_{n-r}]$$

$$+ 2k_p'[W] \sum_{n=1}^{\infty} n^2 \sum_{m=n+1}^{\infty} [P_m] - k_p'[W] \sum_{n=2}^{\infty} n^2(n-1)[P_n]$$

$$= 2k_p[P_1]_0^2 + \frac{k_p'}{3}[W]\{[P_1]_0 - \lambda_3\} \tag{2.9.10}$$

This equation involves λ_3 which implies that λ_2 cannot be determined unless a suitable equation for λ_3 is established. If a similar relation for λ_3 is written, it would be found to involve λ_4, and so on, thus forming a hierarchy of equations to be solved.

To be able to solve for the moments, the usual procedure is to assume a suitable form of MWD (it would be preferred to use the exact MWD form) and relate λ_3 to the lower moments λ_2, λ_1, and λ_0. There have been many efforts in this direction and in the literature, this procedure is generally known as moment-closure approximation. Hulbert and Katz[38,39] have used Laguerre polynomials and after relating λ_3 to λ_2, the moments λ_2 and λ_0 are determined by solving the differential equations numerically. Subsequent works have assumed the MWD given by the Schultz–Zimm equation to relate λ_3 to λ_2, λ_1, and λ_0 as[40,41]

$$\lambda_3 \cong \frac{\lambda_2(2\lambda_2\lambda_0 - \lambda_1^2)}{\lambda_1\lambda_0} \tag{2.9.11}$$

and then integrating Eq. (2.9.12). The computations show that the results are fairly insensitive to the nature of approximation made for λ_3, and almost identical results are obtained[42] irrespective of the relationship used. Equation (2.9.11) is now preferred on account of its simplicity. In these, the concentration of water in the reaction mass is assumed to be given by the stoichiometric relation.

Polymerization of ARB monomers is usually accompanied by several side reactions, the most common being the redistribution reaction

$$P_m + P_n \underset{k_f'}{\overset{k_f'}{\rightleftharpoons}} P_{m+n-q} + P_q \tag{2.9.12}$$

in which there is no generation of the condensation product (Flory has called this the ester interchange reaction). This reaction is known to play an important role in determining the MWD of the polymer formed. This reaction has an equilibrium constant of 1 because the forward and reverse reactions are identical in nature. Mellichamp[43] and Szabo and Leathrun[44] have studied the polymerization in presence of the redistribution reaction. They have shown that the zeroth and first moments of the distribution are not affected, but the second moment is greatly affected by reaction (2.9.12).

There have been very few attempts in the literature to solve the performance of HCSTRs carrying out reversible step growth poly-merizations. The steady state mole balance equations for HCSTRs can be written as[44,45]

$$[P_1]_0 = [P_1] - k_p \bar{\theta} \left\{ -2[P_1] \sum_{m=1}^{\infty} [P_m] + \frac{[W]}{K} \sum_{m=2}^{\infty} [P_m] \right\} \qquad (2.9.13a)$$

$$0 = [P_n] - k_p \bar{\theta} \left\{ -2[P_n] \sum_{m=1}^{\infty} [P_m] + \sum_{r=1}^{n-1} [P_r][P_{n-r}] \right.$$

$$\left. + \frac{2[W]}{K} \sum_{m=n+1}^{\infty} [P_m] - (n-1)[P_n]\frac{[W]}{K} \right\}, \qquad n \geqslant 2 \qquad (2.9.13b)$$

The mole balance equation for water can be similarly derived as

$$[W]_0 = [W] - k_p \bar{\theta} \left\{ \left(\sum_{n=1}^{\infty} [P_n] \right)^2 - \frac{[W]}{K} \sum_{n=2}^{\infty} (n-1)[P_n] \right\} \qquad (2.9.14)$$

where $[P_1]_0$ and $[W]_0$ are the monomer and water concentrations in the feed and it is assumed that the condensation product is not removed from the reaction mass. Equations (2.9.13) and (2.9.14) for various values of n represent an infinite set of coupled, nonlinear algebraic equations which must be solved simultaneously. The usual method employed to solve a large number of such equations is the Gauss–Jordon technique, for which one must provide (a) an initial guess, which is close to the actual solution, and (b) a Jacobian matrix, J, whose elements are the partial derivatives of the function with respect to the various independent variables. To provide a good initial guess is always tricky, whereas to obtain the Jacobian is cumbersome.

In Brown's technique,[45-48] both these difficulties have been overcome. The computer program has a built-in subroutine which generates the necessary partial derivatives. In addition, Brown's method is more efficient than the Gauss–Jordon technique because in the former equations are solved sequentially, in contrast to the latter, where they are solved simultaneously.

For small residence times, the solution for batch reactors can be assumed to be close to those for HCSTRs and can serve as the initial guess. The residence time of the HCSTR is then increased by a small amount $\Delta\bar{\theta}$ with the MWD previously calculated serving as the initial guess for this residence time. In this way, the residence time of the HCSTR can be increased to any desired value and the MWD obtained using Brown's technique. The truncation error can be minimized by increasing the total number of equations to be solved. Results on the computed MWDs are presented in later chapters.

2.10. CONCLUSIONS

In this chapter, step growth polymerizations of ARB monomer has been analyzed. In the equal reactivity hypothesis, it is assumed that all the oligomers in the reaction mass react with the same rate constant independent of their chain length. With this kinetic model, mole balance equations for various oligomers in batch reactors were established, and their solution showed that the molecular weight distribution was completely characterized by a single parameter, the conversion of functional groups. The polydispersity index, Q, of the polymer formed was limited to a maximum value of two for pure monomer feed.

Mole balance equations governing the MWD in batch and plug flow reactors (PFRs) were shown to be identical under the transformation of the length coordinate. However, it is possible to recycle a fraction of the product stream and the analysis of PFRs with recycle shows that the controllability of the reactor improves considerably.

Among the continuous reactors, HCSTRs are preferred over tubular reactors because of the well-mixed conditions existing in the former. Since polymerization is usually exothermic in nature, a better temperature control can be obtained in HCSTRs. However, since the monomer concentration in the product stream is higher, these reactors can at best be used as the beginning reactors in a sequence. The most striking difference between the batch reactor and HCSTRs is that, in the latter, Q is not limited to the maximum value of 2.

Lastly, reversible polymerization of ARB monomers has been analyzed in various types of reactors. Analytical solutions are generally not possible, and recourse to numerical computations must be made. Calculations of MWD in batch reactors is relatively straightforward and can be done using the Runge–Kutta method of the fourth order. To find the MWD for HCSTRs is fairly involved because of a large number of nonlinear algebraic equations which have to be solved simultaneously. For this purpose, Brown's numerical technique is found to be suitable.

APPENDIX 2.1. MOMENT GENERATION FUNCTION

The moment generating function and the Z transform are nothing but the Taylor series component and residue, respectively, of the Laurent series expansion of any complex function. In the context of combining various MWD equations, the moment generating function is defined as

$$G(s, \tau) = \sum_{n=1}^{\infty} s^n [P_n] \qquad (A2.1.1)$$

where s is an arbitrary transform variable such that $|s| < 1$. This technique is found useful in combining a large number of equations into a single one involving G. The various moments of the MWD can easily be determined from $G(s, \tau)$ as shown below.

The zeroth moment, λ_0, is given by taking the limit of $G(s, \tau)$ as $s \to 1$, i.e.,

$$\lambda_0 = \lim_{s \to 1} G(s, \tau) = \sum_{n=1}^{\infty} [P_n] \qquad (A2.1.2)$$

In order to obtain the first moment, λ_1, $G(s, \tau)$ is differentiated with respect to s:

$$\frac{\partial G(s, \tau)}{\partial s} = \sum_{n=1}^{\infty} n[P_n] s^{n-1} \qquad (A2.1.3)$$

On rearranging this, the following equation is obtained:

$$\frac{\partial G(s, \tau)}{\partial (\ln s)} = s \frac{\partial G(s, \tau)}{\partial s} = \sum_{n=1}^{\infty} n[P_n] s^n \qquad (A2.1.4)$$

λ_1 is, therefore, given by

$$\lambda_1 = \lim_{s \to 1} \frac{\partial G(s, \tau)}{\partial (\ln s)} = \sum_{n=1}^{\infty} n[P_n] \qquad (A2.1.5)$$

The above result can easily be generalized to give the following equation for the kth moment:

$$\lambda_k = \lim_{s \to 1} \frac{\partial^k G(s, \tau)}{\partial (\ln s)^k} \qquad (A2.1.6)$$

The parameter y defined in Eq. (2.6.7) as

$$y(s, \tau) = \frac{G(s, \tau)}{G(1, \tau)} = \frac{G(s, \tau)}{[P]} \tag{A2.1.7}$$

has the property that

$$y(1, \tau) = 1 \tag{A2.1.8}$$

A similar differentiation of $y(s, \tau)$ as in Eq. (A2.1.6) leads to normalized moments, λ_k, given by

$$\mu_k \equiv \frac{\lambda_k}{\lambda_0} = \lim_{s \to 1} \frac{\partial^k y(s, \tau)}{\partial (\ln s)^k} \tag{A2.1.9}$$

In terms of normalized moments, μ_n and μ_w can be written as

$$\mu_n = \mu_1 \tag{A2.1.10a}$$

$$\mu_w = \mu_2 / \mu_1 \tag{A2.1.10b}$$

REFERENCES

1. P. J. Flory, *Principles of Polymer Chemistry*, 1st ed., Cornell University Press, Ithaca (1953).
2. A. Kumar and S. K. Gupta, *Fundamentals of Polymer Science and Engineering*, 1st ed., Tata McGraw-Hill, New Delhi (1978).
3. R. Aris, *Introduction to the Analysis of Chemical Reactors*, 1st ed., Prentice-Hall, Englewood Cliffs, New Jersey (1965).
4. S. W. Benson, *Foundations of Chemical Kinetics*, 1st ed., McGraw-Hill, New York (1960).
5. E. Rabinowitch, Collision, coordination, diffusion and reaction velocity in condensed systems, *Trans. Faraday Soc.* **33**, 1225–1233 (1937).
6. S. L. Rosen, *Fundamental Principles of Polymeric Materials*, Wiley, New York (1982).
7. P. J. Flory, Kinetics of polyesterification: A study of the effects of molecular weight and viscosity on reaction rate, *J. Am. Chem. Soc.* **61**, 3334–3340 (1939).
8. K. Ueberreiter and W. Hager, On the kinetics of polycondensation of ethylene glycol and dicarboxylic acid in melt, *Macromol Chem.* **180**, 1697–1706 (1979).
9. A. C. Tang and K. S. Yao, Mechanism of hydrogen ion catalysis in esterification II. Studies on the kinetics of polyesterification reactions between dibasic acids and glycols, *J. Polym. Sci.* **35**, 219–233 (1959).
10. V. V. Korshak and S. V. Vinogradova, *Polyesters*, 1st ed., Pergamon, New York (1965).
11. L. B. Sokolov, *Synthesis of Polymers by Polycondensation*, 1st ed., Israel Program for Scientific Translations, Jerusalem (1968).
12. S. Chen and J. Hsiao, Kinetics of polyesterification: I. Dibasic acid and glycol systems, *J. Appl. Polymer Sci.* **19**, 3123–3136 (1981).
13. D. H. Solomon, in *Step Growth Polymerization*, 1st ed., D. H. Solomon, Ed., Marcell Dekker, New York (1978).

14. C. C. Lin and K. H. Hsieh, The kinetics of polyesterification. I. Science acid and ethylene glycol, *J. Polym. Sci.* **21**, 2711–2719 (1977).

15. C. C. Lin and P. C. Yu, The kinetics of polyesterification. II. Succinic acid and ethylene glycol, *J. Polym. Sci., Polym. Chem. Ed.* **16**, 1005–1016 (1978).

16. C. C. Lin and P. C. Yu, The kinetics of polyesterification III. A mathematical model for quantitative prediction of the apparent rate constants, *J. Appl. Polym. Sci.* **22**, 1797–1803 (1978).

17. H. Kilkson, Effect of reaction path and initial distribution on MWD of irreversible condensation polymers, *Ind. Eng. Chem. Fundam.* **3**, 281–293 (1964).

18. H. Kilkson, Generalization of various polycondensation problems, *Ind. Eng. Chem. Fundam.* **7**, 354–362 (1968).

19. W. H. Ray, On the mathematical modelling of polymerization reactors, **C8**, 1–50 (1972).

20. J. A. Biesenberger, Yield and molecular size distributions in batch and continuous linear condensation polymerizations, *AIChE J.* **11**, 369 (1965).

21. P. J. Flory, Molecular size distribution in linear condensation polymers, *J. Am. Chem. Soc.* **58**, 1877–1885 (1936).

22. P. J. Flory, Fundamental principles of condensation polymerization, *Chem. Rev.* **39**, 137–197 (1946).

23. W. J. M. Rootsaert and J. G. Vande Vurse, Kinetics of an infinite series of consecutive reactions. Studies on the preparation of epoxy esters, *Chem. Eng. Sci.* **21**, 1067–1078 (1966).

24. R. Zeman and N. R. Amundson, Continuous models for polymerization, *AIChE J.* **9**, 297–302 (1963).

25. R. Zeman and N. R. Amundson, Continuous polymerization models I—Polymerization in CSTRs, *Chem. Eng. Sci.* **20**, 331–361 (1965).

26. R. Zeman and N. R. Amundson, Continuous polymerization models II—Batch reactor polymerization, *Chem. Eng. Sci.* **20**, 637–664 (1965).

27. S. Lynn and J. E. Huff, Paper 146 presented in 65 National Meeting of AIChE on Kinetics of Polymerization Symposium, May 1–7 (1969).

28. J. L. Throne, *Plastics Process Engineering,* 1st ed., Marcell Dekker, New York (1979).

29. N. H. Smith and G. A. Sather, Polycondensation in CSTRs, *Chem. Eng. Sci.* **20**, 15–23 (1965).

30. W. H. Abraham, Transient polycondensation calculations—An analytical solution, *Chem. Eng. Sci.* **21**, 327–336 (1966).

31. K. Denbigh, *Chemical Reaction Engineering,* 2nd ed., Wiley, New York (1972).

32. O. Levenspiel, *Chemical Reaction Engineering,* 2nd ed., Wiley, New York (1972).

33. Z. Tadmor and J. A. Biesenberger, Influence of segregation on MWD in continuous linear polymerization, *I & EC Fundam.* **5**, 336–343 (1966).

34. P. J. Flory, Thermodynamics of heterogeneous polymers and their solutions, *J. Chem. Phys.* **12**, 425–438 (1944).

35. A. V. Tobolsky, Equilibrium distribution in sizes for linear polymer molecules, *J. Chem. Phys.* **12**, 402–404 (1944).

36. A. Kumar, P. Rajora, N. L. Agarwalla, and S. K. Gupta, Reversible polycondensation characterized by unequal reactivities, *Polymer* **23**, 222–228 (1982).

37. S. K. Gupta, N. L. Agarwalla, P. Rajora, and A. Kumar, Simulation of reversible polycondensations wih monomer having reactivity different from that of higher homologs, *J. Polym. Sci. Polym. Phys. Ed.* **20**, 933–945 (1982).

38. R. L. Laurence and M. Tirrel, *Polymerization Reaction Engineering,* Academic, New York, in preparation.

39. H. M. Hulbert and S. Katz, Some problems in particle technology. A statistical mechanical formulation, *Chem. Eng. Sci.* **19**, 555–574 (1964).

40. S. K. Gupta, A. Kumar, and K. K. Agarwal, Simulation of three stage nylon 6 reactors with intermediate mass transfer at finite rates, *J. Appl. Polymer Sci.* **27**, 3089–3101 (1982).

41. K. Tai, Y. Arai, H. Teranishi, and T. Tagawa, The kinetics of hydrolytic polymerization of ε-caprolactum. IV. Theoretical aspect of the molecular weight distribution, *J. Appl. Polym. Sci.* **25**, 1789–1782 (1980).
42. S. K. Gupta and A. Kumar, Simulation of step growth polymerization, *Chem. Eng. Commun.* **20**, 1–52, 1983.
43. D. A. Mellichamp, Reversible polycondensation in semibatch reactors, *Chem. Eng. Sci.* **24**, 125–139 (1969).
44. T. T. Szabo and J. F. Leathrum, Analysis of condensation polymerization reactors. II. Batch reactors, *J. Appl. Polym. Sci.* **13**, 487–491 (1969).
45. T. T. Szabo and J. F. Leathrum, Analysis of condensation polymerization III. Continuous reactors, *J. Appl. Polym. Sci.* **13**, 561–570 (1969).
46. A. Kumar, S. Kuruville, A. R. Raman, and S. K. Gupta, Simulation of reversible nylon-66 polymerization, *Polymer* **22**, 387–390 (1981).
47. A. Kumar, R. Agarwal, and S. K. Gupta, Simulation of reversible nylon-66 polymerization in homogeneous continuous flow stirred tank reactors, *J. Appl. Polym. Sci.* **27**, 1759–1769 (1982).
48. K. Brown, in *Numerical Solutions of Systems of Nonlinear Algebraic Equations* (C. D. Byrne and C. A. Hall, Eds.), 1st ed., Academic, New York (1973).

EXERCISES

1. Derive Eq. (2.4.8). This can be done by adding Eqs. (2.4.6) and (2.4.7) for all n and using results of summation given in Chapter 1.

2. Consider the polymerization of the nonequimolar ratio of ARA and BRB and show that it can be represented by Eq. (2.4.13).

3. Integrate Eq. (2.5.8) and derive Eq. (2.5.10).

4. Derive Eq. (2.6.9) and from this obtain expressions for μ_n, μ_w, and Q for the combined distribution.

5. Verify results of Table 2.2. This can be done by determining total moles of various functional groups flowing at all points in Fig. 2.9.

6. Using Eq. (2.7.10), determine μ_n, μ_w, and Q at all points in Fig. 2.9.

7. If $C(\tau)$ from Eq. (2.8.8) [or (2.8.9)] is substituted in Eq. (2.8.2), expressions for C_1, C_2, etc. can be solved easily. Find expressions $C_1(\tau)$ for ICs given in Eqs. (2.8.4) and (2.8.5).

8. Write mole balance relations for ARB polymerization in HCSTRs; show that by having recycle on the reactor, there is no change in MWD.

9. For irreversible ARB polymerization in HCSTRs, assume that the feed consists of $[P_1]_0$, $[P_2]_0$, etc. Modify Eq. (2.8.10) to give the MWD of the polymers.

LINEAR STEP GROWTH POLYMERIZATION VIOLATING THE EQUAL REACTIVITY HYPOTHESIS

3.1. INTRODUCTION

In Chapter 2, it was assumed that the functional groups of the various oligomers have the same reactivity, independent of the chain length, n, of the molecules on which they are located. Mathematical results derived for step growth polymerization based on the equal reactivity hypothesis were subsequently used to explain the gross kinetic features of the poly-esterification of adipic acid with ethylene glycol. Comparison with experimental data sufficiently indicated that the overall polymerization is far more complex and the assumption of the equal reactivity hypothesis is a considerable simplification. This is true not only for polyesterification, but for most step growth polymerization systems.[1] In addition, as the polymerization progresses, the viscosity of the reaction mass increases by severalfold and the overall reaction, at some stage depending upon the reactor used, becomes mass transfer controlled.[2-5] In diffusion controlled reactions, the method of reactor analysis is quite different and will be discussed in Chapter 5.

In the reaction controlled region, unequal reactivity of functional groups is commonly observed. This can arise when the reacting groups are not kinetically equivalent, as, for example, in the polymerization of a diacid with a glycol having primary and secondary OH groups which have different reactivities. These are called asymmetric monomers and are usually represented symbolically as the polymerization of (AA + BC) monomers, where functional groups B and C react with A with different rate constants. The polymerization of cyclic monomers can also be described by a similar mathematical development even though the chemistry of reaction in this case is different from that of (AA + BC) monomers. The ring opening reaction, as found in anhydrides, generates two functional groups which participate in the chain building process, and these two steps generally have different reactivities. The formation of epoxy resins, which will be discussed in detail in Chapter 10, can be expressed kinetically as

$$\text{\textbf{www}ROH} + \overset{\displaystyle O}{\overset{\displaystyle /\backslash}{CH_2-CH-CH_2Cl}} \xrightarrow{k_1} \text{www}R-O-CH_2\overset{\displaystyle O}{\overset{\displaystyle /\backslash}{CH-CH_2}}$$

$$(3.1.1a)$$

$$\text{www}RO\,CH_2\overset{\displaystyle O}{\overset{\displaystyle /\backslash}{CH-CH_2}} + HOR\text{www} \xrightarrow{k_2} \text{www}ROCH_2\overset{\displaystyle O}{\overset{\displaystyle |}{CH}}-CH_2-OR\text{www}$$

$$(3.1.1b)$$

A different type of unequal reactivity is found in the polymerization of divinyl benzene and *p*-cresol and in urethanes. In the former, the dimer and trimer are formed in larger amounts than that predicted by the results derived from the equal reactivity hypothesis. The trimer is predominantly phenol-terminated, implying that the ortho hydrogens of unreacted cresol are more reactive compared to the hydrogens of the monosubstituted products. A similar phenomenon is also exhibited in urethane polymerizations, in which case the unequal reactivity can be explained by examining the

charge density on the monomer, $\text{OCN}-\!\!\!\left\langle\bigcirc\right\rangle\!\!\!-\text{NCO}$

$$\underset{-\delta}{O} = \overset{①}{\underset{+\delta}{C}} = \underset{-\delta}{N}-\!\!\!\left\langle\bigcirc\right\rangle\!\!\!-\underset{-\delta}{N} = \overset{②}{\underset{+\delta}{C}} = \underset{-\delta}{O} \qquad (3.1.2)$$

In the presence of two isocyanate groups as on the monomer, there is a competition for the electrons of the phenyl ring, as shown in Eq. (3.1.2). The nitrogens withdraw electrons from the carbons marked ① and ②, making them positively charged, which, in turn, can attract the OH functional groups of other molecules, in this way leading to the formation of higher oligomers. When one of the NCO groups is reacted, in the unreacted isocyanate group, this competition of the electron cloud from the phenyl ring is reduced and the electropositivity of carbon ① in Eq. (3.1.2) (after carbon ② has reacted) reduces, in this way reducing its ability to interact with OH functional groups. This phenomenon is known as *induced asymmetry* and can be modeled kinetically as

$$BB + AA \xrightarrow{k_a} \text{\scriptsize\textasciitilde}A + \text{\scriptsize\textasciitilde}B \qquad (3.1.3a)$$

$$\text{\scriptsize\textasciitilde}B + AA \xrightarrow{k_a} A\text{\scriptsize\textasciitilde} \qquad (3.1.3b)$$

$$\text{\scriptsize\textasciitilde}B + A\text{\scriptsize\textasciitilde} \xrightarrow{k_b} \text{\scriptsize\textasciitilde}BA\text{\scriptsize\textasciitilde} \qquad (3.1.3c)$$

where AA represents the diisocyanate. Evidently, the MWD derived from the kinetic model of Eq. (3.1.3) would be different from that derived in Chapter 2. Case[6] has analyzed various situations of asymmetry and induced asymmetry and derived the molecular weight distribution in terms of the probabilities of reaction of the various functional groups. The effect of the unequal reactivity has now been established to have considerable ramifications on the MWD of the polymer and cannot be ignored.

There is yet another class of unequal reactivity polymerizations in which the rate constants of the various oligomers are dependent upon their chain lengths. As an example, in the polymerization of sodium-*p*-fluoro thiophenoxide,[7] the reaction mass has more unreacted monomer than that predicted by the equal reactivity hypothesis. This observation suggests that the monomer has a lower reactivity. In the formation of polyimides, on the other hand, polymer molecules have been found to have lower reactivity.[8]

There are several cases of unequal reactivity in step growth polymerization, but in the limited space of this chapter the discussion is limited to the following because of their importance:

(a) polymerization with chain length dependent reactivity and
(b) polymerization with asymmetric and induced asymmetric monomers.

3.2. IRREVERSIBLE ARB POLYMERIZATION WITH OLIGOMERS HAVING CHAIN LENGTH DEPENDENT REACTIVITY[9]

For irreversible step growth polymerization of ARB monomers the overall mechanism of reaction is written as

$$P_m + P_n \xrightarrow{k_{p,mn}} P_{m+n} + W \qquad (3.2.1)$$

where the rate constant $k_{p,mn}$ is assumed to depend on m as well as n. The mole balance equations for various oligomers in a batch reactor can be written as

$$\frac{d[P_1]}{dt} = -[P_1] \sum_{r=1}^{\infty} k_{p,1r}[P_r] \tag{3.2.2a}$$

$$\frac{d[P_n]}{dt} = \frac{1}{2} \sum_{r=1}^{n-1} k_{p,r(n-r)}[P_r][P_{n-r}] - [P_n] \sum_{r=1}^{\infty} k_{p,nr}[P_r], \qquad n \geq 2 \tag{3.2.2b}$$

For a polymerizing system it is very difficult to evaluate $k_{p,mn}$ experimentally and, therefore, it must be modeled. The MWD calculated from Eq. (3.2.2) can then be compared with the experimental molecular weight distribution of the polymer formed to justify the model. To demonstrate this method, the variation of the rate constant is assumed to be a linear function of the chain lengths of the two oligomers reacting as[9]

$$k_{p,mn} = k\{1 - (m + n)\alpha\} \tag{3.2.3}$$

where k and α are curve fit constants to be evaluated later on. Equation (3.2.2) then becomes

$$\frac{d[P_1]}{dt} = -[P_1] \sum_{r=1}^{\infty} k\{1 - (1 + r)\alpha\}[P_r] \tag{3.2.4a}$$

$$\frac{d[P_n]}{dt} = \frac{1}{2} \sum_{r=1}^{n-1} k\{1 - \alpha n\}[P_r][P_{n-r}]$$

$$- [P_n] \sum_{r=1}^{\infty} k\{1 - (n + r)\alpha\}[P_r], \qquad n \geq 2 \tag{3.24b}$$

It is noted that the total number of repeat units in the reaction mass remains unchanged during the course of polymerization. Equation (3.2.4) can be added for all n to give for pure monomer feed

$$\frac{d[P]}{dt} = \alpha k[P_1]_0[P] - (k/2)[P]^2 \tag{3.2.5a}$$

IC: $$[P] = [P_1]_0 \qquad \text{at } t = 0 \tag{3.2.5b}$$

Equation (3.2.5) can easily be integrated to give

$$P = \frac{2[P_1]_0 \alpha \, e^{[P_1]_0 \alpha k t}}{2\alpha - 1 + e^{[P_1]_0 \alpha k t}} \tag{3.2.6}$$

Equation (3.2.4) can now be integrated sequentially to obtain $[P_1]$, $[P_2]$, $[P_3]$, etc., and the MWD can be shown by induction to be given by

$$[P_n] = ab^{2(1-\alpha n)}(1-b)^{n-1}f_n, \qquad n = 1, 2, \ldots \qquad (3.2.7)$$

where

$$a = [P_1]_0\, e^{[P_1]_0 \alpha k t} \qquad (3.2.8a)$$

$$b = \frac{2\alpha}{2\alpha - 1 + e^{[P_1]_0 \alpha k t}} \qquad (3.2.8b)$$

$$f_1 = 1 \qquad (3.2.8c)$$

$$f_n = \frac{1 - \alpha n}{(n-1)} \sum_{i=1}^{n-1} f_{n-i} f_i, \qquad \text{for } n \geqslant 2 \qquad (3.2.8d)$$

In this kinetic model, α is an adjustable parameter to fit the experimental molecular weight distribution. In Fig. 3.1 experimental data for nylon-66[10] have been curve fitted using $\alpha = -10^3$ and Eq. (3.2.7) is found to be a better description of the MWD than Flory's distribution, which has also been plotted in this figure for comparison.

The assumption of the rate constant varying linearly with chain length in Eq. (3.2.3) is relatively arbitrary and has little theoretical or experimental basis. Recent experiments[11] have shown that the change in reactivity of the functional groups with the chain length is described by an S-shaped curve. In the literature, this reactivity change has been approximated by a step function in which the monomer is assumed to react differently from the way other oligomers in the reaction mass react. In the analysis presented

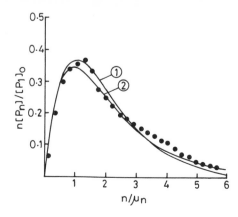

Figure 3.1. Weight fraction distribution versus (n/μ_n).[10] Experimental results (indicated by ○) have been taken from Ref. 10 for nylon-66. Curve ① gives Flory's distribution and curve ② has been calculated for $\alpha = -10^{-3}$ and $\alpha[P_1]_0 kt = -0.1803$. (Reprinted from Ref. 9 with permission of the American Institute of Physics, New York.)

next, it is shown that this has considerable influence on the MWD of the polymer formed.

3.3. REVERSIBLE STEP GROWTH POLYMERIZATION WITH OLIGOMERS HAVING CHAIN LENGTH DEPENDENT REACTIVITY[12-19]

Step growth polymerization of ARB monomer can be written in its most general form as

$$P_m + P_n \underset{k'_{p,m+n}}{\overset{k_{p,mn}}{\rightleftharpoons}} P_{m+n} + W, \qquad m, n = 1, 2, \ldots, \qquad (m + n) = 2, 3, \ldots$$
$$(3.3.1)$$

The chain length dependence of the reactivity of functional groups in the forward step can be idealized into two different limiting cases:

(a) P_1 reacts with P_1 alone with a different rate constant (model I) and
(b) P_1 reacts with all oligomers with a different rate constant (model II).

The unequal reactivity of the forward step in actual systems is expected to behave intermediate to these limiting situations. The reverse reaction involves the reaction between the condensation product W (a small molecule) and a reacted —AB— bond on P_n (a large molecule). The reactivity of —AB— with W would, in general, be dependent upon its position because the local electronic structure near the chain ends is expected to be different from that in the middle. In addition, polymer chains usually exist in a coiled state in the reaction mass with both their ends near the outer surface and the middle —AB— groups near the center of the sphere. There is a possibility of some sort of a molecular shielding effect which may make the reactivity of various —AB— groups different. It is expected that the reactivity of the reacted —AB— bonds would vary gradually from the end of the chain to

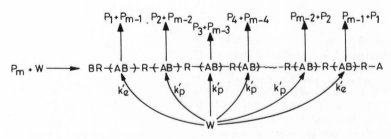

Figure 3.2. Unequal reactivity in the reverse reaction of polymerization.

its middle position, but as a first approximation, the end groups have been modeled to have different reactivity compared to that of the remaining —AB— groups, as shown in Fig. 3.2.

In combination with the unequal reactivity in the forward reaction (models I and II), it is now possible to have the following two kinetic models for reversible polymerization:

Reversible Model I:

$$k_{p,11} = k_{11} \tag{3.3.2a}$$

$$k_{p,mn} = 2k_p; \quad m \neq n; \quad m, n = 1, 2, \ldots \tag{3.3.2b}$$

$$k_{p,mm} = k_p; \quad m = 2, 3, \ldots \tag{3.3.2c}$$

$$k'_{p,m} = k'_e; \quad \text{if } (AB) \text{ is the chain end} \tag{3.3.2d}$$

$$k'_{p,m} = k'_p; \quad \text{if } (AB) \text{ is not the chain end} \tag{3.3.2e}$$

Reversible Model II:

$$k_{p,11} = k_{11} \tag{3.3.3a}$$

$$k_{p,n1} = k_{p,1n} = 2k_{11}; \quad n = 2, 3, \ldots \tag{3.3.3b}$$

$$k_{p,mn} = 2k_p; \quad m \neq n; \quad m, n = 2, 3, \ldots \tag{3.3.3c}$$

$$k_{p,mm} = k_p; \quad m = 2, 3, \ldots \tag{3.3.3d}$$

$$k'_{p,m} = k'_e; \quad \text{if } (AB) \text{ is the chain end} \tag{3.3.3e}$$

$$k'_{p,m} = k'_p; \quad \text{if } (AB) \text{ is not the chain end} \tag{3.3.3f}$$

where k_{11} is the reactivity of the monomer, which is different from that of polymer molecules. With these kinetic models, the mole balance equations for batch reactors can easily be written and are given in Table 3.1.

It may be noted that the mole balance equations for P_n involve the concentrations of P_{n+1}, P_{n+2}, etc., which means that all these equations must be solved simultaneously to obtain the MWD. The balance equations of Table 3.1 for isothermal batch reactors involve three dimensionless parameters R, R', and K, where R and R' give a measure of the degree of unequal reactivity for the forward and reverse steps of step growth polymerization and K is the equilibrium constant representing the majority of

TABLE 3.1. Mole Balance Relations in Batch Reactors under Unequal Reactivities Hypothesis

Reversible model I:

$$\frac{d[P_1]}{dt} = -2(k_{11} - k_p)[P_1]^2 - 2k_p[P_1][P] + 2k'_e[W] \sum_{n=2}^{\infty} [P_n]$$

$$\frac{d[P_2]}{dt} = -2k_p[P_2][P] + k_{11}[P_1]^2 + 2k'_e[P_3][W] + 2k'_p[W] \sum_{n=4}^{\infty} [P_n] - k'_e[W][P_2]$$

$$\frac{d[P_n]}{dt} = k_p \sum_{m=1}^{n-1} [P_m][P_{n-m}] - 2k_p[P_n][P] + 2k'_e[P_{n+1}][W]$$

$$+ 2k'_p[W] \sum_{m=n+2}^{\infty} [P_m] - 2k'_e[P_n][W] - k'_p(n-3)[P_n][W], \qquad n \geq 3$$

and

$$\frac{d[W]}{dt} = k_{11}[P_1]^2 + k_p \sum_{n=3}^{\infty} \sum_{m=1}^{n-1} [P_m][P_{n-m}] - k'_e[P_2][W] - \sum_{n=3}^{\infty} \{2k'_e + (n-3)k'_p\}[W][P_n]$$

Reversible model II:

$$\frac{d[P_1]}{dt} = -2k_{11}[P_1][P] + 2k'_e[W] \sum_{n=2}^{\infty} [P_n]$$

$$\frac{d[P_2]}{dt} = 2(k_p - k_{11})[P_1][P_2] - 2k_p[P_2][P] + k_{11}[P_1]^2$$

$$+ 2k'_e[P_3][W] + 2k'_p[W] \sum_{n=4}^{\infty} [P_n] - k'_e[P_2][W]$$

$$\frac{d[P_n]}{dt} = 2(k_p - k_{11})[P_1][P_n] - 2k_p[P_n][P] + k_p \sum_{m=1}^{n-1} [P_m][P_{n-m}]$$

$$+ 2(k_{11} - k_p)[P_1][P_{n-1}] + 2k'_p[W] \sum_{m=n+2}^{\infty} [P_m] - 2k'_e[P_n][W]$$

$$- k'_p(n-3)[W][P_n] + 2k'_e[W][P_{n+1}], \qquad n \geq 3$$

$$\frac{d[W]}{dt} = k_{11}[P_1]^2 + 2(k_{11} - k_p)[P_1] \sum_{m=2}^{\infty} [P_m] + k_p \sum_{n=3}^{\infty} \sum_{m=1}^{n-1}$$

$$[P_m][P_{n-m}] - k'_e[W][P_2] - [W] \sum_{m=3}^{\infty} \{2k'_e + (m-3)k'_2\}[P_m]$$

Dimensionless groups

$$R = \frac{k_{11}}{k_p} \qquad R' = \frac{k'_e}{k'_p} \qquad K = \frac{k_p}{k'_p}$$

$$C_n = \frac{[P_n]}{[P_1]_0}, \qquad n = 1, 2, \ldots \qquad \tau = k_p[P_1]_0 t$$

reactions and is defined with respect to the equilibrium of functional groups as in Chapter 2. For R and R' equal to one, it can be confirmed that the MWD equations of Table 3.1 reduce to those given in Section 2.9.

The MWD of the polymer can be solved numerically using the Runge-Kutta method of fourth order treating R, R', and K as parameters. Results for irreversible polymerization can be easily derived by substituting k'_p and k'_e equal to zero in Table 3.1.

As pointed out in Chapter 1, it is sometimes desired to find the polydispersity index and number average chain length only instead of the entire MWD of the polymer. It is easy to establish the moment generation relations using Table 3.1, and after appropriately differentiating, one can determine λ_0, λ_1, and λ_2. For irreversible polymerization these are summarized in Table 3.2.

It may be observed that for irreversible polymerization, it is possible to determine the moments, λ_0, λ_1, and λ_2 precisely after solving C and C_1, and no moment closure assumptions need be made. Since it is not possible to get moments analytically, for a given value of R, moments have been solved numerically.

TABLE 3.2. Moment Relations for Irreversible Polymerization with Chain Length Dependent Reactivity in Batch Reactors

---

(1) Model I

$$\frac{d\lambda_0}{dt} = -(k_{11} - k_p)[P_1]^2 - k_p\lambda_0^2$$

$$\frac{d\lambda_1}{dt} = 0$$

$$\frac{d\lambda_2}{dt} = 2k_p\lambda_1^2 + 2(k_{11} - k_p)[P_1]^2$$

I.C.:

$$\lambda_0 = \lambda_1 = \lambda_2 = [P_1]_0$$

(2) Model II

$$\frac{d\lambda_0}{dt} = (k_{11} - k_p)[P_1]^2 - k_p\lambda_0^2 - 2(k_{11} - k_p)[P_1]\lambda_0$$

$$\frac{d\lambda_1}{dt} = 0$$

$$\frac{d\lambda_2}{dt} = 2k_p\lambda_1^2 - 2(k_{11} - k_p)[P_1]^2 + 4(k_{11} - k_p)[P_1]\lambda_1$$

I.C.:

$$\lambda_0 = \lambda_1 = \lambda_2 = [P_1]_0$$

With the kinetic equations derived for models I and II, it is easy to derive the steady state mole balance equations for homogeneous continuous-flow stirred tank reactors. The moments and the molecular weight distributions can be obtained by solving these coupled equations for irreversible polymerization. Results for irreversible polymerization in batch reactors are obtained by using the Runge-Kutta method of fourth order whereas those for HCSTRs are found using the Gauss-Jordon method. Results for both batch reactors and HCSTRs are given in Figs. 3.3–3.8.

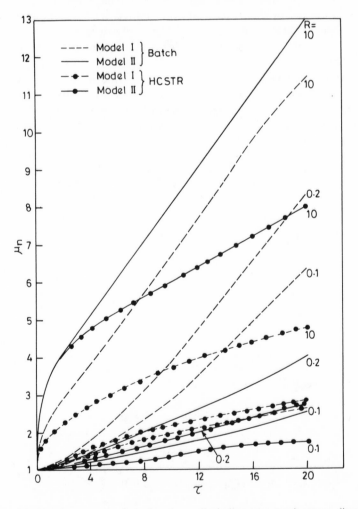

Figure 3.3. Number average chain length μ_n (including monomer) versus dimensionless reaction of residence time for various reactivity ratios R in (irreversible) models I and II in batch reactors and HCSTRs.[15-18]

In Fig. 3.3, the number average chain length μ_n for irreversible polymerization in batch reactors and HCSTRs are given for three values of the reactivity ratio R. As R is increased from 0.1 to 10, for a given dimensionless time, τ, μ_n increases monotonically. For a given kinetic model and τ, μ_n for batch reactors is always higher. However, for a given reactor, μ_n for model I for $R > 1$ is less than those for model II but for $R < 1$, μ_n for model I is higher.

The plots of the weight average chain length μ_w for irreversible polymerization show that for a given residence time, τ, μ_w for HCSTRs compare with those for batch reactors. Therefore, the polydispersity index Q for

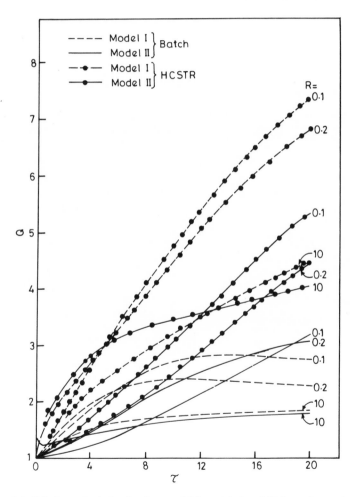

Figure 3.4. Polydispersity index for (irreversible) models I and II for various values of R in batch reactors and HCSTRs.[15-18]

HCSTRs for a given kinetic model is always higher than for batch reactors as observed from Fig. 3.4. The Q of the polymer obtained from HCSTRs for the kinetic model I is considerably higher compared to those for model II. Similar conclusions cannot be drawn for batch reactors.

The molecular weight distributions of the polymer obtained for irreversible polymerization in batch reactors are given in Figs. 3.5 and 3.6. For $R < 1$ (Fig. 3.5), the MWDs for model II are found to split into two curves for odd and even n, whereas no such phenomenon is observed for model

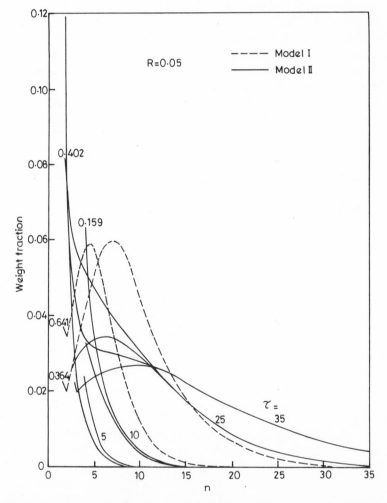

Figure 3.5. MWDs for (irreversible) models I and II in batch reactors for $R = 0.05$. The MWD curves are found to split for model II with the upper curve for even n and the lower one for odd n. Values in the left at the end of the MWD curves give the monomer mass fraction.[15,16]

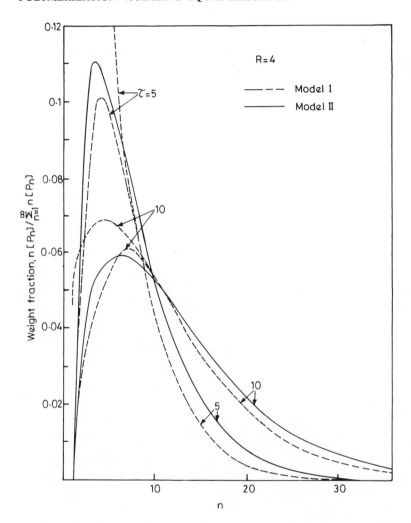

Figure 3.6. MWDs for (irreversible) models I and II in batch reactors for $R = 4$. The MWD curves split for model I, with the upper curves for even n and the lower curves for odd n.[15,16]

I. For $R > 1$ (Fig. 3.6), the MWDs for model I are found to split for odd and even n, whereas for model II, a continuous curve for MWD is obtained. The splitting of the MWDs is explained as follows. For model I, P_1 reacts with P_1 at higher rates when $R > 1$ and gets depleted quickly, forming dimers. Because of the high concentration of P_2 and the lower concentration of P_1, P_4 would be formed in larger amounts compared to P_3, and so on. Model II assumes that monomers reacts at a different rate compared to other species and for $R < 1$ it reacts at a lower rate. In view of this, the

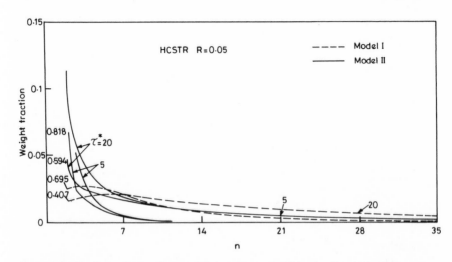

Figure 3.7. MWDs for irreversible models I and II in HCSTRs for $R = 0.05$. MWD curves split for model II, with upper curve for even n and the lower one for odd n. Values in the left at the end of the MWD curves give the monomer mass fraction.[17,18]

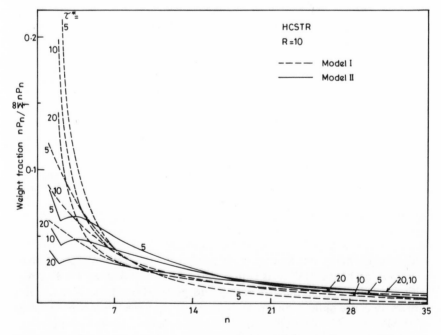

Figure 3.8. MWDs for (irreversible) models I and II in HCSTRs for $R = 10$. MWD curves for model I split, with upper curves for even n and the lower one for odd n.[17,18]

reaction mass contains considerable amounts of unreacted monomer. As soon as dimers are formed, they have a higher tendency to react with themselves than with monomer molecules. Consequently, the relative concentration of tetramers is expected to be higher than that of trimers and so on, and a split in the MWDs results.

The MWDs obtained in HCSTRs for models I and II are compared in Figs. 3.7 and 3.8 for $R = 0.05$ and $R = 10.0$. Similar splitting of MWDs for odd and even n is obtained for HCSTRs. For $R < 1$ (as in Fig. 3.7), MWDs for model II split and the concentration of various oligomers (except

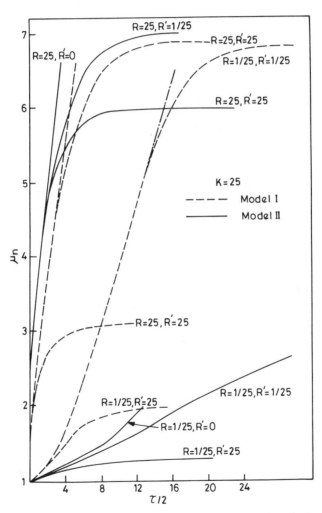

Figure 3.9. Number average chain length versus τ for reversible polymerization for models I and II for various values of R and R'.[12,13]

for monomer) is generally lower than those for model I because of lower conversions. The converse is found to hold for model I when $R > 1$ as seen in Fig. 3.8. The split of the MWDs is commonly observed for step growth polymerizations characterized by the unequal reactivity of functional groups. Also, it has been observed that for such cases a one-parameter representation of the MWD in terms of the conversion of functional groups is not possible.

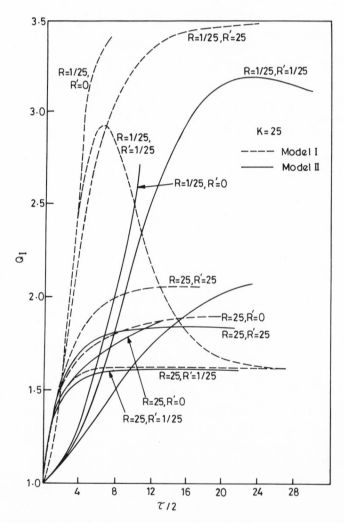

Figure 3.10. Polydispersity index Q versus τ for reversible polymerization[12,13] for various values of R and R'. $K = 25$.

For reversible polymerization, the MWD of the polymer formed in batch reactors can be determined by numerically solving equations of Table 3.1 for specified values of R, R', and K, and some of the results are given in Figs. 3.9 and 3.10. In Fig. 3.9, the number average chain length, μ_n, has been plotted for both models I and II. For any given model, the effect of reversibility is that μ_n approaches an asymptotic value at large times. This asymptotic value decreases as R' is increased, which is because the reverse reactions are being speeded up. For $R > 1$ (results for $R = 25$ shown in the figure), μ_n for model II is higher than that for model I, and this is because the forward reactions are speeded up more in model I than in model II. For $R < 1$ (results for $R = 1/25$ shown), the reverse is observed for similar reasons. In Fig. 3.10, the polydispersity index, Q, has been plotted. For model I, Q increases monotonically to an asymptotic value for some cases, and for some others it attains the asymptotic value after exhibiting a maximum. A study of the MWD shows that the maximum occurs in those cases where there is a preponderance of the monomer in the reaction mass. Similar results are observed for reversible model I as shown in Fig. 3.10.

3.4. STEP GROWTH POLYMERIZATION INVOLVING ASYMMETRIC FUNCTIONAL GROUPS[6,20,21]

In Section 3.1, various cases of unequal reactivity were discussed and it was argued that polymerization involving asymmetric monomers cannot be described by the results based on the equal reactivity hypothesis. In this section, the polymerization of (AA + BC) monomers has been analyzed using a probabilistic approach. Kilkson[22] has suggested that for batch reactors, at times it is advantageous to combine the kinetic and probabilistic approaches to obtain the reactor performance, and in this section, this is the approach adopted. Using the latter approach, the molecular weight distribution and the number and weight average molecular weights have been related to the conversions of functional groups. The conversions of functional groups can be easily determined using the kinetic approach by using Eq. (3.1.3). It may be added that the same results can also be obtained using the kinetic approach alone, as presented earlier. However, the probabilistic technique is presented here deliberately since it gives a simultaneous introduction to a new technique which is particularly useful for handling network polymer formation, discussed in Chapter 4.

It is assumed that the reaction mass has initially $[A]_0$, $[B]_0$, and $[C]_0$ concentrations of functional groups. Since in an AA monomer molecule there are two A groups, and in a BC molecule there is one B and one C group, the total number of molecules, N_0, per unit volume at $t = 0$ is given by

At $t = 0$:

$$N_0 = [A]_0/2 + [B]_0 \tag{3.4.1}$$

$$[B_0] = [C]_0 \tag{3.4.2}$$

However, as the polymerization progresses, [B] need not be equal to [C] because if B has a higher reactivity compared to that of C, B would be preferentially reacted. It is in view of this, that the following three probabilities are defined:

$$p_A = \text{probability of finding a reacted A group at time } t \tag{3.4.3a}$$

$$p_B = \text{probability of finding a reacted B group at time } t \tag{3.4.3b}$$

$$p_C = \text{probability of finding a reacted C group at time } t \tag{3.4.3c}$$

Since these are equal to the fraction of reacted groups in the reaction mass at this particular time, one has

$$p_A = \frac{[A]_0 - [A]}{[A]_0} = \frac{\text{number of A groups reacted}}{\text{total number of A groups}} \tag{3.4.4a}$$

$$p_B = \frac{[B]_0 - [B]}{[B]_0} = \frac{\text{number of B groups reacted}}{\text{total number of B groups}} \tag{3.4.4b}$$

$$p_C = \frac{[C]_0 - [C]}{[C]_0} = \frac{\text{number of C groups reacted}}{\text{total number of C groups}} \tag{3.4.4c}$$

where [A], [B], and [C] are the molar concentrations of A, B, and C groups at time t. Because of the stoichiometry of the reaction, the number of A groups reacted must be equal to the sum of B and C reacted, which means that

$$p_A = \frac{[B]_0}{[A]_0}(p_B + p_C) \tag{3.4.5a}$$

$$[A]_0 = 2N_0 \Big/ \left(1 + \frac{2p_A}{p_B + p_C}\right) \tag{3.4.5b}$$

$$[B]_0 = N_0 \left\{ 1 - \frac{1}{1 + 2p_A/(p_B + p_C)} \right\} \tag{3.4.5c}$$

It is thus seen that among p_A, p_B, and p_C, only two are independent variables because of the stoichiometry.

By definition, the number average molecular weight, $\bar{M}_n$; at time t is given by the ratio of total weight and total number of molecules within the reactor. As the polymerization progresses, the loss in the weight of the reaction mass occurs only by the removal of the condensation product, which is always small. If this is neglected, the total weight is time invariant and is therefore given by the value at $t = 0$

$$W_t = \frac{\text{total weight}}{\text{volume}}$$

$$= (\text{total number of molecules of AA monomer/volume at } t = 0)M_{AA}$$

$$+ (\text{total molecules of BC monomer/volume at } t = 0)M_{BC}$$

$$= \frac{[A]_0}{2} M_{AA} + [B]_0 M_{BC}$$

$$= \frac{N_0}{\left(1 + \frac{2p_A}{p_B + p_C}\right)} \left\{ M_{AA} + \frac{2p_A}{(p_B + p_C)} M_{BC} \right\} \qquad (3.4.6)$$

where M_{AA} and M_{BC} are the molecular weights of the repeat units *formed* through the monomers AA and BC, respectively. In defining M_{AA} and M_{BC} in this way, the error introduced in assuming the weight of the reaction mass as time-invariant is further reduced.

The total concentration of molecules, N_t, in the reaction mass at time t can be determined by observing that the total number of molecules is reduced by one whenever an A functional group reacts with either a B or a C group (assuming no intramolecular reactions). This leads to

$$N_t = N_0 \left\{ 1 - \frac{2p_A}{1 + \frac{2p_A}{p_B + p_C}} \right\} \qquad (3.4.7)$$

The number average molecular weight, $\bar{M}_n$, is therefore, obtained as

$$\bar{M}_n = \frac{W_t}{N_t} = \frac{(p_B + p_C)M_{AA} + 2p_A M_{BC}}{2p_A + p_B + p_C - 2p_A(p_B + p_C)} \qquad (3.4.8)$$

It is necessary to analyze the molecular weight distribution to determine the weight-average molecular weight. To find the MWD, the various poly-meric oligomers in the reaction mass are classified according to their end functional groups into the following:

$$A\text{\large$\sim\sim$}A_n \quad \text{denoting (AABC)}\underset{n-1}{\text{------}}AA \tag{3.4.9a}$$

$$A\text{\large$\sim\sim$}B_n \quad \text{denoting (BCAA)}_n \tag{3.4.9b}$$

$$A\text{\large$\sim\sim$}C_n \quad \text{denoting (AABC)}_n \tag{3.4.9c}$$

$$B\text{\large$\sim\sim$}B_n \quad \text{denoting (BCAA)}\underset{n}{\text{---}}CB \tag{3.4.9d}$$

$$B\text{\large$\sim\sim$}C_n \quad \text{denoting (BCAA)}\underset{n}{\text{)---}}BC \tag{3.4.9e}$$

$$C\text{\large$\sim\sim$}C_n \quad \text{denoting (CBAA)}\underset{n}{\text{)---}}CB \quad n \geqslant 1 \tag{3.4.9f}$$

In addition to these species, unreacted BC monomer is also present in the system. In molecule (a), there are nAA and $(n-1)$BC units; in molecules (b) and (c), there are nAA and nBC units; and in (d), (e), and (f), there are nAA and $(n+1)$BC units. The MWD is obtained by first determining the probability of finding any particular chain in the reaction mass and then summing up appropriate terms as given below.

The probability of finding an $\text{(AABC)}_{\overline{n-1}}$AA molecule in the reaction mass is equal to the probability of finding an unreacted A group from among various end groups, multiplied by the probability of finding the sequence $-A-B-C(AABC)\overline{_n}\text{------}A-$, multiplied by the probability of finding an unreacted A group. The probability of finding an unreacted A, B, or C group from all the end groups can be found by dividing the total number of unreacted groups of a particular type by the total number of *end* groups in the reaction mass. The latter is twice the total number of molecules $\{=2N_t\}$. Thus,

Probability of finding an unreacted end A group

$$= \frac{1 - p_A}{1 + \dfrac{2p_A}{p_B + p_C} - 2p_A} \tag{3.4.10a}$$

Probability of finding an unreacted end B group

$$= \frac{p_A}{p_B + p_C} \frac{1 - p_B}{1 + \dfrac{2p_A}{p_B + p_C} - 2p_A} \tag{3.4.10b}$$

In the above equations, $[A]_0/N_0$ and $[B]_0/N_0$ have been eliminated using Eq. (3.4.5). Using a similar approach, one can determine the probability of an unreacted end C group. With this, the probability of finding the unreacted BC monomer can be found as

Prob(BC) = probability of finding unreacted BC monomer

   = (probability of randomly picking B end)(probability
      of finding unreacted C)

   + (probability of randomly picking C end)(probability
      of finding unreacted B)

$$= \frac{2p_A}{p_B + p_C} \frac{(1 - p_B)(1 - p_C)}{1 + \dfrac{2p_A}{p_B + p_C} - 2p_A} \tag{3.4.11}$$

The probability of the sequence $\{AA-BC\}_{n-1}$ is obtained by successive multiplication of the probability of finding the repeat unit $-AABC- (n - 1)$ times. In $-AABC-$, the occurrence of the first reacted A group has a probability of p_A; the subsequent A occurs with probability 1 (since it is already connected to the previous A); the probability of finding a reacted B out of B and C reacted functional groups is given by $p_B/(p_B + p_C)$, whereas finding a reacted C group (reaction with A) has a probability of p_C. Since $\{AA-BC\}$ and $\{AA-CB\}$ both have the probability of $p_A[p_B/(p_B + p_C)]p_C$, the probability of finding either of these two sequences is given by $2p_Ap_Bp_C/(p_B + p_C)$. Therefore,

Prob($A\text{\textasciitilde\textasciitilde}A_n$) $\equiv$ probability of $A\text{\textasciitilde\textasciitilde}A_n$

$$= \left[\frac{1 - p_A}{1 + \dfrac{2p_A}{p_B + p_C} - 2p_A} \right] \left(\frac{2p_Ap_Bp_C}{p_B + p_C} \right)^{n-1} (1 - p_A) \tag{3.4.12}$$

To find the probability of polymer oligomers with dissimilar ends, it is observed that the probability of finding *both* B and C reacted is equal to $2p_Bp_C/(p_B + p_C)$. Therefore, the probability of finding *either* an unreacted B *or* C is given by

$$\text{Probability of either an unreacted B or C} = \left(1 - \frac{2p_Bp_C}{p_B + p_C} \right) \tag{3.4.13}$$

Thus, the probability of molecular structures (b) and (c) can be combined and written as

$\text{Prob}(A \wedge C_n, A \wedge B_n) \equiv \text{Prob}\{A \wedge C_n \text{ or } A \wedge B_n\}$

$\qquad = \text{Prob(randomly picking unreacted A)}$

$\qquad \times \text{Prob}(-AABC-)^{n-1} \times \text{Prob(unreacted B or C)}$

$\qquad + \text{Prob(randonly picking unreacted B or C)}$

$\qquad \times \text{Prob}(-AABC)^{n-1} \times \text{Prob(unreacted A)}$

$$= 2 \frac{1 - p_A}{1 + \dfrac{2p_A}{p_B + p_C} - 2p_A} \left(\frac{2 p_A p_B p_C}{p_B + p_B} \right)^{n-1} p_A \left(1 - \frac{2 p_B p_C}{p_B + p_C} \right)$$

$$(3.4.14)$$

The oligomeric structures $B \wedge B_n$, $B \wedge C_n$, and $C \wedge C_n$ can be similarly considered together as follows. The probability of finding these can be found to be

$\text{Prob}(B \wedge B_n, B \wedge C_n, C \wedge C_n)$

$\qquad \equiv \text{probability of } B \wedge B_n, B \wedge C_n, \text{ or } C \wedge C_n$

$\qquad = \text{(probability of B } or \text{ C at the chain end)} p_A \text{(probability of } \{AABC\}^{n-1})$

$\qquad \times p_A \text{(probability of finding unreacted B or C)}$

$$= \frac{1 - \dfrac{2 p_B p_C}{p_B + p_C}}{1 + \dfrac{2p_A}{p_B + p_C} - 2p_A} p_A \left(\frac{2 p_A p_B p_C}{p_B + p_C} \right)^{n-1} p_A \left(1 - \frac{2 p_B p_C}{p_B + p_C} \right) \qquad (3.4.15)$$

The number or mole fraction distribution of a molecule of size n is, therefore, given by the sum of Eqs. (3.4.12), (3.4.14), and (3.4.15) and is given by

$$\frac{[P_n]}{\sum_{n=1}^{\infty} [P_n]} = \text{Prob}(A \wedge A_n) + \text{Prob}(A \wedge C_n, A \wedge B_n)$$

$$+ \text{Prob}(B \wedge B_n, B \wedge C_n, C \wedge C_n) \qquad (3.4.16)$$

To evaluate the weight average molecular weight, $\bar{M}_w$, it is required to determine {(weight of molecule)^2 (number of that kind of molecule)}. The number of specific kinds of molecules can be written as the product of (probability of that kind of molecule) and the (total number of molecules in the reaction mass). Therefore,

$$M_{n1}^2[\text{A}\text{\tiny\char`\~}\text{A}_n] = \frac{N_0}{1 + \dfrac{2p_A}{p_B + p_C}}\{(1 - p_A)^2\alpha^{n-1}[nM_{AA} + (n - 1)M_{BC}]^2\}$$

$$(3.4.17)$$

where

$$\alpha = \frac{2p_Ap_Bp_C}{p_B + p_C} \qquad (3.4.18)$$

Similarly,

$$M_{n2}^2\{[\text{A}\text{\tiny\char`\~}\text{B}_n] + [\text{A}\text{\tiny\char`\~}\text{C}_n]\}$$

$$= \frac{N_0}{1 + \dfrac{2p_A}{p_B + p_C}}\{2p_A(1 - p_A)\beta\alpha^{n-1}[nM_{AA} + nM_{BC}]^2\}$$

$$(3.4.19a)$$

and

$$M_{n3}^2\{[\text{B}\text{\tiny\char`\~}\text{B}_n] + [\text{B}\text{\tiny\char`\~}\text{C}_n] + [\text{C}\text{\tiny\char`\~}\text{C}_n]\}$$

$$= \frac{N_0}{1 + \left(\dfrac{2p_A}{p_B + p_C}\right)}\{p_A^2\beta^2\alpha^{n-1}[nM_{AA} + (n + 1)M_{BC}]^2\} \qquad (3.4.19b)$$

where

$$\beta = 1 - \frac{2p_Bp_C}{p_B + p_C} \qquad (3.4.20)$$

The weight of unreacted BC is given by

$$M_{BC}^2[\text{BC}] = \frac{N_0}{1 + \dfrac{2p_A}{p_B + p_C}}\frac{2p_A(1 - p_B)(1 - p_C)M_{BC}^2}{(p_B + p_C)} \qquad (3.4.21)$$

Therefore,

$$\sum M_n^2[P_n] = \frac{N_0}{1 + \dfrac{2p_A}{(p_B + p_C)}} \left\{ M_{AA}^2[(1 - p_A)^2 + 2p_A(1 - p_A)\beta \right.$$

$$+ \beta^2 p_A^2] \sum n^2 \alpha^{n-1} + 2M_{AA}M_{BC}[(1 - p_A)^2 \sum n(n - 1)\alpha^{n-1}$$

$$+ 2p_A(1 - p_A)\beta \sum n^2 \alpha^{n-1} + p_A^2 \beta^2 \sum n(n + 1)\alpha^{n-1}]$$

$$+ M_{BC}^2 \left[(1 - p_A)^2 \sum (n - 1)^2 \alpha^{n-1} + 2p_A(1 - p_A)\beta \sum n^2 \alpha^{n-1} \right.$$

$$\left. \left. + p_A^2 \beta^2 \sum n(n + 1)\alpha^{n-1} + \frac{2p_A(1 - p_B)(1 - p_C)}{(p_B + p_C)} \right] \right\}$$

$$= \frac{N_0}{\left(1 + \dfrac{2p_A}{p_B + p_C}\right)(1 - \alpha)} \left\{ \left[M_{AA}^2(1 + \alpha) + 4p_A M_{AA}M_{BC} \right] \right.$$

$$\left. + M_{BC}^2 \left[\frac{2p_A}{(p_B + p_C)}(p_B + p_C + p_A p_B^2 + p_A p_C^2) \right] \right\} \qquad (3.4.22)$$

The weight average molecular weight is therefore given by

$$\bar{M}_w = \frac{(1 + \alpha)M_{AA}^2 + 4p_A M_{AA}M_{BC} + M_{BC}^2 \left[\dfrac{2p_A}{P_B + P_C}(p_B + p_C + p_A p_B^2 + p_A p_C^2) \right]}{(1 - \alpha)\left(M_{AA} + \dfrac{2p_A}{p_B + p_C}M_{BC} \right)}$$

$$\qquad (3.4.23)$$

Consequently the polydispersity index Q is equal to

$$Q \equiv \frac{\bar{M}_w}{\bar{M}_n}$$

$$= [2p_A + (1 - 2p_A)(p_B + p_C)]\left\{ (1 + \alpha)M_{AA}^2 + 4p_A M_{AA}M_{BC} \right.$$

$$\left. + M_{BC}^2 \left[\frac{2p_A}{p_B + p_C}(p_B + p_C + p_A p_B^2 + p_A p_C^2) \right] \right\}$$

$$\overline{\left\{ (p_B + p_C - 2p_A p_B p_C)\left[M_{AA} + \frac{2p_{AA}}{p_B + p_C}M_{BC}^2 \right]^2 \right\}^{-1}} \qquad (3.4.24)$$

From the stoichiometry it follows that all of A functional groups reacted would be equal to all B and C functional groups, i.e.,

$$p_A = \frac{r}{2}(p_B + p_C) \qquad (3.4.25)$$

where $r(=2[B]_0/[A]_0)$ represents the initial ratio of moles of BC and AA. In terms of r, Eqs. (3.4.8) and (3.4.24) can be rewritten as

$$\bar{M}_n = \frac{M_{AA} + rM_{BC}}{r + 1 - 2p_A} \qquad (3.4.26a)$$

$$Q = \frac{M_{AA}^2(1 + rp_B p_C) + 4p_A M_{AA} M_{BC} + M_{BC}^2\left[r + r^2\left(\dfrac{p_B^2 + p_C^2}{2}\right)\right]}{(1 - rp_B p_C)(rM_{BC} + M_{AA})^2}$$

$$(3.4.26b)$$

Thus, once p_A, p_B, and p_C are known as a function of time, $\bar{M}_n$ and Q can be determined. The kinetic approach is now presented for evaluating the conversions.

To determine p_A, p_B, and p_C as a function of time, it is necessary that the balance equations for the functional groups A, B, or C be solved. The reaction mechanism for the polymerization of asymmetric monomer BC can be written as

$$\text{\small ⋀⋀}AA + \text{\small ⋀⋀}CB \xrightarrow{k_1} \text{\small ⋀⋀}AABC\text{\small ⋀⋀} \qquad (3.4.27a)$$

$$\text{\small ⋀⋀}AA + \text{\small ⋀⋀}BC \xrightarrow{k_2} \text{\small ⋀⋀}AACB\text{\small ⋀⋀} \qquad (3.4.27b)$$

In this mechanism, the functional groups B and C are shown to have different reactivities. The time dependence of p_A, p_B, and p_C defined in Eq. (3.4.4) can be easily written from the kinetic mechanism given above as

$$\frac{dp_A}{dt} = \frac{[A]_0}{2}(1 - p_A)\{k_1(1 - p_B) + k_2(1 - p_C)\} \qquad (3.4.28a)$$

$$\frac{dp_B}{dt} = \frac{2k_1[A]_0}{r}(1 - p_B)(1 - p_A) \qquad (3.4.28b)$$

$$\frac{dp_C}{dt} = \frac{2k_2[A]_0}{r}(1 - p_C)(1 - p_A) \qquad (3.4.28c)$$

Dividing Eq. (3.4.28b) by Eq. (3.4.28c) and on integration, one obtains

$$p_B = 1 - (1 - p_C)^{1/R} \tag{3.4.29}$$

where

$$R = (k_2/k_1) \tag{3.4.30}$$

Thus, with the knowledge of p_C, p_B can be calculated from Eq. (3.4.29) and p_A is determined by integrating Eqs. (3.4.28a) numerically. $\bar{M}_n$, $\bar{M}_w$, and Q are then evaluated from Eqs. (3.4.26).

For various values of R and the feed ratio, r, of AA and BC monomers, p_C is numerically obtained and has been plotted in Fig. 3.11 as a function of time. For a given R, p_C rises quickly reaching an asymptotic value corresponding to the residual functional groups. For $r = 1$, all the functional groups would react and therefore p_C for large times should be *one*, which is indeed found to be so in Fig. 3.11.

In Fig. 3.12, the effect of the feed ratio r of AA and BC monomers on the polydispersity index Q of the polymer formed at the end of the reaction (i.e., $t \to \infty$) has been examined. For $r < 1$, AA is in excess and at the end of the reaction, p_B as well as p_C are equal to one, and p_A, according to Eq. (3.4.25) is equal to r. Therefore, Eq. (3.4.26) reduces to

$$\lim_{t \to \infty} Q_I = \frac{r^2 + 6r + 1}{(1 + r)^2} \quad \text{for } M_{AA} = M_{BC}, \quad r < 1 \tag{3.4.31}$$

which is independent of R. For $r > 1$, BC is in excess and for large times

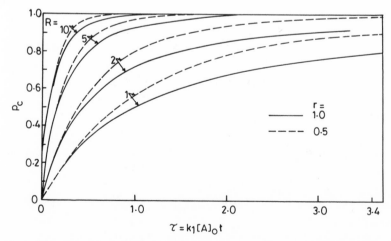

Figure 3.11. Conversion of C functional group, p_c, versus time in (AA + BC) polymerization for various values of R and r.[21]

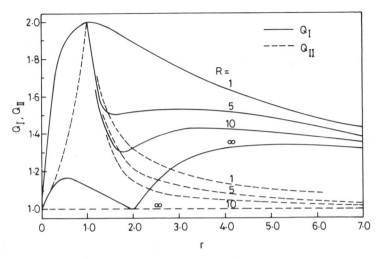

Figure 3.12. Q_I (including monomer) and Q_{II} (PDI excluding the monomer) versus r for various values of R in (AA + BC) polymerization[21] at complete conversion of limiting functional group when $M_{AA} = M_{BC}$. (Reprinted from Ref. 21 with permission of the American Institute of Chemical Engineers, New York.)

(i.e., $t \to \infty$), all of AA would be reacted, thus giving $p_A = 1$. The values of p_B and p_C would then be governed by R. As r is increased beyond 1, Q_I undergoes a short minimum followed by a broad maximum. In the limit of $R \to \infty$ (i.e., $k_1 = 0$), the monomer BC acts like a monofunctional compound and for $r = 2$, only trimers of the type CB—AA—BC are formed. The polymer thus formed is completely monodisperse (i.e., $Q_I = 1$). Any further addition of BC (i.e., $r > 2$) would not lead to any further reaction and would cause dilution of the monodisperse reaction mass. In view of this, a modified polydispersity index, Q_{II}, and a modified number average molecular weight, $\bar{M}_{n,1}$, are sometimes defined from which the monomers are excluded. For $M_{AA} = M_{BC}$, they can be found to be

$$\bar{M}_{n,1} = \frac{4p_A - rp_B p_C - p_A^2}{2p_A - rp_B p_C - p_A^2} M_{AA} \tag{3.4.32a}$$

$$Q_{II} = \frac{(p_A - rp_B p_C)^2 + rp_B p_C (1 - p_A)^2 + 8p_A}{(1 - rp_B p_C)(4p_A - rp_B p_C - p_A^2)}(2p_A - rp_B p_C - p_A^2) \tag{3.4.32b}$$

The Q_{II} calculated from the above equation has also been plotted in Fig. 3.12 which, as expected, falls monotonically from the value of 2 for $r > 1$. Evidently, this reaction scheme reduces identically to that given in Chapter 2 for $R = 1$.

3.5. CONCLUSIONS

There are various situations in step growth polymerizations where the equal reactivity hypothesis does not hold. The unequal reactivities are generally of two types, one arising due to the chain length dependent rate constant and the other due to the presence of asymmetric functional groups. In this chapter, both these have been discussed in detail.

Chain length dependent reactivities of polymer molecules have been modeled in two ways, called models I and II, and for these the MWDs have been obtained. When the reactivity ratio $R > 1$, for model I, the MWD is found to split for odd and even chain length but for model II no such phenomenon is observed. Similarly for $R < 1$, the split in the MWD is found for model II but no such observation is made for model I. The chain length dependent unequal reactivity in the reverse reaction has also been modeled and the effects of various reaction parameters have been carefully examined.

The step growth polymerizations involving asymmetric monomers in batch reactors have been modeled adopting the probabilistic approach. The MWD, the number and weight average molecular weights, and polydispersity index have been derived in terms of the conversions of various functional groups. This probabilistic approach has been adopted primarily to introduce a new approach because the same results can also be derived through the kinetic approach as done earlier in this chapter. The former is found to be advantageous in modeling network formation.

REFERENCES

1. D. H. Solomon, ed. *Step Growth Polymerization*, Marcel Dekker, New York (1972).
2. S. K. Gupta and A. Kumar, Simulation of step growth polymerization, *Chem. Eng. Commun.* **20**, 1-52 (1983).
3. M. Amon and C. D. Denson, Simplified analysis of the performance of wiped-film polycondensation reactors, *Ind. Eng. Chem. Fundam.* **19**, 415-420 (1980).
4. S. K. Gupta, N. L. Agarwalla, and A. Kumar, Mass transfer effects in polycondensation reactors wherein functional groups are not equally reactive, *J. Appl. Polym. Sci.* **27**, 1217-1231 (1982).
5. S. K. Gupta, A. Kumar, and K. K. Agarwal, Simulation of AA′ + B′B″ reversible polymerization with mass transfer of condensation products, *Polymer* **23**, 1367-1371 (1982).
6. L. C. Case, Molecular distributions in polycondensations involving unlike reactants. II. Linear distributions, *J. Polym. Sci.* **29**, 455-495 (1958).
7. R. W. Lenz, C. E. Handlovitz, and H. A. Smith, Phenylene sulfide polymers. III. The synthesis of linear polyphenylene sulfide, *J. Polym. Sci.* **58**, 351-367 (1962).
8. J. H. Hodkin, Reactivity changes during polyimide formation, *J. Polym. Sci. Polym. Chem. Ed.* **14**, 409-431 (1976).
9. V. S. Nanda and S. C. Jain, Effect of variation of the bimolecular rate constant with chain length on the statistical character of condensation polymers, *J. Chem. Phys.* **49**, 1318-1320 (1968).

10. G. B. Taylor, The distribution of the molecular weight of nylon as determined by fractionation in a phenol-water system, *J. Am. Chem. Soc.* **69**, 638–644 (1947).
11. S. I. Kuchanov, M. L. Keshtov, P. G. Halatur, V. A. Vasnev, S. V. Vinogradova, and V. V. Korshak, On the principle of equal reactivity in solution polycondensation, *Macromol. Chem.* **184**, 105–111 (1983).
12. S. K. Gupta, N. L. Agarwalla, P. Rajora, and A. Kumar, Simulation of reversible polycondensations with monomers having reactivities different from that of higher homologs, *J. Polym. Sci. Polym. Phys. Ed.* **20**, 933–945 (1982).
13. A. Kumar, P. Rajora, N. L. Agarwalla, and S. K. Gupta, Reversible condensation characterized by unequal reactivities, *Polymer* **23**, 222–228 (1982).
14. R. Goel, S. K. Gupta, and A. Kumar, Rate of condensation polymerization for monomers having reactivities different from their polymers, *Polymer* **18**, 851–852 (1977).
15. S. K. Gupta, A. Kumar, and A. Bhargava, Molecular weight distributions and moments for condensation polymerizations characterized by two rate constants, *Europ. Polym. J.* **15**, 557–564 (1979).
16. S. K. Gupta, A. Kumar, and A. Bhargava, Molecular weight distribution and moments for condensation polymerization of monomers having reactivities different from their homologues, *Polymer* **20**, 305–310 (1979).
17. A. Kumar, S. K. Gupta, and R. Saraf, Condensation polymerization of ARB type monomers in CSTRs wherein the monomer is R times more reactive than other homologues, *Polymer* **21**, 1323–1326 (1980).
18. S. K. Gupta, A. Kumar, and R. Saraf, Condensation polymerization in ideal continuous flow-stirred tank reactors of monomers violating the equal reactivity hypothesis, *J. Appl. Polym. Sci.* **25**, 1049–1058 (1980).
19. A. S. Gupta, A. Kumar, and S. K. Gupta, Condensation polymerization with unequal reactivity in segregated HCSTRs, *British Polymer J.* **13**, 76–81 (1981).
20. L. H. Peebles, *Molecular Weight Distribution in Polymers*, 1st ed., Interscience, New York (1971).
21. K. S. Gandhi and S. V. Babu, Kinetics of step polymerization with unequal reactivities, *AIChE J.* **25**, 266–272 (1979).
22. H. Kilkson, Generalization of various polycondensation problems, *Ind. Eng. Chem. Fundam.* **7**, 354–363 (1968).

EXERCISES

1. Derive the mole balance equations of Table 3.1. From this derive the moment generation relations for λ_0, λ_1 and λ_2.

2. Assume that in the irreversible polymerization of ARB, P_1 and P_2 exhibit unequal reactivity. Extend the analysis of Section 3.3 for this case. Derive expressions for λ_0, λ_1, and λ_2 in batch reactors.

3. Derive moment generation relations for irreversible polymerization in HCSTRs for the special case of Problem 3.2.

4. Assuming that $k_{p,mn}$ is given by Eq. (3.2.3), derive expressions for moment generation of λ_0, λ_1, and λ_2. Do you think you need any moment closure relation? Extend this analysis to HCSTRs.

5. Consider the polymerization of glycerine with adipic acid. The end OH groups of glycerine have different reactivity compared to the middle one. Analyze this system for batch reactors.

6. Extend the analysis of Sections 3.4 and 3.5 to the polymerization of (AA + BC) monomers in HCSTRs.

NONLINEAR STEP GROWTH POLYMERIZATION

4.1. INTRODUCTION

The discussion in Chapters 2 and 3 has been limited to the polymerization of bifunctional monomers, which give linear polymer chains. There are several industrially important systems, e.g., polyesters from adipic acid (or phthalic anhydride) and glycerol (or pentaerythritol), curing of epoxy prepolymers with diamines, curing of phenol-formaldehyde prepolymers with hexamethylene tetramine, etc., which involve the use of compounds with functionalities more than two. In these polymerizations, branched molecules are formed at low conversions of functional groups. At some definite conversion some of these branched molecules convert into an infinite network structure of macroscopic dimensions, called gel. This phenomenon occurs much before the functional groups are completely consumed, and the point at which it occurs is referred to as the critical or gel point. Experimentally, the gel point is recognized as the state when the viscosity of the reaction mass becomes infinite and gas bubbles fail to rise up through it. Thereafter, the reaction mass is characterized by the presence of two "phases"—the infinite network or gel, interspersed randomly within which are finite polymer molecules, called the sol. If a solvent is added to this system, the sol fraction dissolves and can be extracted. The effect of the solvent on the infinite network is only to swell it—without dissolution. As the polymerization progresses, the larger chains of the sol are incorporated chemically onto the infinite network. The weight fraction as well as the average molecular weight of the sol decreases continuously till at complete conversion of functional groups the entire reaction mass is a single, giant (network) molecule.

Nonlinear step growth polymerizations are more complex than the linear cases. Several alternative approaches have been taken by different workers to model them. Flory[1-3] and Stockmayer[4-7] have approached this problem by determining the probabilities of finding various branched molecular structures in the reaction mass in much the same manner as discussed in Sections 2.6 and 3.4. Thereafter, they have used these probability distributions to compute the number and weight average molecular weights of the polymer before gelation. Their approach, however, becomes exceedingly complex for systems of industrial importance. Other workers[8-20] have attempted to derive the average molecular weights *directly*, without first obtaining the detailed distributions. These techniques are easier to use compared to the earlier theories of Flory and Stockmayer, particularly for more complex polymerization systems. One of these techniques, discussed by Macosko, Miller, and co-workers,[16-20] is presented here, since it is easier both conceptually as well as mathematically, and yields results which are identical to those obtained using the other methods. Only idealized systems are discussed in this chapter in order to present the concepts involved. More complex industrial systems have been treated in later chapters, e.g., polyurethanes and epoxies.

4.2. AVERAGE MOLECULAR WEIGHTS AND THE ONSET OF GELATION IN BATCH REACTORS

The polymerization of $R_1A_f + R_2B_2$ monomers is considered in detail, where f is greater than 2 and where the reaction is occurring between A and B functional groups. An example is urethane formation from pentaerythritol and 1,6 hexane diisocyanate. After some time of polymerization, in addition to the unreacted monomers, there are large and branched polymer chains in the reaction mass. The reaction can take place between functional groups on different polymer molecules (intermolecular reactions) as well as between functional groups on the same molecule (intramolecular reactions). The latter lead to the formation of ring or cyclic structures. It is anticipated that intramolecular reactions would gain in importance as larger molecules get formed near the gel point. However, these reactions are neglected in the present analysis to simplify the mathematical description of this complex process.

Two commonly used laws in probability theory are presented first, since these form the basis of the entire modeling of the polymerization of multifunctional monomers. The probability of an event Y can be written in terms of *conditional probabilities*[21] as

$$\text{Prob}(Y) = \sum_{i=1}^{n} \text{Prob}(Y/A_i)\,\text{Prob}(A_i) + \text{Prob}(Y/\bar{A})\,\text{Prob}(\bar{A}) \quad (4.2.1)$$

where Y/A_i represents the occurrence of event Y *given that* event A_i

$(i = 1, 2, \ldots, n)$ has occurred, $\mathrm{Prob}(Y/A_i)$ is the conditional probability of event Y occurring given that A_i has occurred and $\bar{A}$ is the event that none of the A_i events have occurred. Alternatively, it is possible to include $\bar{A}$ in the sample space of A_i, but this is not done to emphasize the importance of this term. Equation (4.2.1) is referred to as the law of total probability. Similarly, the law of total probability for expectations (average values) states that

$$E(Y) = \sum_{i=1}^{n} E(Y/A_i)\,\mathrm{Prob}(A_i) + E(Y/\bar{A})\,\mathrm{Prob}(\bar{A}) \qquad (4.2.2)$$

where $E(Y/A_i)$ is the expectation or average value of Y given that event A_i has occurred, and $\bar{A}$ is defined as above.

In the batch polymerization of $R_1A_f + R_2B_2$, the two monomers are assumed to have concentrations $[R_1A_f]_0$ and $[R_2B_2]_0$ initially. At any later time, a typical molecule in the reaction mass would have structures of the type shown in Fig. 4.1. At this stage, a fraction p_A of the A groups and a fraction p_B of the B groups are reacted, where the conversions p_A and p_B are defined in the usual manner $\{p_A = ([A]_0 - [A])/[A]_0; p_B = ([B]_0 - [B])/[B]_0\}$. The initial concentrations of the functional groups, $[A]_0$ and $[B]_0$, with $[A]_0 \equiv r[B]_0$, are given in terms of the concentrations of the molecular species as $[A]_0 = f[R_1A_f]_0$ and $[B]_0 = 2[R_2B_2]_0$. Evidently, the two conversions p_A and p_B are not independent, since stoichiometry implies that one A group reacts with only one B group. This leads to $p_B = rp_A$.

We now focus attention on an A group chosen randomly (e.g., group A' in Fig. 4.1). The expected value, $E(M_A^{\mathrm{out}})$, of the mass, M_A^{out}, that is found to be attached to this A group as one looks *out* in direction 1 (i.e., in a direction *outwards* of the monomeric unit of which A' is a part) is then computed. This can be obtained by realizing that there are two possibilities, viz., the randomly chosen A' group can either be reacted with an R_2B_2 unit or it can be unreacted. Equation (4.2.2) can then be written as

$$E(M_A^{\mathrm{out}}) = E(M_A^{\mathrm{out}}/\text{A reacted})\,\mathrm{Prob}(\text{A reacted})$$

$$+ E(M_A^{\mathrm{out}}/\text{A unreacted})\,\mathrm{Prob}(\text{A unreacted}) \qquad (4.2.3)$$

Since M_A^{out} is zero if A is unreacted, and $\mathrm{Prob}(\text{A reacted})$ is the same as

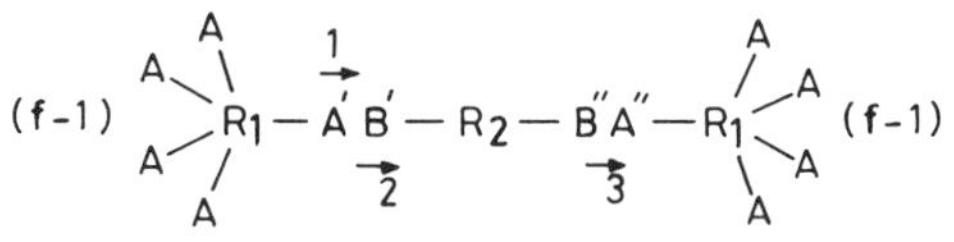

Figure 4.1. A typical structure in $R_1A_f + R_2B_2$ polymerization.

p_A, Eq. (4.2.3) simplifies to

$$E(M_A^{out}) = p_A E(M_A^{out}/A \text{ reacted}) \tag{4.2.4}$$

The mass attached to A' looking out, M_A^{out}, where A' is reacted, is exactly equal to M_B^{in}, the mass attached to group B' as one looks into the monomeric unit of which the latter is a part (in direction 2 in Fig. 4.1).[†] Thus, Eq. (4.2.4) can be rewritten as

$$E(M_A^{out}) = p_A E(M_B^{in}) \tag{4.2.5}$$

The expected value of M_B^{in} can be written as the sum of the molecular weight, $M_{R_2B_2}$, of R_2B_2 and the expected value of the mass attached to group B'', looking outwards in direction 3 (Fig. 4.1):

$$E(M_B^{in}) = M_{R_2B_2} + E(M_B^{out}) \tag{4.2.6}$$

where it is assumed that there is no byproduct formed by reaction. $E(M_B^{out})$, in turn, can be written using Eq. (4.2.2) as

$$E(M_B^{out}) = E(M_B^{out}/B \text{ reacted}) \text{ Prob(B reacted)}$$

$$+ E(M_B^{out}/B \text{ unreacted}) \text{ Prob(B unreacted)}$$

$$= p_B E(M_B^{out}/B \text{ reacted}) = p_B E(M_A^{in}) = rp_A E(M_A^{in}) \tag{4.2.7}$$

Finally, $E(M_A^{in})$ can be written in analogy with Eq. (4.2.6) as

$$E(M_A^{in}) = M_{R_1A_f} + (f-1)E(M_A^{out}) \tag{4.2.8}$$

where $M_{R_1A_f}$ is the molecular weight of R_1A_f. Equations (4.2.5)-(4.2.8) form a series of recursive relations which can be solved to give

$$E(M_A^{out}) = \frac{p_A M_{R_2B_2} + rp_A^2 M_{R_1A_f}}{1 + rp_A^2(f-1)} \tag{4.2.9a}$$

$$E(M_A^{in}) = \frac{M_{R_1A_f} + p_A(f-1)M_{R_2B_2}}{1 - rp_A^2(f-1)} \tag{4.2.9b}$$

[†] Note that even though $M_A^{out} = M_B^{in}$ for *any particular reacted* A group, $E(M_A^{out}) \neq E(M_B^{in})$. This is because one must appropriately sum up over both reacted as well as unreacted A groups (for which $M_B^{in} = 0$) in the reaction mass to get $E(M_A^{out})$. Similar arguments hold for $E(M_B^{out})$.

$$E(M_B^{out}) = \frac{rp_A\{M_{R_1A_f} + p_A(f-1)M_{R_2B_2}\}}{1 - rp_A^2(f-1)} \qquad (4.2.9c)$$

$$E(M_B^{in}) = \frac{M_{R_2B_2} + rp_A M_{R_1A_f}}{1 - rp_A^2(f-1)} \qquad (4.2.9d)$$

If $E(M_{R_1A_f}^*)$ is the expected mass of the entire molecule to which a randomly chosen R_1A_f unit belongs, and $E(M_{R_2B_2}^*)$ is the expected mass of the entire molecule to which a randomly chosen R_2B_2 unit belongs, then

$$E(M_{R_1A_f}^*) = E(M_A^{in}) + E(M_A^{out})$$

$$E(M_{R_2B_2}^*) = E(M_B^{in}) + E(M_B^{out}) \qquad (4.2.10)$$

The weight average molecular weight is then given by

$$\bar{M}_w = (\text{mass fraction of } R_1A_f \text{ units})E(M_{R_1A_f}^*)$$

$$+ (\text{mass fraction of } R_2B_2 \text{ units})E(M_{R_2B_2}^*) \qquad (4.2.11)$$

The final equation is given in Table 4.1 [Eq. (a)]. It may be mentioned that the use of mass fraction as a weighting factor in Eq. (4.2.11) means physically that a *unit of mass* is chosen at random and the expected mass of that molecule is computed, of which this "unit mass" is a part. This is similar to what has been done in Chapter 1 in defining the weight average chain length or molecular weight for ARB polymerization, for which [Eq. (1.3.4)]

$$\mu_w = \sum_{n=1}^{\infty} \left(\frac{n[P_n]}{\sum_m m[P_m]}\right)n \propto \sum_{n=1}^{\infty} (\text{mass frac. of } P_n)(\text{mass of } P_n) \quad (4.2.12)$$

Equation (4.2.11), therefore, gives the weight average molecular weight and not the number average value.

Equation (a) of Table 4.1 is identical to the expression derived by Stockmayer[6,7] using combinatorial arguments. It is seen that $\bar{M}_w$ is a function of the initial reactant ratio, r, and the conversion, p_A, of A functional groups (or alternatively, the time of reaction) for a given system (i.e., with f, $M_{R_1A_f}$, and $M_{R_2B_2}$ specified). Figure 4.2 shows[16] the weight average *chain length* for some systems (including the linear polymerization of $AR_1A + BR_2B$) when stoichiometric amounts of functional groups are taken initially. It is observed that $\bar{M}_w$ rises sharply to infinity at values of $p_A < 1$ for multifunctional systems. In fact, Eq. (a) of Table 4.1 gives $\bar{M}_w \to \infty$ when

$$p_{A,c}^2 = \frac{1}{r(f-1)} \qquad (4.2.13)$$

where the subscript c indicates the critical point or the onset of gelation.

TABLE 4.1. Final Equations for $\bar{M}_w$ and $\text{Prob}(F_A^{\text{out}})$, for Nonlinear Polymerization

---

$R_1A_f + R_2B_2$

$\bar{M}_w$:

$$\bar{M}_w = \frac{(2r/f)(1 + rp_A^2)M_{R_1A_f}^2 + \{1 + (f-1)rp_A^2\}M_{R_2B_2}^2 + 4p_A r M_{R_1A_f} M_{R_2B_2}}{[(2r/f)M_{R_1A_f} + M_{R_2B_2}]\{1 - r(f-1)p_A^2\}} \qquad (a)$$

$R_{1i}A_{fi} \; (i = 1, 2, \ldots, n) + R_{2j}B_{gj} \; (j = 1, 2, \ldots, l)$

$\bar{M}_w$:

$$\bar{M}_w = \frac{p_B m_a' + p_A m_b'}{p_B m_a + p_A m_b} + \frac{p_A p_B \{p_A(f_e - 1)M_b^2 + p_B(g_e - 1)M_a^2 + 2M_a M_b\}}{(p_B m_a + p_A m_b)\{1 - p_A p_B(f_e - 1)(g_e - 1)\}} \qquad (b)$$

where $\quad p_B = rp_A \qquad (c)$

$$m_a = \sum_{i=1}^{n} M_{R_{1i}A_{fi}}[R_{1i}A_{fi}]_0 \bigg/ \sum_{i=1}^{n} f_i[R_{1i}A_{fi}]_0 \qquad (d)$$

$$m_a' = \sum_{i=1}^{n} M_{R_{1i}A_{fi}}^2[R_{1i}A_{fi}]_0 \bigg/ \sum_{i=1}^{n} f_i[R_{1i}A_{fi}]_0 \qquad (e)$$

$$a_{fi} = \frac{f_i[R_{1i}A_{fi}]_0}{\sum_{i=1}^{n} f_i[R_{1i}A_{fi}]_0} \qquad (f)$$

$$f_e = \sum_{i=1}^{n} f_i a_{fi} \qquad (g)$$

$$M_a = \sum_{i=1}^{n} M_{R_{1i}A_{fi}} a_{fi} \qquad (h)$$

with similar expressions for m_b, m_b', g_e, b_{gj}, and M_b in terms of the $R_{2j}B_{gj}$.

$\text{Prob}(F_A^{\text{out}}) \equiv x; \; 0 \leqslant x \leqslant 1$:

$$p_A \sum_{j=1}^{l} b_{gj} \left\{ 1 - p_B + p_B \sum_{i=1}^{n} a_{fi} x^{(f_i - 1)} \right\}^{(g_j - 1)} - x - p_A + 1 = 0 \qquad (i)$$

R_1A_f (self-condensation of A groups)

$\bar{M}_w$:

$$\bar{M}_w = M_{R_1A_f} \frac{1 + p_A}{1 - p_A(f-1)} \qquad (j)$$

$\text{Prob}(F_A^{\text{out}}) \equiv x$:

$$p_A x^{(f-1)} - x - p_A + 1 = 0 \qquad (k)$$

Macosko and Miller[16] have extended the above analysis to the polymerization of mixtures of $R_{1i}A_{fi}$ $(i = 1, 2, \ldots, n)$ and $R_{2j}B_{gj}$ $(j = 1, 2, \ldots, l)$ monomers, as well as to the self-polycondensation of R_1A_f monomers (A reacting with A). The corresponding expressions are also given in Table 4.1.

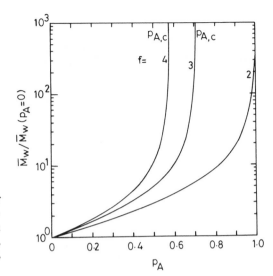

Figure 4.2. Polymerization[16] of $R_1A_f + R_2B_2$ systems. $r = 1$, $p_A = p_B$. (Reprinted with permission from Ref. 16. Copyright 1976, American Chemical Society, Washington, DC.)

These equations give the following expressions for the gel point:

$$p_{A,c}p_{B,c} = \frac{1}{(f_e - 1)(g_e - 1)}; \quad R_{1i}A_{fi} + R_{2j}B_{gj} \qquad (4.2.14a)$$

$$p_{A,c} = \frac{1}{f - 1}; \quad R_1A_f \qquad (4.2.14b)$$

These expressions are also consistent with those obtained by Stockmayer[4-7] and Flory.[1-3]

The number average molecular weight, $\bar{M}_n$, can be derived easily for the polymerization of $R_{1i}A_{fi} + R_{2j}B_{gj}$ as the ratio of the total mass and total moles per unit volume, and is given by

$$\bar{M}_n = \frac{\sum_{i=1}^{n} M_{R_{1i}A_{fi}}[R_{1i}A_{fi}]_0 + \sum_{j=1}^{l} M_{R_{2j}B_{gj}}[R_{2j}B_{gj}]_0}{\sum_{i=1}^{n} [R_{1i}A_{fi}]_0 + \sum_{j=1}^{l} [R_{2j}B_{gj}]_0 + p_A \sum_{i=1}^{n} f_i[R_{1i}A_{fi}]_0} \qquad (4.2.15)$$

assuming no by-product formation and no intramolecular reactions. Figure 4.3 shows[16] how $\bar{M}_n$ remains finite while $\bar{M}_w \rightarrow \infty$ at the gel point.

Early experimental results[1,3] on some systems reveal that the experimental values of $p_{A,c}$ are slightly above theoretically predicted values. This discrepancy has been attributed to the presence of intramolecular reactions, particularly near the gel point, as well as to the unequal reactivity of functional groups. Results[22] on the polymerization of pentaerythritol and

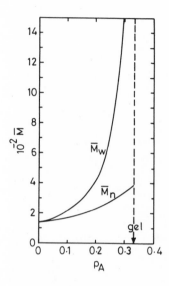

Figure 4.3. $\bar{M}_w$ and $\bar{M}_n$ for polymerization of pentaerythritol,[16] $C(CH_2OH)_4$. (Reprinted with permission from Ref. 16. Copyright 1976, American Chemical Society, Washington, DC.)

adipic acid $(R_1A_4 + R_2B_2)$ with $r = 1$, for example, give the experimentally determined critical conversion as 0.630 while the value predicted by Eq. (4.2.13) is 0.577. However, the value of the critical conversion, $p_{A,c}$, in the presence of diluents is 0.578 ± 0.005 when extrapolated such that $\{1/(\text{reactant concentration})\} \to 0$ (so that there are no intramolecular reactions). This is in excellent agreement with the theoretical value. A more recent compilation of gel point data,[23,24] however, indicates that there is still considerable disagreement among different workers in this area.

The results derived in this section using the expectation theory approach of Macosko and Miller have also been derived using other techniques. A short discussion of these is now presented. Flory[1-3] was the first to study nonlinear polymerizations to determine the gel point by evaluating a *branching coefficient*. For the polymerization of $R_1A_3 + R_2A_2 + R_3B_2$ monomers, with A reacting with B only, it is observed that a typical substructure of a polymer molecule is

$$
\begin{array}{c}
A \\
\diagdown \\
R_1-A\!\!\left[\!\!BR_3B-AR_2A\right]_i\!\!-BR_3B-A-R_1\!\!\diagup \\
\diagup \diagdown \\
A
\end{array}
$$

with i $-B-R_3B-AR_2A-$ repeat units $(i = 0, 1, 2, \ldots)$ between the trifunctional groups. If p_A is the fraction of (all) A groups reacted and p_B is the fraction of B groups reacted, the probability of finding the above substructure can be written as [see Section 3.4 or discussion preceding Eq. (2.6.13)]

$$\text{Prob}(i) = p_A\{p_B(1 - a_f)p_A\}^i\{p_B a_f\}$$

$$= p_A p_B a_f \{p_A p_B (1 - a_f)\}^i \qquad (4.2.16)$$

where a_f is the initial fraction of *all* As on R_1A_3. The term $p_B(1 - a_f)p_A$ gives the probability of the sequence $-B-AR_2-AB-$, which is the product of the probabilities that a B group has reacted with an A of R_2A_2, and that the other A of this R_2A_2 has reacted with B of R_3B_2. The summation of Prob(i) for all values of i (from 0 to ∞) gives the probability that an A group of R_1A_3 ends up in another trifunctional group, and is called the branching coefficient, α. It can be easily derived as

$$\alpha = \frac{p_A p_B a_f}{1 - p_A p_B (1 - a_f)} = \frac{r p_A^2 a_f}{1 - (1 - a_f) r p_A^2} \qquad (4.2.17)$$

The two forms of α in this equation are obtained using stoichiometry. If $\alpha = 0.5$, there is a 100% likelihood that a chain leads to an infinite network (since each A group of R_1A_3 has a 50% probability of leading to *two* other chains on another R_1A_3), and so, for gelation in this system,

$$\left.\frac{r p_A^2 a_f}{1 - (1 - a_f) r p_A^2}\right|_c = \frac{1}{2} \qquad (4.2.18)$$

This is identical to Eq. (4.2.14a) since $g_e = 2$, $a_{f1} = a_f$, and $f_e = a_f + 2$ for this system.

Stockmayer[4-7] has analyzed several other systems of multifunctional polymerizations using this probabilistic technique. However, his method becomes very tedious for systems of practical importance wherein unequal reactivity of functional groups, etc., may be involved. Gordon and co-workers[8-14] have used the theory of stochastic branching processes,[15] also called the cascade theory, with vectorial probability generating functions to compute molecular weight averages directly, without computing the distributions. Their technique, though general, is conceptually quite complex, and is once again tedious to apply to systems of practical importance. The kinetic approach has also been used recently[23,25-31] to predict chain length distributions and the onset of gelation, both for batch reactors[23,25-28] as well as for homogeneous continuous-flow stirred tank reactors (HCSTRs).[29-31] The kinetic approach is particularly well suited for modeling nonlinear polymerizations in HCSTRs. In fact, probabilistic techniques have not yet been applied for this purpose. For batch reactors, the kinetic as well as probabilistic methods give identical results for most cases. Statistically speaking, however, these approaches are not equivalent because in the kinetic approach, the integrity of the existing structures and the information on the formation history is preserved, whereas in the approach of Gordon and co-workers, where the *final* assembly of molecules is generated from building blocks, it is not. In fact, Mikeš and Dušek[32] have shown recently using a Monte Carlo method for simulating these systems in batch reactors that even though in most cases these two techniques give virtually the same results, there are *some* special cases where they differ.

4.3. POSTGEL PROPERTIES FOR NONLINEAR POLYMERIZATIONS IN BATCH REACTORS

In the previous section, attention was focused on the pregel region of nonlinear step growth polymerizations in batch reactors. The most important information was the conversion at which the gel point occurred, which places an upper limit on the conversion that can be achieved in flow reactors. Beyond the gel point, the material behaves more like a "solid" and in this region, *in situ* polymerization in molds is required. In this section, a few important properties for the postgel region are being considered, as for example, the weight fraction of the sol, the concentration of elastically effective network or cross-link points, weight fraction of pendant chains on the gel, etc. These molecular characteristics determine the mechanical properties of the gelled mass and so are important for reactor design. Once again, the expectation theory of Macosko and Miller[16-20] is presented because of its conceptual simplicity.

The polymerization of $R_1A_f + R_2B_2$ monomers is analyzed once again since this illustrates the major features of the procedure to be followed. In Fig. 4.1, an A group, say A', can be selected randomly, and then the expected value of the molecular weight looking *out* in direction 1 can be computed. There can be either a finite chain attached to A', or an infinite chain leading to the walls of the container. F_A^{out} is defined as the event that a finite chain is attached to A' in direction 1. According to Eq. (4.2.1), the probability of F_A^{out} can be written as

$$\text{Prob}(F_A^{out}) = \text{Prob}(F_A^{out}/A \text{ reacted}) \text{ Prob}(A \text{ reacted})$$

$$+ \text{Prob}(F_A^{out}/A \text{ unreacted}) \text{ Prob}(A \text{ unreacted}) \quad (4.3.1)$$

As in the previous section, the probability of finding a reacted A group is p_A, and that of an unreacted A group is $1 - p_A$. The probability that the chain is finite when A is unreacted is unity. Hence

$$\text{Prob}(F_A^{out}) = p_A \text{ Prob}(F_A^{out}/A \text{ reacted}) + (1 - p_A) \quad (4.3.2)$$

The probability of finding a finite chain from A', when A' is reacted, is the same as $\text{Prob}(F_B^{in})$, where F_B^{in} is the event that there is a finite chain starting from B', looking *inwards* in direction 2. Hence,

$$\text{Prob}(F_A^{out}) = p_A \, \text{Prob}(F_B^{in}) + (1 - p_A) \qquad (4.3.3)$$

The probability of finding a finite chain from B', looking inwards, is the same as the probability of finding a finite chain from B", looking out in direction 3 in Fig. 4.1. Thus,

$$\text{Prob}(F_B^{in}) = \text{Prob}(F_B^{out}) \qquad (4.3.4)$$

Using Eq. (4.2.1) again, one has

$$\text{Prob}(F_B^{out}) = \text{Prob}(F_B^{out}/B \text{ reacted}) \, \text{Prob}(B \text{ reacted})$$

$$+ \text{Prob}(F_B^{out}/B \text{ unreacted}) \, \text{Prob}(B \text{ unreacted})$$

$$= p_B \, \text{Prob}(F_A^{in}) + (1 - p_B) \qquad (4.3.5)$$

Finally, to complete the recursive relations, it is observed that finite chains can be obtained looking inwards from A" only when each of the other $(f - 1)$ A's have finite chains on looking outwards, i.e.,

$$\text{Prob}(F_A^{in}) = \{\text{Prob}(F_A^{out})\}^{f-1} \qquad (4.3.6)$$

Equations (4.3.3)–(4.3.6) give the following equation for $\text{Prob}(F_A^{out})$:

$$rp_A^2 \{\text{Prob}(F_A^{out})\}^{f-1} - \{\text{Prob}(F_A^{out})\} + 1 - rp_A^2 = 0 \qquad (4.3.7)$$

Since $\text{Prob}(F_A^{out}) = 1$ is one of the roots of this equation, it can be factored out to give

$$\text{for } f = 3: \quad (x - 1)\left(x - \frac{1 - rp_A^2}{rp_A^2}\right) rp_A^2 = 0 \qquad (4.3.8a)$$

$$\text{for } f = 4: \quad (x - 1)\left\{x - \left[\left(\frac{1}{rp_A^2} - \frac{3}{4}\right)^{1/2} - \frac{1}{2}\right]\right\}$$

$$\times \left\{x - \left[-\left(\frac{1}{rp_A^2} - \frac{3}{4}\right)^{1/2} - \frac{1}{2}\right]\right\} rp_A^2 = 0 \qquad (4.3.8b)$$

where

$$x \equiv \text{Prob}(F_A^{out}) \qquad (4.3.8c)$$

Roots for higher values of f can be determined numerically. The third root in Eq. (4.3.8b) is unrealistic and so the two feasible roots are

$$\text{for } f = 3: \quad x = 1, \qquad \left\{ \frac{1}{rp_A^2} - 1 \right\} \tag{4.3.9a}$$

$$\text{for } f = 4: \quad x = 1, \qquad \left\{ + \left(\frac{1}{rp_A^2} - \frac{3}{4} \right)^{1/2} - \frac{1}{2} \right\} \tag{4.3.9b}$$

The interpretation of Eq. (4.3.9) is very interesting for the following reason. For $f = 3$, Fig. 4.4 shows plots of x vs. p_A for some values of r. The function is monotonically decreasing and x lies between 0 and 1 only in the physically meaningful range $[1/(2r)^{1/2}] \leq p_A \leq 1$ or $[1/(2r)^{1/2}] \leq p_A \leq (1/r^{1/2})$, depending on whether $r \geq 1$ or $r \leq 1$. In this range of p_A, $x = 1/rp_A^2 - 1$ is one of the roots of Eq. (4.3.8a) and $\text{Prob}(F_A^{\text{out}}) \leq 1$, i.e., there is a possibility that a randomly chosen A group is part of an infinite network. Alternatively speaking, part of the reaction mass is a gel if $[1/(2r)^{1/2}] \leq p_A \leq 1$ or $[1/(2r)^{1/2}] \leq p_A \leq (1/r^{1/2})$. For $p_A \leq [1/(2r)^{1/2}]$, $x = (1 - rp_A^2)/(rp_A^2)$ lies beyond the physically meaningful range of $0 \leq x \leq 1$ and therefore, the only meaningful solution of Eq. (4.3.8a) is $x = \text{Prob}(F_A^{\text{out}}) = 1$. This indicates that for $p_A \leq [1/(2r)^{1/2}]$, the probability of finding any A group as part of

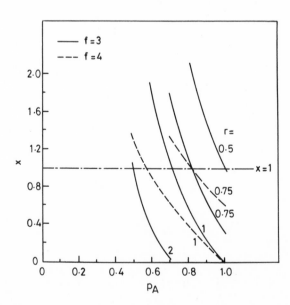

Figure 4.4. Plots of $x\{\equiv \text{Prob}(F_A^{\text{out}})\}$ vs. p_A as a function of r. x is given by the second of the roots in Eq. (4.3.9).

an infinite network is zero, i.e., gelation has not yet started. $p_A = 1/(2r)^{1/2}$ thus represents the gel point for this system. This is also consistent with Eq. (4.2.13).

The case of $f = 4$ is similar to that for $f = 3$. Once again, the second root $x = [(1/rp_A^2 - \frac{3}{4})^{1/2} - \frac{1}{2}]$ in Eq. (4.3.9b) decreases monotonically with p_A. x has physically meaningful values of $0 \leq x \leq 1$ only for p_A lying between $1/(3r)^{1/2}$ [the gel point, consistent with Eq. (4.2.13)] and $p_A = 1$ (for $r \leq 1$) or $p_A = 1/r^{1/2}$ (for $r > 1$, at which point A is completely reacted).

The mass fraction of the sol can be obtained as follows. For the $R_1A_f + R_2B_2$ system being discussed, a randomly chosen R_1A_f unit is a part of the sol only if each of its f A groups are linked to finite chains. Similarly, a randomly chosen R_2B_2 unit is a part of the sol if each of its two B groups is linked to a finite chain. Hence the mass fraction of the sol, w_s, can be easily written as the weighted sum (using mass fractions) of these two probabilities as

$$w_s = 1 - w_g = \frac{[R_1A_f]_0 M_{R_1A_f} \{Prob(F_A^{out})\}^f + [R_2B_2]_0 M_{R_2B_2} \{Prob(F_B^{out})\}^2}{M_{R_1A_f} [R_1A_f]_0 + [R_2B_2]_0 M_{R_2B_2}}$$

(4.3.10)

where w_g is the mass fraction of the gel in the reaction mass. Equation (4.3.10) can be used with Eq. (4.3.9) (or its generalization for higher f) and the following equation for $Prob(F_B^{out})$ [derived from Eqs. (4.3.5) and (4.3.6)]

$$Prob(F_B^{out}) = 1 - rp_A(1 - x^{f-1})$$

(4.3.11)

The above derivations can easily be generalized[17] for the polymerization of mixtures of $R_{1i}A_{fi}$ ($i = 1, 2, \ldots, n$) and $R_{2j}B_{gj}$ ($j = 1, 2, \ldots, l$) and for the self-polycondensation of R_1A_f monomers. Equations similar to Eq. (4.3.7) for these systems are given in Table 4.1. The results obtained match well with those derived by Flory[1-3] and Gordon[8] using different approaches.

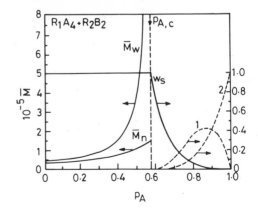

Figure 4.5. Various properties[17] for the polymerization of R_1A_4 {(HCH$_3$—⟨O⟩—Si—O)$_4$—Si} with R_2B_2 {vinyl terminated monodisperse polydimethylsiloxane, $M_{R_2B_2} = 5000$}. $r = 1$. Curve 1 is $Prob(X_{3,4})$ and Curve 2 is $Prob(X_{4,4})$. (Reprinted with permission from Ref. 17. Copyright 1976, American Chemical Society, Washington, DC.)

Figure 4.5 shows results for an $R_1A_4 + R_2B_2$ system.[17] It is observed that w_s is almost negligible after about 90% conversion of A groups. It may be noted that $w_s = 1$ before the gel point.

The weight fraction of the sol is a good characteristic of the progress of polymerization and can be measured experimentally. Another molecular parameter, which is of importance in determining the mechanical strength of the gel, is the concentration, μ_e, of elastically "effective" network *chains or strands*. This, in turn, is related to the concentration, ν_e, of elastically effective network *junctions*[1,33] or cross-link *points*. These are shown schematically in Fig. 4.6. In fact, the tensile or shear modulus G, of the gelled material, as obtained by creep experiments, is related to μ_e by the Flory[1,33] equation

$$G = \mu_e RT \qquad (4.3.12)$$

where R is the universal gas constant and T is the absolute temperature. An effective cross-link *point* in an $R_1A_f + R_2B_2$ system is formed when *at least* three of the A's of an R_1A_f unit are connected to infinite chains (Fig. 4.6). The probability of $X_{n,f}$, defined as the event that *any n* of the f A groups in $R_1A_f(n \le f)$ lead to infinite chains, is given by

$$\text{Prob}(X_{n,f}) = \{^fC_n\}\{\text{Prob}(F_A^{\text{out}})\}^{(f-n)}\{1 - \text{Prob}(F_A^{\text{out}})\}^n \qquad (4.3.13)$$

where $\{^fC_n\}$ is the total number of distinct ways in which n A groups can be chosen out of f, and is equal to $f!/n!(f-n)!$. $\text{Prob}(X_{3,4})$ and $\text{Prob}(X_{4,4})$, as computed from Eq. (4.3.13), are shown in Fig. 4.5 for the $R_1A_4 + R_2B_2$ system. It is observed that beyond the gel point, trifunctional cross-links are formed first. After some time, these decrease gradually, with tetrafunctional cross-links increasing. The concentration, ν_e, of elastically effective network *junctions*, can then be written as

$$\nu_e = [R_1A_f]_0 \sum_{n=3}^{f} \text{Prob}(X_{n,f}) \qquad (4.3.14)$$

The concentration, ν_e, of network junctions increases from zero at the gel point to the asymptotic value of $[R_1A_f]_0$ when the A groups are completely reacted. In a mixture of multifunctional $R_{2j}B_{gj}$'s, additional terms must be included in Eq. (4.3.14) to account for effective cross-links of the B type, as discussed in Ref. 17.

An expression for μ_e for the $R_1A_f + R_2B_2$ system can now be derived. It can be seen from Fig. 4.6 that a network junction of "strength" n (having n A groups leading to infinite chains) contributes $n/2$ strands to the network structure, since each elastically effective strand lies between two elastically

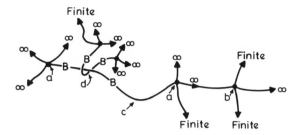

Figure 4.6. Elastically effective (a) and ineffective (b) network *junctions*, elastically effective network strand (c), and a trapped entanglement (d) shown schematically.

effective junctions. Hence,

$$\mu_e = [R_1A_f]_0 \sum_{n=3}^{f} \frac{n}{2} \text{Prob}(X_{n,f}) \tag{4.3.15}$$

In the above discussion, only chemical cross-links were considered. In addition to these, physical entanglements may be trapped[33] in between consecutive, elastically effective network junctions, as shown schematically in Fig. 4.6. These trapped entanglements increase the resistance to chain motion, and so contribute to the tensile or shear modulus, G. Equation (4.3.12) must be modified to account for this effect. If G_e is the effective contribution to the modulus due to each physical entanglement, Eq. (4.3.12) can be written as

$$G = \mu_e RT + G_e T_e \tag{4.3.16}$$

where T_e is the probability of forming a trapped entanglement. G_e is usually a curve-fit parameter, and can be approximated as G_N^0, the modulus of the rubbery plateau obtained from experiments on *linear* polymers having similar backbones as the gel. In the polymerization of R_1A_f with long molecules of R_2B_2, an entanglement is described by the fact that when one looks inside from any of the four B's around the entanglement (Fig. 4.6), one finds an infinite chain, and so

$$T_e = \{1 - \text{Prob}(F_B^{in})\}^4 \tag{4.3.17}$$

Further refinements of rubber elasticity theory[1,33] account for the fluctuations of the junctions and give

$$G = (\mu_e - \nu_e)RT + G_e T_e \tag{4.3.18}$$

The last postgel property discussed herein is the mass fraction of pendant (dangling) chains in the gel.[19] This property is related to another mechanical property, viz., the dynamic loss modulus,[33] G'', of the rubbery network. The method of computing this molecular parameter is again illustrated for $R_1A_f + R_2B_2$ polymerization. The probability that an R_2B_2 unit selected at random is part of a pendant chain (see Fig. 4.7) is given by

$$\text{Prob}(P_{R_2B_2}) = \{^2C_1\} \, \text{Prob}(F_B^{out})\{1 - \text{Prob}(F_B^{out})\} \qquad (4.3.19)$$

where $\{^2C_1\}$ accounts for the two distinct ways in which an R_2B_2 unit can lie in a pendant chain, with one B connected to a finite chain and the other B to an infinite chain. In order to obtain similar expressions for randomly selected R_1A_f units, it is observed that there are two ways in which these units can be present in a gel, as shown in Fig. 4.7. One of these is marked R_1', and represents R_1A_f units lying in the intermediate positions in pendant chains. Such units have only one A connected to an infinite network, while the remaining $(f-1)$ A groups are connected to finite chains. The probability of finding such R_1A_f groups is given by

$$\text{Prob}(P_{R_1'A_f}) = \{^fC_1\}\{\text{Prob}(F_A^{out})\}^{(f-1)}\{1 - \text{Prob}(F_A^{out})\} \qquad (4.3.20)$$

The multiplying factor of $\{^fC_1\}$ and the exponent $f-1$ appear because any one of the f A groups can be connected to an infinite network. It is to be noted that the entire $R_1'A_f$ unit is part of the pendant chain. The other type of R_1A_f units are those marked R_1^* in Fig. 4.7. These are the R_1A_f units in which two or more A groups are connected to infinite networks and only a *part* of the entire unit contributes to the pendant chain. If i of the A

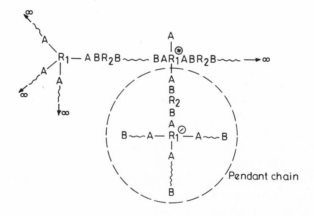

Figure 4.7. Placing of R_1A_4 units in pendant chains in a network.

groups ($i = 1, 2, \ldots, f - 2$) in $R_1^* A_f$ are finite, the probability of the occurrence of such units is

$$\text{Prob}(P_{R_1^* A_{f,i}}) = \{^f C_i\} \{\text{Prob}(F_A^{\text{out}})\}^i \{1 - \text{Prob}(F_A^{\text{out}})\}^{(f-i)} \quad (4.3.21)$$

The mass fraction of pendant material, w_P, is obtained by appropriate summation as

$$w_P = \left\{ [R_2 B_2]_0 M_{R_2 B_2} \text{Prob}(P_{R_2 B_2}) + M_{R_1 A_f} [R_1 A_f]_0 \text{Prob}(P_{R_1 A_f}) \right.$$

$$\left. + M_{R_1 A_f} [R_1 A_f]_0 \sum_{i=1}^{f-2} \frac{i M_A}{M_{R_1 A_f}} \text{Prob}(P_{R_1^* A_{fi}}) \right\}$$

$$\times \{ [R_2 B_2]_0 M_{R_2 B_2} + M_{R_1 A_f} [R_1 A_f]_0 \}^{-1} \quad (4.3.22)$$

where M_A is the molecular weight of one branch of $R_1 A_f$. The term $i M_A / M_{R_1 A_f}$ is chosen to be the weighting factor for $\text{Prob}(P_{R_1^* A_{fi}})$ so that the entire mass of the $R_1^* A_f$ unit is not added to the pendant chain. Figure 4.8 shows how w_P varies with conversion[19] for an $R_1 A_3 + R_2 B_2$ system. It is to be noted that the mass fraction of elastically effective networks, w_e, will be given by

$$w_e = 1 - w_s - w_P \quad (4.3.23)$$

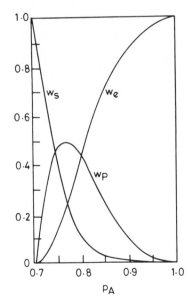

Figure 4.8. Mass fractions of sol, pendant groups, and elastically effective networks as a function of conversion for the $R_1 A_3 + R_2 B_2$ system, beyond the gel point. $r = 1$. $M_{R_1 A_3} = 329$: [$HSi(CH_3)_2 O \!\!\!+_3 \!\!\!-Si-C_6 H_5$; $M_{R_2 B_2} = 11600$ (vinyl terminated PDMS: $Vi-C_6 H_5-CH_3- SiO\!\!\!+\!Si(CH_3)_2 O\!\!\!+_n Si-CH_3-C_6 H_5-Vi$). (Reprinted from Ref. 19 with permission of Soc. Plast. Engr., Brookfield Center, Connecticut.)

The value of w_P goes through a maximum at some intermediate conversion, which matches the qualitative behavior of the loss modulus, G'', vs. conversion[34] plot. The peaks in both these plots are found to occur at about the same conversion.

Macosko and co-workers have used their technique to derive expressions for many other molecular characteristics of the product from network polymerizations, which are related to their physical and mechanical properties. These include[19] the weight average number of branches per molecule, weight average number of chains per molecule, and the weight average molecular weight of the longest chain passing through a branched molecule, all in the pregel region. These molecular parameters are important because they determine the flow properties of the reaction mass before it gels. Similarly, expressions for the extent of reaction *in the sol*, fraction of trapped entanglements and the weight average molecular weight of network chains, have been derived for the postgel region, since these are related to rheological properties like G', G'', G, etc.[3] Many other systems, e.g., polymerization of oligomeric mixtures,[19,35] the cross-linking or vulcanization of polymers including those with unsaturation on the backbone, and network formation by chain (addition) polymerization have also been studied. It is evident that nonlinear polymerizations present a more fascinating field of study than linear polymerizations in view of the wide range of important molecular characteristics that can be used to describe the product formed. Very little work is available on how to tinker with reactor operating variables in nonlinear polymerizations in order to manufacture a product having a specified range of these characteristics, even though one has some idea of the effect of each of these variables. As in the polymerization of bifunctional monomers, the effect of unequal reactivity of functional groups is important in multifunctional polymerization, and is presented next.[18,20]

4.4. EFFECT OF UNEQUAL REACTIVITY[18,20] OF FUNCTIONAL GROUPS

As discussed in Chapter 3, several polymerizations of industrial interest involve functional groups which are characterized by unequal reactivity. Macosko and Miller[18,20] have extended their technique to study systems exhibiting several types of unequal reactivity of functional groups. In this section, however, only one typical case is presented and it is left to the reader to derive results for the other cases as an exercise. The system chosen is the polymerization of asymmetric monomers, $R_1A_f + R_3A_2 + R_2BC$, with A reacting with B or C at different rates (rate constants k_1 and k_2, respectively). The conversions, p_A, p_B, and p_C, are defined in the usual

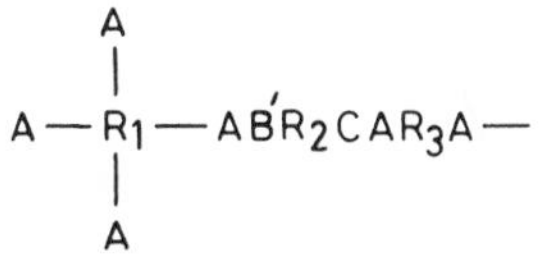

$$
\begin{array}{c}
\mathsf{A} \\
| \\
\mathsf{A}-\mathsf{R_1}-\mathsf{A}\overset{'}{\mathsf{B}}\mathsf{R_2}\,\mathsf{C}\,\mathsf{A}\mathsf{R_3}\mathsf{A}- \\
| \\
\mathsf{A}
\end{array}
$$

Figure 4.9. Structure for study of $R_1A_f + R_3A_2 + R_2BC$ system.

manner with $p_{BC} \equiv \{[B]_0 + [C]_0 - ([B] + [C])\}/([B]_0 + [C]_0) = rp_A$, where $r = [A]_0/\{[B]_0 + [C]_0\}$.

In order to apply the expectation theory to this system the structure shown in Fig. 4.9 is considered. It is possible to derive the following expressions using concepts developed in earlier sections:

$$E(M_B^{out}) = E(M_{A_f}^{in})p_B a_f + E(M_{A_2}^{in})p_B(1 - a_f) \tag{4.4.1a}$$

$$E(M_C^{out}) = E(M_{A_f}^{in})p_C a_f + E(M_{A_2}^{in})p_C(1 - a_f) \tag{4.4.1b}$$

$$E(M_{A_f}^{in}) = M_{R_1A_f} + (f - 1)E(M_A^{out}) \tag{4.4.1c}$$

$$E(M_{A_2}^{in}) = M_{R_3A_2} + E(M_A^{out}) \tag{4.4.1d}$$

$$E(M_A^{out}) = 0 \times (1 - p_A) + E(M_B^{in})p_A\frac{p_B}{p_B + p_C} + E(M_C^{in})p_A\frac{p_C}{p_B + p_C} \tag{4.4.1e}$$

$$E(M_B^{in}) = M_{R_2B_2} + E(M_C^{out}) \tag{4.4.1f}$$

$$E(M_C^{in}) = M_{R_2B_2} + E(M_B^{out}) \tag{4.4.1g}$$

where a_f is the initial fraction of A groups on $R_1A_f\{=f[R_1A_f]_0/(f[R_1A_f]_0 + 2[R_3A_2]_0)\}$. The expected mass attached to a randomly chosen R_1A_f unit is $M_{R_1A_f} + fE(M_A^{out})$; that on an R_3A_2 is $M_{R_3A_2} + 2E(M_A^{out})$; on an R_2BC it is $M_{R_2BC} + E(M_B^{out}) + E(M_C^{out})$. These, along with the expressions for the mass fractions, can be used to give an expression for $\bar{M}_w$ in a manner similar to Eq. (4.2.11). At the gel point, $\bar{M}_w$ goes to infinity. The only other information required to solve the entire set of equations developed above is the relationship between p_B and p_C. This is obtained from stoichiometry. This is the same as Eq. (3.4.29). It can be shown easily that the gel point is given by

$$1 - \frac{p_{B,c}p_{C,c}}{r}\{1 + (f - 2)a_f\} = 0 \tag{4.4.2}$$

The postgel properties are similarly evaluated. A set of recursive relations for $\text{Prob}(F_A^{out})$, $\text{Prob}(F_B^{out})$, etc., can now be written. Some of the equations are simple, while the others are given below:

$$\text{Prob}(F_A^{out}) = \{(1 - p_A) \times 1\} + p_A \left\{ \frac{p_B}{p_B + p_C} \text{Prob}(F_B^{in}) + \frac{p_C}{p_B + p_C} \text{Prob}(F_C^{in}) \right\}$$

(4.4.3a)

$$\text{Prob}(F_B^{out}) = \{(1 - p_B) \times 1\} + p_B\{a_f \text{Prob}(F_{A_f}^{in}) + (1 - a_f) \text{Prob}(F_{A_2}^{in})\}$$

(4.4.3b)

$$\text{Prob}(F_C^{out}) = (1 - p_C) + p_C\{a_f \text{Prob}(F_{A_f}^{in}) + (1 - a_f) \text{Prob}(F_{A_2}^{in})\} \qquad (4.4.3c)$$

The mass fraction of the sol is then written in a manner similar to that of Eq. (4.3.10) to give w_s as a function of the conversions.

The expression for $\text{Prob}(X_{n,f})$ is the same as Eq. (4.3.13), with $\text{Prob}(F_A^{out})$ obtained from the above arguments. Figure 4.10 shows[18,36] some typical results. The gel point is delayed owing to the effect of unequal reactivity (note that results are symmetrical if $k_1/k_2 < 1$). The effects of unequal reactivity are observed to be most significant around the gel point and must be considered in the design of reactors for some multifunctional polymerizations.

Macosko and Miller[20] have also considered the effect of induced asymmetry on nonlinear polymerizations. This is not presented here since a similar derivation[37] involving induced asymmetry is presented in Chapter 10 on the curing of epoxy resins.

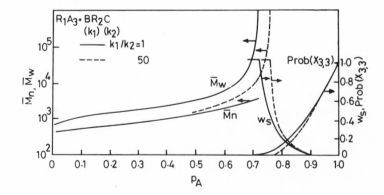

Figure 4.10. Results on $R_1A_3 + R_2BC$ system,[18,36] A reacting with B and C with rate constants k_1 and k_2, respectively. $r = 1$. $M_{R_1A_3} = 1000$, $M_{R_2BC} = 174$.

4.5. EXPERIMENTAL RESULTS

The experimental verification of the critical conversion for the onset of gelation in nonlinear polymerizations has been a subject of intensive research activity. Recently, Stafford[23] has compiled a considerable body of the available information and has compared the results with theoretical predictions. It has been found that even though in some systems the onset of gelation corresponds to the theoretically predicted conditions when $\bar{M}_w \rightarrow \infty$, results on several other systems are better explained if the critical point is defined *empirically* as that where $\bar{M}_n \rightarrow \infty$. In this section, however, some results on one particular system which has been studied extensively,[38-40] viz., the cross-linking of vinyl terminated polydimethyl siloxanes with hydrosilanes, is presented.

Table 4.2 shows the comparison between theoretical and experimental results for some conditions.[38] The conversion is measured by IR spectroscopy, rheological properties (viscosity and normal stress) by a mechanical spectrometer, and $\bar{M}_w$ by light scattering (of quenched samples). The gel time is estimated as that when the normal force just exceeds 1000 g force/cm^2. Equation (4.2.13) is used to compute $p_{A,c}$ and Eq. (4.2.11) (with appropriate average molecular weights[19] of the polydisperse R_2B_2) to obtain $\bar{M}_w$. The agreement between experimental and theoretical results is seen to be satisfactory. These workers also find that the zero-shear viscosity

TABLE 4.2. Theoretical and Experimental Results[a] for $R_1A_f + R_2B_2$[b]

System	r	$p_{A,c}$ (expt)	$p_{A,c}$ (theor)	p_A (expt)	$10^{-3}\,\bar{M}_w$ (expt)	$10^{-3}\,\bar{M}_w$ (theor)
$R_1A_3 + R_2B_2$	0.999	0.703	0.708	—	—	—
	1.002	0.712	0.706	—	—	—
	1.008	0.710	0.701	—	—	—
	1.008	—	—	0.641	179	134
	1.008	—	—	0.665	232	207
	1.008	—	—	0.679	321	305
$R_1A_4 + R_2B_2$	0.999	0.581	0.578	—	—	—
	1.003	0.588	0.577	—	—	—
	1.008	0.583	0.573	—	—	—
	1.008	—	—	0.559	410	380

[a] Reference 38.

[b] R_2B_2: $H_2C = HC-C_6H_5-CH_3SiO\{Si(CH_3)_2-O\}_n-SiCH_3C_6H_5-CH=CH_2$:

$\bar{M}_n = 11{,}600$

$Q = 2.21$ (with 5% low molecular weight species)

R_1A_f: $[H-Si(CH_3)_2O\}_f-Si-(C_6H_5)_{4-f}$

(Reprinted with permission from Ref. 38. Copyright 1979, American Chemical Society, Washington, DC.)

TABLE 4.3. Comparison of Theoretical and Experimental Results[39] on w_s for $R_1A_f + R_2B_2$ at Complete Conversion[a,b]

System	r	w_s (expt)	w_s (theor)[c]
$R_1A_3 + R_2B_2$	0.907	0.055	0.011
	0.960	0.016	0.002
	0.977	0.010	0.001
	0.983	0.000	0.000
$R_1A_4 + R_2B_2$	0.516	0.347	0.348
	0.712	0.110	0.095
	1.150	0.010	0.000
	1.560	0.013	0.004

[a] Same system as in Table 4.2.
[b] (Reprinted with permission from Ref. 39. Copyright 1979, American Chemical Society, Washington, DC.)
[c] Computed using Eq. (4.3.10) with $p_A = 1$ for $r < 1$ and $p_A = 1/r$, $p_B = 1$ for $r > 1$.

data for all the systems fall on a single, smooth curve when plotted against the theoretically computed weight-average molecular weight of the longest chain in the pregel region, indicating that the latter molecular property should be used for the correlation of experimental data on the viscosity of branched systems. Table 4.3 compares the experimental and theoretical results for w_s at complete conversion. The agreement is satisfactory, par-

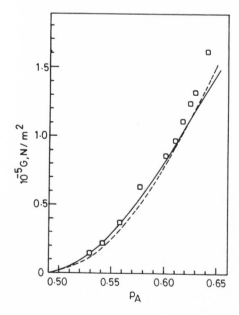

Figure 4.11. Dynamic storage modulus vs. extent of reaction for $R_1A_4 + R_2B_2$ (Table 4.2). $r = 1.56$. Solid line: Eq. (4.3.12). Dotted line: Eq. (4.3.18) with $G_e = \frac{3}{4}G_N^0 = 1.6 \times 10^5$ Pa, $T = 310$ K. (Reprinted with permission from Ref. 39. Copyright 1979 American Chemical Society, Washington DC.)

ticularly if it is noted that the experimental errors in these measurements can be substantial. The interpretation of experimental data on the tensile modulus (creep measurements) or the dynamic storage modulus (mechanical spectrometer) of the gelled material is more difficult since the effect of trapped entanglements is not quantitatively well established. Figure 4.11 shows some experimental data on the modulus ($G \simeq G'$, obtained from dynamic measurements), vs. extent of reaction (from IR) for the $R_1A_f + R_2B_2$ system of Table 4.2. The solid line is the Flory equation (4.3.12), and it is observed that the agreement with this equation is fairly good. Figure 4.11 also shows good agreement of the experimental data with Eq. (4.3.18) if G_e is arbitrarily assumed to be equal to $0.75G_N^0$. Recently,[40] a more comprehensive study of 12 sets of experimental results on PDMS networks revealed similar agreement with theory.

4.6. KINETIC APPROACH IN THE POLYMERIZATION OF RA_f IN BATCH REACTORS

The kinetic approach presented in this section has two advantages: firstly, it can be used easily to predict the behavior of nonlinear polymerizations in HCSTRs, and secondly, it can be more *easily* adapted to account for intramolecular reactions than the probabilistic models discussed in earlier sections. Moreover, this technique is even easier to understand than that of Macosko and Miller.

The analysis of the polymerization of R_1A_f monomers, with A reacting with A, is presented to illustrate this technique. Let P_n represent an *n*-mer in the reaction mass. *If there are no intramolecular reactions,* there will be $\{nf - 2(n - 1)\}$ unreacted A groups on P_n since every bond formed consumes two A groups. Mole balance equations can easily be written for a batch reactor and are summarized in Table 4.4. It may be pointed out that a factor of $1/2$ is included in these because reactions between like functional groups are involved. It can be confirmed that the following equation[1] satisfies these mole balance equations:

$$\frac{[P_n]}{[P_1]_0} = \frac{\{n(f - 1)\}!f}{n!\{n(f - 2) + 2\}!}p_A^{n-1}(1 - p_A)^{n(f-2)+2} \qquad (4.6.1)$$

when a pure monomer feed is used, where p_A is the conversion of A groups $\{=1 - [A]/[A]_0\}$. Equation (4.6.1) can also be derived using probabilistic arguments[1,4-7,13,25] since there are $n(f - 2) + 2$ unreacted A groups (with probability $= 1 - p_A$) and $(n - 1)$ —AA— bonds (the probability for each *bond* being p_A) in a molecule of P_n. In order to obtain p_A as a function of time, an equation for $\sum_{n=1}^{\infty} [P_n]$ must first be obtained by summing up the

TABLE 4.4. Balance Equations for RA_f Polymerization in Batch Reactors

Mole balance

$$\frac{d[P_1]}{dt} = -k\frac{f}{2}[P_1] \sum_{m=1}^{\infty} (mf + 2 - 2m)[P_m] \tag{a}$$

$$\frac{d[P_n]}{dt} = \frac{k}{2} \sum_{m=1}^{n-1} (mf + 2 - 2m)\{(n-m)f + 2 - 2(n-m)\}[P_m]\frac{[P_{n-m}]}{2}$$

$$- k(nf + 2 - 2n)[P_n] \sum_{m=1}^{\infty} \frac{(mf + 2 - 2m)}{2}[P_m] \qquad n = 2, 3, 4 \ldots \tag{b}$$

Moment equations

$$\frac{d\lambda_0}{dt} = -\frac{k}{4}(f-2)^2\lambda_1^2 - k\lambda_0\{\lambda_0 + \lambda_1(f-2)\} \tag{c}$$

$$\frac{d\lambda_1}{dt} = 0 \tag{d}$$

$$\frac{d\lambda_2}{dt} = \frac{k}{2}(f-2)^2\lambda_2^2 + 2k\lambda_1\{\lambda_2(f-2) + \lambda_1\} \tag{e}$$

$$\lambda_0 = [P_1]_0\left(1 - \frac{fp_A}{2}\right) \tag{f}$$

$$\lambda_1 = [P_1]_0 \tag{g}$$

$$\lambda_2 = [P_1]_0\frac{1 + p_A}{1 - (f-1)p_A} \tag{h}$$

equations in Table 4.4 over all n. This leads to

$$\frac{d}{dt} \sum_{n=1}^{\infty} [P_n] = -\frac{k}{4}\left\{ \sum_{n=1}^{\infty} (nf + 2 - 2n)[P_n]\right\}^2$$

$$= -k\left\{\frac{f-2}{2}[P_1]_0 + \sum_{n=1}^{\infty} [P_n]\right\}^2 \tag{4.6.2}$$

The fact that $\sum_{n=1}^{\infty} n[P_n]$ is the initial concentration, $[P_1]_0$, of R_1A_f molecules in the system, has been used in Eq. (4.6.2). Integration of Eq. (4.6.2) leads to

$$\sum_{n=1}^{\infty} [P_n] = \frac{(f/2)[P_1]_0}{1 + (f/2)[P_1]_0kt} - \left(\frac{f}{2} - 1\right)[P_1]_0 \tag{4.6.3}$$

The conversion is now obtained as

$$p_A = \frac{(f/2)[P_1]_0kt}{1 + (f/2)[P_1]_0kt} \tag{4.6.4}$$

The various moments of the MWD can easily be obtained by appropriate summation of Eq. (4.6.1) [the binomial expansion of $(1 - p_A)^{n(f-2)+2}$ has to be written]. This leads to[23] Eqs. (f)-(h) in Table 4.4. These equations give $\bar{M}_n$ and $\bar{M}_w$ values identical to Eq. (4.2.15) (with $p_A/2$ replacing p_A, since one molecule is removed from the reaction mass for every two A groups reacted) and Eq. (j) of Table 4.1, respectively. Alternatively, moment equations may be written,[43] as given in Table 4.4, and then integrated to give their time variations. The behavior must be consistent with that predicted by Eqs. (4.6.4) and (f)-(h) of Table 4.4.

Stafford[23,27] has found that for other systems like the polymerization of mixtures of $R_{1i}A_{fi}$ and $R_{2j}B_{gj}$ monomers, the same equations are obtained provided appropriately defined average functionalities, $\bar{f}$, are used in place of f. The concept of average functionalities has also been utilized by Durand and Bruneau[41,42] to explain the polymerization of several nonlinear monomer systems.

Equation (h) of Table 4.4 can easily be used to predict the gel point ($\bar{M}_w$ or $\lambda_2 \to \infty$). Alternatively, the detailed equations for the individual polymeric species P_1, P_2, ... can be solved numerically, and the moments computed by appropriate summations.[26] It is found that $\bar{M}_w$ and Q increase very sharply near the gel point, and even extremely small values of the increment Δt used lead to numerical instabilities very near the gel point. In fact, this numerical integration technique cannot really predict the *exact* gel point, though it can approach it fairly closely. The advantage of this numerical approach is that it can handle more complicated problems, e.g., the presence of intramolecular reactions (cyclization).

4.7. INTRAMOLECULAR REACTIONS IN NONLINEAR POLYMERIZATION

The deviations of experimental results from theoretical predictions in the pregel region have been attributed to the presence of intramolecular reactions (cyclization).[44] Experimental results on an $R_1A_3 + R_2B_2$ system (polyoxypropylene triol + hexamethylene diisocyanate in bulk, and in benzene at 70°C)[45] reveal that the average number of ring structures per molecule increases with conversion till it is almost 0.3 near the gel point. Any meaningful modeling exercise must, therefore, account for intramolecular reactions. In fact, Stepto and co-workers[46,47] find that the properties of the network material formed can be markedly affected by these reactions.

Several attempts have been made to study the effect of intramolecular reactions theoretically, and these have been reviewed recently by Stepto.[44] Most of these studies[12,14,48] use the cascade theory. Experimental confirmation of these theoretical results both for the gel point, as well as for the variation of the average number of ring structures per molecule with the conversion in the pregel region, is still to be made.[44,46] The kinetic approach has also been extended to account for intramolecular reactions in nonlinear polymerization.[26,49] Initial results of Temple[49] on $R_1A_3 + R_2B_2$ systems, which were limited to conversions below about 30% due to the complexity of the equations, reveal that the cascade theory underestimates the effect of intramolecular reactions, a discrepancy which has also been referred to by Mikes and Dušek[32] using Monte Carlo generation of polymer chains on the computer. In this section, the incorporation of the intramolecular reactions in the kinetic scheme of Section 4.6 is presented.

One method of studying the effects of intramolecular reactions in R_1A_f polymerization is to define several species, $P_n, C_{n1}, C_{n2}, \ldots$. P_n represents a molecule with n R_1A_f units without any rings due to intramolecular reactions and C_{nj} represents a molecule with n R_1A_f units, but having j ($j = 1, 2, \ldots$) rings (irrespective of where they are placed) caused by intramolecular reactions. Mole balance equations can then be written for each of these species, keeping in mind that P_n can cyclize (intramolecularly) to give C_{n1}, it can react with C_{mj} (intermolecularly) to give $C_{m+n,j}$ in addition to reacting (intermolecularly) with P_m to give P_{n+m}. Similarly, the various $C_{n,j}$ can cyclize (intramolecularly) to give $C_{n,j+1}$, or react (intermolecularly) with $C_{m,i}$ to give $C_{n+m,i+j}$. The rate constant for the intermolecular reactions is chosen as k (per reactive site), as in Section 4.6. The rate constants for the cyclization reactions depend on intramolecular configurations and need to be modeled properly.

Two approaches have been taken in the literature to model the rate constants for intramolecular reactions.[44] In one approach[50,51] the rate of reaction, $\mathcal{R}_{i,P_n}$, between two given functional groups located i monomeric units apart, in a molecule of say, P_n, is written as[44]

$$\mathcal{R}_{i,P_n} = k[P_n]\left\{\frac{1}{i^{3/2}}\left(\frac{3}{2\pi\nu l^2}\right)^{3/2}\frac{1}{N_{Av}}\right\} \equiv \frac{k[P_n]}{i^{3/2}}a \qquad (4.7.1)$$

where k is the intrinsic reactivity of functional groups (and is the same as that used for intermolecular reactions) and l is the backbone-bond length. In Eq. (4.7.1), ν is the number of backbone bonds per monomeric unit (so that νi is approximately the number of backbone bonds between the reacting functional groups) and N_{Av} is the Avogadro number. Equation (4.7.1) follows from the fact that for a random flight chain model[1,33] of the molecule of P_n, the probability of locating the two A groups at a distance r and within a small volume dV is $\{3/(2\pi\nu i l^2)\}^{3/2} \exp\{-r^2/\nu i l^2\} dV$. Two functional groups can react only if they come within a volume dV_0 (at $r \to 0$) of each other. In writing mole balance equations for P_n, $C_{n,1}$, $C_{n,2}$, etc., intermolecular terms can easily be written. The depletion and generation of these species by intramolecular reactions occur at rates

$$\tfrac{1}{2}ka[P_n] \sum_{\substack{j=1}}^{n} \sum_{\substack{i=1 \\ i \neq j}}^{n} (f-2)^2/|i-j|^{3/2};$$

and

$$\tfrac{1}{2}ka[C_{n,1}] \sum_{\substack{j=1}}^{n} \sum_{\substack{i=1 \\ i \neq j}}^{n} (f-2)^2/|i-j|^{3/2}$$

respectively. In writing these terms, it is assumed that functional groups on the same monomeric unit cannot react and that no account has been taken of the intramolecular bond already present in $C_{n,1}$. The error introduced because of the latter is small if the chain lengths are large.

Two important assumptions have been made in the above analysis. The first is that *any* two unreacted A groups on a molecule can undergo intramolecular reaction. This would not necessarily be so, particularly when $|i-j|$ is small, and it becomes difficult for a given A group to come close to another one nearby due to chain inflexibility, i.e., the random flight chain model breaks down. The second assumption is that the same expression applies for the rate of intramolecular reaction between two A groups equally spaced on a P_n as on a molecule of $C_{n,j}$. This also need not be so because a highly cyclized molecule is more compact in size, and the configurations become constrained. The random flight chain model would once again break down. Platé and Noah[52,53] have written the rate of removal of species $C_{n,j}$ by cyclization as $k[C_{n,j}]\bar{Z}_{n,j}$ and have generated chains with intramolecular rings on a computer using the Monte Carlo technique. This method overcomes some of the limitations of Eq. (4.7.1) as discussed above. It was found that their results could be fitted empirically by the following equation:

$$\bar{Z}_{n,j} = a(n)j + b(n) \tag{4.7.2}$$

up to a certain value of j, where a and b depend on n.

In either of the two models discussed above, the cyclization rate can be written in the form $k[C_{n,j}]\phi(n,j)$ where $\phi(n,j)$ is $\bar{Z}_{n,j}$ for the second model, and can easily be written (by comparison) for the first model. Mole balance equations for the various oligomers in the reaction mass can now be written[26] and are given in Table 4.5. These equations can be further simplified by using two assumptions:

(a) To prevent the existence of highly "strained" species like $C_{1,1}$, $C_{2,1}$, etc., species above a certain value of n are allowed for any value of j in $C_{n,j}$.

(b) In species $C_{m,j}$, likewise, j can, at best, go to a maximum value of $\{m(f-2)+2\}/2$, and even these may be unlikely to occur. It is, thus, necessary to avoid species above a certain value of j.

TABLE 4.5. Mole Balance Equations for Multifunctional Polymerizations with Intramolecular Reactions[a]

$$\frac{d[P_n]}{dt} = \frac{k}{2} \sum_{m=1}^{n-1} (mf + 2 - 2m)\{(n-m)(f-2) + 2\}[P_m][P_{n-m}]\delta_{n \neq 1}/2$$

$$- k\{n(f-2) + 2\}[P_n]\left(\sum_{m=1}^{\infty} \{m(f-2) + 2\}[P_m]/2 \right.$$

$$+ \sum_{m=1}^{\infty} \{m(f-2)\}[C_{m,1}]/2 + \sum_{m=1}^{\infty} \{m(f-2) - 2\}[C_{m,2}]/2 + \cdots \left. \right) - k[P_n]\phi(n,0)$$

($\delta_{n \neq 1}$ is zero when $n = 1$ and unity otherwise)

$$\frac{d[C_{n,j}]}{dt} = k[C_{n,j-1}]\phi(n, j-1) - k[C_{n,j}]\phi(n,j)$$

$$+ \frac{k}{2} \sum_{m=1}^{n-1} \{m(f-2) + 2 - 2j\}\{(n-m)(f-2) + 2\}[C_{m,j}][P_{n-m}]\delta_{n \neq 1}/2$$

$$+ \frac{1}{2} \frac{k}{2} \sum_{r=1}^{j-1} \sum_{m=1}^{n-1} \{m(f-2) + 2 - 2(j-r)\}\{(n-m)(f-2) + 2 - 2r\}$$

$$\times [C_{m,j-r}][C_{n-m,r}]\delta_{n \neq 1} - k\{n(f-2) + 2 - 2j\}\frac{[C_{n,j}]}{2}$$

$$\times \left(\sum_{m=1}^{\infty} \{m(f-2) + 2\}[P_m] + \sum_{r=1}^{\infty} \sum_{m=1}^{\infty} \{m(f-2) + 2 - 2r\}[C_{m,r}] \right)$$

[a] Reference 26.

Appropriate modifications are necessary in the equations to account for these simplifications.

These equations seem formidable, but recently some numerical computations have been performed.[26] Moment closure techniques have also been established to solve equations derived from these equations for the various moments. Results have been found to be relatively insensitive to the closure approximations used, as was observed in the case of linear step growth polymerizations.

4.8. NONLINEAR STEP GROWTH POLYMERIZATION IN HCSTRs[29–31,43,54]

Several synthetic rubbers and large volume nonlinear polymers are now produced in homogeneous, continuous flow, stirred tank reactors (HCSTRs) in order to exploit their excellent heat transfer characteristics. Obviously, only conversions below the critical value can be attained in such reactors. The study of the performance of such polymerizations is most conveniently done using the kinetic approach since the probabilistic approach has not yet been applied for HCSTRs. For an HCSTR operating, in general, in the unsteady mode, as for example, during start-up, the equations for RA_f monomers are similar to those given in Table 4.4, with inflow and outflow terms, $([P_n]/\bar{\theta} - [P_n]_0/\bar{\theta})$, included on the left-hand sides of Eqs. (a) and (b) in the table. Here, $\bar{\theta}$ is the mean residence time $(= V/\mathring{Q})$, $[P_1]$, $[P_2]$, ..., are the concentrations in the reactor (which are the same as those in the output stream) at time t, and subscript 0 refers to the input-stream concentrations. It is assumed that the feed may contain a mixture of several oligomers and so the analysis is applicable to a cascade of HCSTRs. Cozewith et al.[31] have neglected the constant 2 in the term $(mf + 2 - 2m)$, and have replaced $(f - 2)$ by f in order to simplify these equations. They have, thus, solved the set of equations† given in Table 4.6, with k^* defined by

$$k^* \equiv k(f - 2)^2/2 \qquad (4.8.1)$$

The various moments, $\lambda_i(= \sum_{n=1}^{\infty} n^i[P_n])$, $i = 0, 1, 2, \ldots$, can then be written by appropriate summations, in a manner similar to that in Table 4.4 to give the corresponding equations, again listed in Table 4.6. It is interesting to observe that the equation for the ith moment (except for $i = 0$) does not involve higher-order moments. Hence, moment closure conditions are not required.

A commonly encountered industrial problem is to fill the reactor with the feed material ($\lambda_{0,0}$, $\lambda_{1,0}$, $\lambda_{2,0}$, etc.) and assume that the reaction starts thereafter at $t = 0$. Equation (d) in Table 4.6 can then be integrated to give $\lambda_1 = \lambda_{1,0}$ for all times. Equation (e) in this table can be rewritten in terms of $\mu_w/\mu_{w,0}$, where $\mu_w = \lambda_2/\lambda_1$ and $\mu_{w,0}$ is the feed value of μ_w, and integrated to give[55]

$$\int_1^{\mu_w/\mu_{w,0}} \frac{d(\mu_w/\mu_{w,0})}{\left(\dfrac{\mu_w}{\mu_{w,0}}\right)^2 \tau^* - \left(\dfrac{\mu_w}{\mu_{w,0}}\right) + 1} = \int_{t=0}^{t} \frac{dt}{\bar{\theta}} \qquad (4.8.2)$$

where the dimensionless time, τ, is defined by

$$\tau^* \equiv k^* \mu_{w,0}\lambda_{1,0}\bar{\theta} = k^*\lambda_{2,0}\bar{\theta} \qquad (4.8.3)$$

The integral in Eq. (4.8.2) depends on whether the roots of the denominator on the left-hand side are real, imaginary, or equal. The results are given in

† In fact, Cozewith et al. simulated polymer cross-linking in HCSTRs where f is indeed large, but their equation (4.8.1) applies to polymerization of RA_f monomers too with the assumptions mentioned.[43]

TABLE 4.6. Equations for R_1A_f Polymerization in HCSTRs

Balance

$$\frac{d[P_1]}{dt} - \frac{[P_1]_0 - [P_1]}{\bar{\theta}} = -k^*[P_1] \sum_{m=1}^{\infty} m[P_m] \tag{a}$$

$$\frac{d[P_n]}{dt} - \frac{[P_n]_0 - [P_n]}{\bar{\theta}} = \frac{k^*}{2} \sum_{m=1}^{n-1} m(n-m)[P_m][P_{n-m}] - k^*n[P_n] \sum_{m=1}^{\infty} m[P_m] \tag{b}$$

$$\frac{d\lambda_0}{dt} - \frac{\lambda_{0,0} - \lambda_0}{\bar{\theta}} = -\frac{k^*}{2}\lambda_1^2 \tag{c}$$

$$\frac{d\lambda_1}{dt} - \frac{\lambda_{1,0} - \lambda_1}{\bar{\theta}} = 0 \tag{d}$$

$$\frac{d\lambda_2}{dt} - \frac{\lambda_{2,0} - \lambda_2}{\bar{\theta}} = k^*\lambda_2^2 \tag{e}$$

Results on μ_w

for $\tau^* < 1/4$:

$$\frac{\mu_w}{\mu_{w,0}} = \frac{(1-F) - G(1+F)\exp(-Ft/\bar{\theta})}{2\tau^*\{1 - G\exp(-Ft/\bar{\theta})\}} \tag{f}$$

where

$$F = (1 - 4\tau^*)^{1/2} \tag{g}$$

$$G = \frac{2\tau^* - 1 + F}{2\tau^* - 1 - F} \tag{h}$$

for $\tau^* = 1/4$:

$$\frac{\mu_w}{\mu_{w,0}} = 2\left(1 - \frac{2}{4 + t/\bar{\theta}}\right) \tag{i}$$

for $\tau^* > 1/4$:

$$\frac{\mu_w}{\mu_{w,0}} = \frac{1}{2\tau^*} + \frac{(4\tau^* - 1)^{1/2}}{2\tau^*} \tan\left\{\frac{(4\tau^* - 1)^{1/2}}{2}\frac{t}{\bar{\theta}} + \tan^{-1}\frac{2\tau^* - 1}{(4\tau^* - 1)^{1/2}}\right\} \tag{j}$$

Table 4.6 for various ranges of τ^*. As $t \to \infty$, i.e., at large times beyond the start up, these equations give

$$\tau^* < \tfrac{1}{4}: \quad \frac{\mu_w}{\mu_{w,0}} = \frac{1-F}{2\tau^*} = \frac{1 - \{1 - 4\tau^*\}^{1/2}}{2\tau^*} \tag{4.8.4a}$$

$$\tau^* = \tfrac{1}{4}: \quad \frac{\mu_w}{\mu_{w,0}} = 2 \tag{4.8.4b}$$

$$\tau^* > \tfrac{1}{4}: \quad \mu_w/\mu_{w,0} \to \infty \tag{4.8.4c}$$

It can be observed that for $\tau^* \leq \frac{1}{4}$, the value of μ_w is finite. However, for $\tau^* > \frac{1}{4}$, Eq. (j) in Table 4.6 predicts finite values of $\mu_w/\mu_{w,0}$, only for some time $t < t_g$, where

$$t_g = \left\{ \frac{\pi}{2} - \tan^{-1} \frac{2\tau^* - 1}{(4\tau^* - 1)^{1/2}} \right\} \frac{2\bar{\theta}}{(4\tau^* - 1)^{1/2}} \qquad (4.8.5)$$

Thereafter, gelation occurs. HCSTRs, therefore, cannot be profitably used if $\tau^* > 0.25$.

In summary, then, for an HCSTR, steady state operation without gelation is possible only when $\tau^* = k^* \mu_{w,0} \lambda_{1,0} \bar{\theta} \leq \frac{1}{4}$ or, when

$$\bar{\theta} \leq \frac{1}{4k^* \mu_{w,0} \lambda_{1,0}} = \frac{1}{2k(f-2)^2 \mu_{w,0} \lambda_{1,0}} \equiv \bar{\theta}_c \qquad (4.8.6)$$

For values of $\bar{\theta}$ above this critical value, the reaction mass would gel at a time, t_g, given by Eq. (4.8.5). In contrast, the gelation time for a batch reactor for a pure monomer feed is obtained from Eq. (j) of Table 4.1 and Eq. (4.6.5) as

$$t_{c,\text{batch}} = \frac{2}{k[P_1]_0 f(f-2)} \qquad (4.8.7)$$

Thus, it is observed that the mean critical residence time for an HCSTR with pure monomer feed ($\mu_{w,0} = 1$, $\lambda_{1,0} = [P_1]_0$) is almost one-fourth that for batch reactors. This places a severe restriction on the use of HCSTRs for nonlinear polymerizations and necessitates that these reactors can, at best, be used as beginning reactors in a sequence.

The equations for the *steady state operation* of an HCSTR (i.e., when $\tau^* \leq 0.25$) can easily be obtained from Eqs. (c)–(e) of Table 4.6 by putting the time variation terms as zero, giving

$$\lambda_{0,s} = \lambda_{0,0} - k^*(\lambda_{1,0})^2 \bar{\theta}/2 \qquad (4.8.8a)$$

$$\lambda_{1,s} = \lambda_{1,0} \qquad (4.8.8b)$$

$$\lambda_{2,s} = \frac{1 - (1 - 4\tau^*)^{1/2}}{2k^* \bar{\theta}} \qquad (4.8.8c)$$

$$Q_s = \frac{\lambda_{2,s} \lambda_{0,s}}{\lambda_{1,s}^2} = Q_0 \frac{1 - (1 - 4\tau^*)^{1/2}}{2\tau^*} \left(1 - \frac{\tau^*}{2Q_0}\right) \qquad (4.8.8d)$$

Figure 4.12 shows how the polydispersity index Q_s varies with the

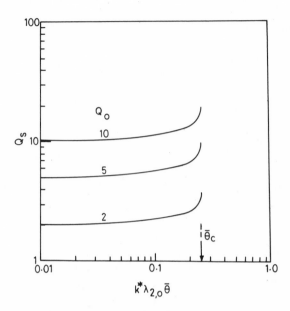

Figure 4.12. Performance of HCSTRs operating at steady state.[31]

dimensionless residence time τ^*. Equation (4.8.8) shows that the Q_s vs. τ^* graph depends on the PDI of the feed. The value of Q_s increases gradually with $\bar{\theta}$ and then rises sharply near $\bar{\theta} = \bar{\theta}_c$ to a limiting value of $2Q_0 - 0.25$. This is in sharp contrast to the Q vs. t behavior in batch reactors, where $Q \to \infty$ as $t \to t_{c,\text{batch}}$.

It is thus found that the performance of HCSTRs differs significantly from that of batch reactors when multifunctional monomers are used. Steady state operation is assured only when mean residence times are below a critical value $\bar{\theta}_c$, which is almost a fourth of the critical time, $t_{c,\text{batch}}$, of batch reactors. When $\bar{\theta} > \bar{\theta}_c$, the operation of the HCSTR is unsteady, and gelation is attained after a time t_g [Eq. (4.8.5)] of the start-up. When $\bar{\theta} \leqslant \bar{\theta}_c$, the HCSTR attains a steady state after some time, and gives a product characterized by finite values of the PDI (note: even at $\bar{\theta} = \bar{\theta}_c$). Another important characteristic of the steady-state HCSTR operation (for $\bar{\theta} < \bar{\theta}_c$) is that the higher moments (than λ_2) of the product tend to infinity before $\bar{\theta}_c$. This is in marked contrast to the behavior of batch reactors in which λ_2 as well as higher moments diverge simultaneously at the critical time.

More recently, balance equations without the Cozewith assumption of large f have been solved,[43] and it is found that even more severe restrictions are placed on the use of HCSTRs (gelation is predicted to occur earlier than at $\tau^* = 0.25$). Periodic operation, with feed concentration changing with time sinusoidally, has been explored[54] as a means of overcoming this restriction and a region of operation of HCSTRs in this mode has been obtained.

4.9. CONCLUSIONS

In this chapter, step growth polymerizations for typical multifunctional monomers have been analyzed, both in batch reactors as well as in HCSTRs. The most important characteristic of such systems is the gelation time, where, *for batch reactors*, the second as well as the higher moments of the chain length distribution diverge. The effects of cyclization have also been focused upon and it is found that the gel point shifts to higher times because of this effect. Gelation is encountered in HCSTRs also and occurs at much lower conversions than in batch reactors. For steady state nongelling operation of HCSTRs, the mean residence time must be below a critical value, and there is a limit to the extent that Q can be increased. However, higher moments of the MWD of the product diverge, even for steady operation in an HCSTR. Predictions for the postgelation properties of the polymer from batch reactors have been found to compare well with experimental data and can be used to predict *in situ* polymerization in molds.

REFERENCES

1. P. J. Flory, *Principles of Polymer Chemistry*, 1st ed., Cornell University Press, Ithaca, New York (1953).
2. P. J. Flory, Molecular size distribution in three-dimensional polymers. I. Gelation; II. Trifunction branching units; III. Tetrafunctional branching units, *J. Am. Chem. Soc.* 63, 3083-3090; 3091-3096; 3096-3100 (1941).
3. P. J. Flory, Fundamental principles of condensation polymerization, *Chem. Rev.* 39, 137-197 (1949).
4. W. H. Stockmayer, Theory of molecular size distribution and gel formation in branched-chain polymers, *J. Chem. Phys.* 11, 45-58 (1943).
5. W. H. Stockmayer, Theory of molecular size distribution and gel formation in branched-chain polymers. II. General cross linking, *J. Chem. Phys.* 12, 125-131 (1944).
6. W. H. Stockmayer, Molecular distribution in condensation polymers, *J. Polym. Sci.* 9, 69-71 (1952).
7. W. H. Stockmayer, Errata: Molecular distribution in condensation polymers, *J. Polym. Sci.* 11, 424 (1953).
8. M. Gordon, Good's theory of cascade processes applied to the statistics of polymer distributions, *Proc. R. Soc. London A* 268, 240-256 (1962).
9. D. S. Butler, G. N. Malcolm, and M. Gordon, Configurational statistics of copolymer systems, *Proc. R. Soc. London A* 295, 29-54 (1966).
10. M. Gordon and T. G. Parker, The graph-like state of matter. I. Statistical effects of correlations due to substitution effects, including steric hindrance, on polymer distributions, *Proc. R. Soc. Edinburgh, A* 69, 181-198 (1970-71).

11. M. Gordon, T. C. Ward, and R. S. Whitney, in *Polymer Networks* (A. J. Chompf and S. Newman, Eds.), 1st ed., pp. 1-21, Plenum, New York (1971).
12. M. Gordon and G. R. Scantlebury, Statistical kinetics of polyesterification of adipic acid with pentaerythritol or trimethylol ethane, *J. Chem. Soc. London B* 1-13 (1967).
13. M. Gordon and M. Judd, Statistical mechanics and the critically branched state, *Nature* **234**, 96-97 (1971).
14. K. Dušek, M. Gordon, and S. B. Ross-Murphy, Graphlike state of matter. 10. Cyclization and concentration of elastically active network chains in polymer networks, *Macromolecules* **11**, 236-245 (1978).
15. T. E. Harris, *The Theory of Branching Processes*, 1st ed., Chap. 1, Springer-Verlag, Berlin (1963).
16. C. W. Macosko and D. R. Miller, A new derivation of average molecular weights of nonlinear polymers, *Macromolecules* **9**, 199-205 (1976).
17. D. R. Miller and C. W. Macosko, A new derivation of post gel properties of network polymers, *Macromolecules* **9**, 206-211 (1976).
18. D. R. Miller and C. W. Macosko, Average property relations for nonlinear polymerization with unequal reactivity, *Macromolecules* **11**, 656-662 (1978).
19. D. R. Miller, E. M. Valles, and C. W. Macosko, Calculation of molecular parameters for stepwise polyfunctional polymerization, *Polym. Eng. Sci.* **19**, 272-283 (1979).
20. D. R. Miller and C. W. Macosko, Substitution effects in property relations for stepwise polyfunctional polymerization, *Macromolecules* **13**, 1063-1069 (1980).
21. W. Feller, *An Introduction to Probability Theory and Its Applications*, 3rd ed., Vol. 1, Wiley, New York (1968).
22. W. H. Stockmayer and L. L. Weil, in *Advancing Fronts in Chemstry* (S. B. Twiss, Ed.), 1st ed., Chap. 6, Reinhold, New York (1945).
23. J. W. Stafford, Multifunctional polycondensation and gelation: A kinetic approach, *J. Polym. Sci., Polym. Chem. Ed.* **19**, 3219-3236 (1981).
24. Y. Tanaka and H. Kakuichi, Critical conditions for formation of infinite networks in branched-chain polymers, *J. Polym. Sci. A* **3**, 3279-3300 (1965).
25. L. M. Pis'men and S. I. Kuchanov, Multifunctional polycondensation and gelation, *Vysokomol. Soedin A* **13**, 791-802 (1971).
26. A. Kumar, S. Wahal, S. Sastri, and S. K. Gupta, Modeling of intramolecular reactions in the step growth polymerization of multifunctional monomers, *Polymer* **27**, 583-591 (1986).
27. J. W. Stafford, Multifunctional polycondensation distributions: A kinetic approach to mixed monomer systems, *J. Polym. Sci., Polym. Chem. Ed.* **22**, 365-381 (1984).
28. K. Dušek, Correspondence between the theory of branching process and the kinetic theory for random crosslinking in the post-gel stage, *Polym. Bull.* **1**, 523-528 (1979).
29. R. Jackson, P. Small, and R. Whiteley, Prediction of molecular weight distributions in branched polymers, *J. Polym. Sci., Polym. Chem. Ed.* **11**, 1781-1809 (1973).
30. A. Chatterjee, W. S. Park, and W. W. Graessley, Free radical polymerization with long-chain branching: Continuous polymerization of vinyl acetate in t-butanol, *Chem. Eng. Sci.* **32**, 167-178 (1977).
31. C. Cozewith, W. W. Graessley, and G. ver Strate, Polymer crosslinking in continuous flow stirred reactors, *Chem. Eng. Sci.* **34**, 245-248 (1979).
32. J. Mikeš and K. Dušek, Simulation of polymer network formation by the Monte Carlo method, *Macromolecules* **15**, 93-99 (1982).
33. A. Kumar and S. K. Gupta, *Fundamentals of Polymer Science and Engineering*, 1st ed., Tata McGraw-Hill, New Delhi, India (1978).
34. F. G. Mussatti and C. W. Macosko, Rheology of network forming systems, *Polym. Eng. Sci.* **13**, 236-240 (1973).
35. K. D. Ziegel, A. W. Fogiel, and R. Pariser, Prediction of molecular weights for condensation polymerizations of polyfunctional monomers, *Macromolecules* **5**, 95-98 (1972).

36. S. K. Gupta and A. Kumar, Simulation of step-growth polymerizations, *Chem. Eng. Commun.* **20**, 1-52 (1983).
37. U. M. Bokare and K. S. Gandhi, Effect of simultaneous polyaddition reaction on the curing of epoxides, *J. Polym. Sci., Polym. Chem. Ed.* **18**, 857-870 (1980).
38. E. M. Valles and C. W. Macosko, Structure and viscosity of poly (dimethylsiloxanes) with random branches, *Macromolecules* **12**, 521-526 (1979).
39. E. M. Valles and C. W. Macosko, Properties of networks formed by end linking of poly (dimethylsiloxane), *Macromolecules* **12**, 673-679 (1979).
40. M. Gottlieb, C. W. Macosko, G. S. Benjamin, K. O. Meyers, and E. W. Merrill, Equilibrium modulus of model poly (dimethylsiloxane) networks, *Macromolecules* **14**, 1039-1046 (1981).
41. D. Durand and C. M. Bruneau, General expressions of average molecular weights in condensation polymerization of polyfunctional monomers, *Brit. Polym. J.* **11**, 194-198 (1979).
42. D. Durand and C. M. Bruneau, Average functionalities of macromolecules in stepwise polyfunctional polymerization, *Polymer* **23**, 69-72 (1982).
43. S. K. Gupta, S. S. Bafna, and A. Kumar, Multifunctional step growth polymerizations in cascades of isothermal CSTRs, *Polym. Eng. Sci.* **25**, 332-338 (1985).
44. R. F. T. Stepto, in *Developments in Polymerization* (R. N. Haward, Ed.), 1st ed., Vol. 3, pp. 81-141, Applied Science Pub., Barking, UK (1982).
45. J. L. Stanford and R. F. T. Stepto, A study of intramolecular reaction and gelation during non-linear polyurethane formation, *Brit. Polym. J.* **9**, 124-132 (1977).
46. R. F. T. Stepto, Theoretical and experimental studies of network formation and properties, *Polymer* **20**, 1324-1326 (1979).
47. A. B. Fasina and R. F. T. Stepto, Formation and properties of triol-based polyester networks, *Makromol. Chem.* **182**, 2479-2493 (1981).
48. K. Dušek and W. Prins, Structure and elasticity of non-crystalline polymer networks, *Adv. Polym. Sci.* **6**, 1-102 (1969).
49. W. B. Temple, The graphlike state of matter IV. Ring-chain competition kinetics in a branched polymerization reaction, *Makromol. Chem.* **160**, 277-289 (1972).
50. H. Jacobson and W. H. Stockmayer, Intramolecular reaction in polycondensations. I. The theory of linear systems, *J. Chem. Phys.* **18**, 1600-1606 (1950).
51. M. Gordon and W. B. Temple, The graph-like state of molecules. III. Ring-chain competition kinetics in linear polymerization reactions, *Makromol. Chem.* **160**, 263-276 (1972).
52. N. A. Platé and O. V. Noah, Theoretical consideration of kinetics and statistics of reactions of functional groups of macromolecules, *Adv. Polym. Sci.* **31**, 133-173 (1979).
53. I. I. Romantsova, Yu. A. Taran, O. V. Noah, A. M. Yel'yashevich, Yu. Ya. Gotlib, and N. A. Platé, Study of intramolecular crosslinking of polymer chains using the Monte Carlo method, *Vysokomol. Soed. A* **19**, 2800-2807 (1977).
54. S. K. Gupta, S. Nath, and A. Kumar, Forced oscillations in continuous flow stirred tank reactors with nonlinear step growth polymerizations, *J. Appl. Polym. Sci.* **30**, 557-569 (1985).
55. M. Abramowitz and I. A. Stegun, *Handbook of Mathematical Functions*, Dover, New York (1965).

EXERCISES

1. Obtain the final equations for $\bar{M}_w$ [Eq. (a) of Table 4.1] for $R_1A_f + R_2B_2$ polymerization by completing the intermediate steps.
2. Derive Eq. (b) of Table 4.1 for $R_{1i}A_{fi} + R_{2j}B_{gj}$ mixtures, with A groups reacting with B. Give physical interpretations of the variables defined in Eqs. (c)-(h) of this table, wherever possible.

3. Derive Eq. (i) of Table 4.1 for the self-polycondensation of R_1A_f monomer (e.g., etherification of pentaerythritol).

4. Obtain equations for w_s for $R_{1i}A_{fi} + R_{2j}B_{gj}$ and R_1A_f systems from first principles. In particular, obtain an equation for $\text{Prob}(F_B^{out})$ for the first of these two systems, since that will be required in the equation equivalent to Eq. (4.3.10) to determine w_s.

5. Set up the problem of estimating the (weight average) number of branch points per molecule for a $R_1A_f + R_2B_2$ system.[19] This molecular characteristic is important in rheological studies.

6. Complete all the intermediate steps in the derivation of Eqs. (4.4.3) for the system $R_1A_f + R_3A_2 + R_2BC$. Also obtain the complete set of equations to obtain w_s.

7. Confirm the equations in Table 4.4.

8. For RA_f polymerization in batch reactors, show that $d[A]/dt = -(k/2)[A]^2$. Show that this is consistent with Eq. (4.6.4).

9. Justify the equations in Table 4.5 and list the approximations assumed.

10. Integrate the equations for RA_f polymerization in HCSTRs without using the small-f approximation.[43]

11. Explore, after reading Section 6.5 and Ref. 54, the periodic operation of RA_f polymerization in HCSTRs, with the feed concentration varying sinusoidally around some average value.

12. Transform Eqs. (a) and (b) in Table 4.4 into the s domain using generating functions (see Exercise 1.1e). Does this indicate the existence of a "shock" phenomenon?

5

MASS TRANSFER IN STEP GROWTH POLYMERIZATION

5.1. INTRODUCTION

The progress of reversible step growth polymerization reactions of ARB monomers (in a closed reactor), represented schematically by

$$P_n + P_m \underset{k_p' = k_p/K}{\overset{k_p}{\rightleftharpoons}} P_{n+m} + W \tag{5.1.1}$$

has been discussed in Section 2.9. The variation of the conversion, p_A, of functional group A with time was derived and given by Eq. (2.9.8). As the polymerization time t [or the dimensionless time τ defined in Eq. (2.9.9b)] approaches infinity, the overall reaction approaches equilibrium and the equilibrium conversion, $p_{A,e}$, is given by Eq. (2.9.8) as[1]

$$p_{A,e} = 1 - \frac{1}{\mu_{n,e}} = \frac{K}{K-1} - \frac{K^{1/2}}{K-1} \tag{5.1.2}$$

where $\mu_{n,e}$ is the equilibrium degree of polymerization. The values of the equilibrium constant, K, for some polymerizations of commercial interest are extremely unfavorable. For example, for the polymerization of polyethylene terephthalate (PET) from bis hydroxyethyl terephthalate, the value of K is about 0.5, independent of temperature.[2] For this value of K, Eq. (5.1.2) predicts the equilibrium degree of polymerization, $\mu_{n,e}$, as 1.71, which is too low a value for the product to be of any interest. It is, therefore, evident that in order to obtain values of $\mu_{n,e}$ of about a hundred, which is of commercial importance, one must drive reaction (5.1.1) in the forward direction. One way of doing this is to remove the low molecular weight by-product W by application of high vacuum (open reactors).

Several examples of commercial importance may be mentioned in which the by-product W must be removed by some means. In nylon 6 polymerization, even though the value of K is severalfold higher than for PET, optimal operation of the reactor requires that the condensation by-product, water, be removed at some intermediate stage of polymerization.[3] One way of achieving this is to bubble inert gas through the reaction mass.[4] Another example where the diffusion of low molecular weight compounds through the reaction mass is important is in the hot water leaching of the solid nylon 6 chips (below its melting point) produced from the reactors. This is done to remove the 8%–10% unreacted caprolactam. As the monomer diffuses out towards the solid–liquid interface, further polymerization occurs in the chips. In nylon-6,6 polymerization, too, the equilibrium constant is low, and vacuum is applied in order to obtain high molecular weight polymer.[5,6] In all these cases, the condensation by-product diffuses towards a vapor–liquid, solid–liquid, or liquid–liquid interface, where its concentration is lower and is determined by appropriate equilibrium relations. A concentration gradient is established in the reaction mass, and concepts of mass transfer with simultaneous step growth polymerization are required to explain the performance of the reactor.

During the polymerization of PET, the viscosity of the reaction mass undergoes a significant increase from a few centipoise to over about [2] 8000 poise. The diffusivity[7,8] of the small molecule W through the reaction mass towards the vapor–liquid interface thus varies over an extremely wide range and can, at times, be very low. This necessitates the use of special equipment which can enhance mass transfer rates. Also, depending upon the relative resistances of the mass transfer and the polymerization steps, different analytical models must be used to account for the mass transfer occurring in the specified geometry of the reactor. At one end of the spectrum lies the situation where the viscosity of the reaction mass is low (for stirred batch PET reactors,[2] till μ_n of about 30, when the viscosity is below about 40 poise) and the diffusivity and the corresponding mass transfer coefficient are high compared to the rate of chemical reaction. It can then be easily assumed that there is no resistance to mass transfer. In fact, Ravindranath and Mashelkar[2] claim that the value of $k_L s$ (the product of the mass transfer coefficient, k_L, and the interfacial area per unit volume, s) for a turbine-agitated PET reactor with $\mu_n \simeq 30$ is about $10^{-2}\ \mathrm{sec}^{-1}$, while the pseudo-first-order reaction rate constant is about $10^{-3}\ \mathrm{sec}^{-1}$, an order of magnitude lower. The rates of mixing under such situations are also very large compared to the rates of reaction. It can, therefore, be assumed that there are no spatial gradients present in the reaction mass in the concentrations of W and the other species. The value of [W] at any time, t, in the entire reaction mass is thus governed solely by vapor–liquid equilibrium conditions existing at the interface, which, in turn, are determined by the pressure and temperature applied. Turbine-agitated semibatch reactors are usually employed under such conditions wherein the mass transfer resistance is negligible. The removal of the volatile by-product as well as the consequent reduction in the volume of the reaction mass must be correctly accounted for in the mole balance equations for such reactors.

At the other end of the spectrum lies the situation where μ_n is about 100, the viscosity is of the order of 8000 poise (e.g., for PET), and the mass transfer coefficient in stirred batch reactors is extremely small compared to the reaction rate constant. The mass transfer resistance cannot, then, be neglected and spatial variations of the concentration of W must be accounted for. Obviously, it is desirable to reduce the mass transfer resistance as much as possible by using special thin film reactors, in order to reduce the length of the diffusion path. Different equipment have been employed for carrying out polymerizations under these conditions.[9-13] Some common equipment employed under such conditions is described below.

(a) *Interfacial Surface Generators.* In these, a thin film of the reaction medium is generated mechanically by means of rotating disks dipped partially in the reaction mass as shown in Fig. 5.1a.[5,14-16] Murukami et al.[16]

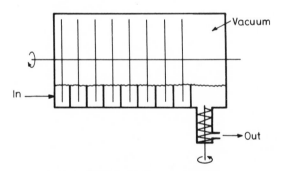

(a) Interfacial surface generator

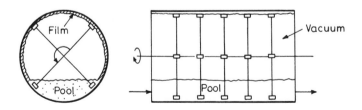

(b) Mechanically agitated thin film contactors

Figure 5.1. Some typical reactors used for finishing stages of step growth polymerizations.

have carried out experimental studies on the fluid mechanical and mass transfer aspects of such equipment in the absence of chemical reaction.

(b) *Mechanically Agitated Thin Film Contactors.* In this equipment, rotor blades apply a thin film of the reaction mass of about 1 mm thickness from a pool on the inside walls of a cylindrical reactor (Fig. 5.1b). The rotating blades may be straight or pitched. Owing to the high shear rates $(10^3-10^4 \, \text{sec}^{-1})$ present in the thin gaps between the inside wall and the blade surface, the viscosity in the film is lower than the low-shear value, and this further accelerates the heat and mass transfer rates. Because of the large number of blades wiping the film surface at a given axial position, film exposure times are very short.[10] Pitched blade rotors can also be used and are usually better for high viscosity fluids. Windmer[10] reports Peclet numbers in the range of 25-1000 and heat transfer coefficients around 100-300 kcal/hr m^2 °C for such reactors.

(c) *Partially Filled Screw Extruders.* Operating at about 100-500 RPM,[6,11,12,17,18] single screw[6,11,18] and intermeshing twin screw[12,17] type extruders have much deeper screws in the central sections so that the channels therein are only partially filled with the reaction mass. The central section of the barrel is connected to a vacuum pump. The fluid mechanical and mass transfer aspects of such equipment are not very well modeled, though some experimental studies have been reported.[11,12,18]

In this chapter, an attempt is made to model ARB polymerizations in a few common reactors wherein mass transfer and step growth polymerization reactions are simultaneously involved. In addition, the range of applicability of the various models currently in use is pointed out.

5.2. POLYMERIZATION WITH NEGLIGIBLE MASS TRANSFER RESISTANCE (SEMIBATCH REACTORS)

In this section, a mathematical model is presented for a semibatch reactor having negligible mass transfer resistance and short mixing times. Such a model is useful to characterize the behavior of reactors used in the first two stages of PET manufacture[2,19] or at the beginning in batch nylon 6 reactors. In addition, the analysis can easily be modified to apply to continuous reactors, e.g., the first stage of nylon 6 production[20] in industrial VK columns. The kinetic scheme used here, however, is that given by Eq. (5.1.1), but more complex kinetic schemes can easily be accounted for and are, indeed, discussed in later chapters on industrially important systems. The reactor is shown schematically in Fig. 5.2. It is assumed that the monomer P_1 and the volatile condensation by-product, W, both vaporize and are present above the reaction mass at a concentration governed by vapor-liquid equilibrium. $P_2, P_3, \ldots$, are assumed not to volatilize. The

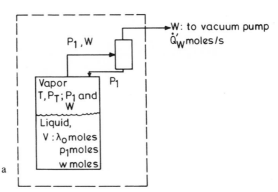

Figure 5.2. Notation used for a semibatch reactor.

vapor consisting of P_1 and W is continuously removed and the monomer is separated and refluxed back to the reactor completely. The total pressure above the reactor is assumed to be P_T, the temperature, T, and the volume of the liquid phase (reactor volume) is V, all these being in general, functions of time. The analysis given below[21] can easily be extended to the case where more or fewer volatile components are present. Since the reactor volume V varies with time, mainly owing to the vaporization of the condensation by-product, it is more convenient to work in terms of the *total moles* of the various components than with their concentrations. In this section, lower case symbols are used to represent these;† for example, p_1 represents the total moles of species P_1 in the liquid phase at time t, and w, the total moles of W in the liquid phase.

Mole balance equations on the control volume shown by dotted lines in Fig. 5.2 can easily be written as[21]

$$\frac{dp_1}{dt} = -V \times 2k_p \left\{ \frac{p_1 \lambda_0}{V^2} - \frac{w(\lambda_0 - p_1)}{KV^2} \right\} \qquad (5.2.1a)$$

$$\frac{d\lambda_0}{dt} = -Vk_p \left\{ \frac{\lambda_0^2}{V^2} - \frac{w(\lambda_1 - \lambda_0)}{KV^2} \right\} \qquad (5.2.1b)$$

$$\frac{dw}{dt} = Vk_p \left\{ \frac{\lambda_0^2}{V^2} - \frac{w(\lambda_1 - \lambda_0)}{KV^2} \right\} - \mathring{Q}'_w \qquad (5.2.1c)$$

where it has been assumed that the reaction takes place only in the liquid phase. In Eq. (5.2.1), $\lambda_0 \equiv V[P] = \sum_{n=1}^{\infty} p_n$, $\lambda_1 = \sum_{n=1}^{\infty} np_n = \lambda_{1,0}$ (since the total number of repeat units in the control volume remains constant) and

---

† Since the conversion of functional groups is not discussed in this section, there need not be any confusion in the use of p_1 as the moles of monomer. Also, the λ's as used here have different units (moles) than in Chapter 2.

$\mathring{Q}'_W$ is the molar rate of removal of the by-product W from the control volume at time t. The rate of reaction terms for P_1 and P are of the same form as derived earlier in Eqs. (2.9.2) and (2.9.5), while that for W can be easily derived. The volume, V, of the reactor can be written as

$$\frac{dV}{dt} = -v_W \mathring{Q}'_W \tag{5.2.2a}$$

or

$$V = V_0 - \int_{t=0}^{t} v_W \mathring{Q}'_W(t)\, dt \tag{5.2.2b}$$

where V_0 is the volume at $t = 0$ and v_W is the molar volume of W, which is, in general, a function of temperature.

Equations (5.2.1) and (5.2.2a) constitute four simultaneous, ordinary differential equations for the five unknown variables p_1, λ_0, w, V, and $\mathring{Q}'_W$ (since $\lambda_1 = \lambda_{1,0}$). Thus, one more equation is required. This is found by using appropriate vapor–liquid equilibrium conditions. If y_W and y_{P_1} represent the mole fractions of W and P_1 in the vapor phase and x_W, x_{P_1}, and x_P, the mole fractions of W, P_1, and P in the liquid phase, the Flory–Huggins theory[22,23] can be used as a good first approximation to give

$$y_{P_1} P_T = P^0_{P_1} v_{P_1} \exp\{v_{P_2 \to P_\infty}(1 + \chi v_{P_2 \to P_\infty} - 1/\mu_n)\} \simeq P^0_{P_1} v_{P_1} e^{(1+\chi)} \tag{5.2.3a}$$

$$y_W P_T = P^0_W v_W \exp\{v_{P_2 \to P_\infty}(1 + \chi v_{P_2 \to P_\infty} - 1/\mu_n)\} \simeq P^0_W v_W e^{(1+\chi)} \tag{5.2.3b}$$

where v_{P_1}, v_W, and $v_{P_2 \to P_\infty}$ are the *volume fractions* of P_1, W, and the nonvolatile species $(P_2, P_3, \ldots)$, respectively, in the liquid phase, χ is the Flory–Huggins interaction parameter (these equations are based on the Flory–Huggins theory for a single "solvent" or a single volatile species, and the same χ has been used for P_1 as well as W), and $P^0_{P_1}$ and P^0_W are the vapor pressures of pure P_1 and W at the temperature T of the reaction mass.[24] The approximate equations are valid for $v_{P_2 \to P_\infty} \simeq 1$ and when μ_n becomes large.

The volume fractions of W and P_1 in the liquid phase can be related to the total moles, w and p_1, in it using

$$v_{P_1} = \frac{p_1 v_{P_1}}{V} \tag{5.2.4a}$$

$$v_W = \frac{w v_W}{V} \tag{5.2.4b}$$

In these equations, v_{P_1} is the molar volume of the monomer. It is assumed that the volume change of mixing is negligible. Equation (5.2.4), when used with Eq. (5.2.3), leads to the following relationships for the partition coefficients:[25]

$$K_{P_1} = \frac{y_{P_1}}{x_{P_1}} = \frac{P_{P_1}^0}{P_T}\left(\frac{w+\lambda_0}{V}\right)v_{P_1}\, e^{1+\chi} \qquad (5.2.5a)$$

$$K_W = \frac{y_W}{x_W} = \frac{P_W^0}{P_T}\left(\frac{w+\lambda_0}{V}\right)v_W\, e^{1+\chi} \qquad (5.2.5b)$$

The partition coefficients are observed to be functions of composition, as well as of T and P_T.

The mole fractions in the vapor phase must add up to unity in the absence of any air, and must be less than unity if air is present along with monomer and W. Thus

$$\left(\frac{p_1 v_{P_1}}{V}\frac{P_{P_1}^0}{P_T} + \frac{w v_W}{V}\frac{P_W^0}{P_T}\right)\exp(1+\chi) \leqslant 1 \qquad (5.2.6)$$

Equation (5.2.6) gives the fifth equation which must be solved simultaneously with Eqs. (5.2.1) and (5.2.2a) in order to obtain the values of p_1, λ_0, w, V, and $\mathring{Q}'_W$ as a function of time.

Several numerical procedures[19,21,25-27] have been developed to solve this set of equations involving both simultaneous ordinary differential equations as well as algebraic inequalities. At any time, if the left-hand side of Eq. 5.2.6 is less than unity, it means that there is some air present in the vapor phase. Once λ_0 is known as a function of time, the number average chain length μ_n can be computed as $\lambda_{1,0}/\lambda_0$. If λ_2 or Q is desired, an equation for it [Eq. (2.9.10)], along with closure conditions [Eq. (2.9.11)] is needed in the set of equations to be solved. Figures 5.3 and 5.4 show[21]

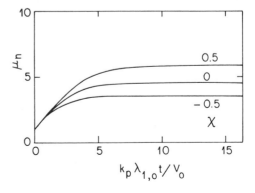

Figure 5.3. Variation[21] of μ_n with time for a semibatch reactor using the Flory–Huggins theory and starting with pure ARB monomer. $P_{P_1}^0/P_T = 0.2161$, $P_W^0/P_T = 7.6965$, $v_{P_1}\lambda_{1,0}/V_0 = 1.0$, $v_W\lambda_{1,0}/V_0 = 0.5$, $K = 1.0$. Values of χ indicated.

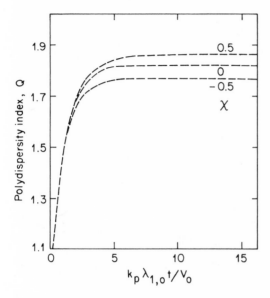

Figure 5.4. Variation[21] of Q with time for a semibatch reactor. Conditions same as in Fig. 5.3.

the variations of μ_n and Q with time for a typically encountered set of dimensionless parameters, for P_T constant with time. It is observed that as χ decreases, i.e., the "solvent" becomes thermodynamically better, there is lesser vaporization and the value of μ_n is lower. The lower values of Q are associated with the lower conversions. The extremely low value of the equilibrium degree of polymerization is because of the low value selected for K. The vaporization of the monomer and W does lead to higher values of μ_n than would have been possible [final $\mu_n = 2$ from Eq. (5.1.2)] in the absence of any evaporation.

Extensive experimentation has shown[23] that for actual systems, χ can be treated as a parameter, taking on values far above the theta-solvent value of 0.5. Experimental vapor–liquid equilibrium data on *model* (nonreacting) systems must, therefore, be taken in order to estimate χ for use in the design of semibatch reactors, or pilot plant data may be used to estimate this parameter. Alternatively, better equilibrium correlations could be used in place of the Flory–Huggins equation to improve the model. Some examples of the use of experimentally derived correlations for the partition coefficients to predict reactor behavior are given in the chapters on nylons and polyesters. It must be emphasized that one must account for the vaporization of volatile components in a manner similar to that described above while interpreting pilot-plant and plant data. Overlooking this aspect can lead to serious errors in the estimation of rate constants from such experimental data or in the prediction of the performance of plants.

As mentioned before, the analysis presented above can either be simplified (e.g., monomer not volatile) or extended (e.g., more complex kinetics, more volatile species necessitating the use of multivariate search techniques along with the Runge–Kutta algorithm, some noncondensables produced by reaction, etc.).[26,27] Some such cases are discussed in appropriate chapters later in this book. Also, the development presented here is completely general, and instead of taking P_T and T as constants, one may assume them as functions of time, $P_T(t)$ and $T(t)$ (as determined by using the characteristics of a vacuum pump, or using heat transfer considerations).

5.3. POLYMERIZATION WITH FINITE MASS TRANSFER RESISTANCE

In this section, attention is focused on modeling polymerizations with desorption of the low molecular weight by-product occurring simultaneously, when these two phenomena have comparable resistances. The desorption of the monomer is neglected since its concentration is usually extremely low for reaction conditions under which this analysis is applicable. A general formulation[28-31] for mass transfer with step growth polymerization of ARB monomers is presented using some simple geometries, so that a basic understanding of the physical phenomena involved in this complex area is developed. The results of this model can be directly applied to explain the behavior of some industrial reactors, while for others this model can be extended by incorporating side reactions, etc.

Two common geometries of interest are shown in Fig. 5.5. In one of these, the reaction mass is a thin film of thickness y_0 applied on an impervious wall at $y = 0$. At $t = 0$, the entire reaction mass has uniform concentrations

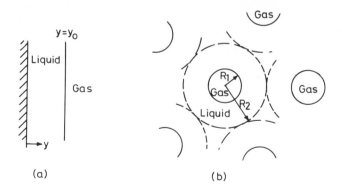

Figure 5.5. Two common reaction-cum-diffusion models.

(usually close to thermodynamic equilibrium with respect to some gas phase pressure) $[P_1]_0$, $[P]_0$, $[W]_0$, etc. At $t \geq 0$, the concentration of the volatile by-product W is lowered in the gas phase usually by application of a vacuum. Because of this, the layer of liquid *just* adjacent to the gas phase attains some new equilibrium concentration of W, say $[W]^*$, given by a vapor–liquid equilibrium relationship similar to Eq. (5.2.3b). Since the value of $[W]^*$ for $t > 0$ is lower than that for $t < 0$, chemical equilibrium is disturbed. In addition, a concentration gradient of $[W]$ is set up, leading to simultaneous mass transfer with polymerization. This continues till a new polymerization equilibrium in the film is attained, with the concentration of W equal to $[W]^*$ throughout its thickness. It is this transition from one equilibrium state towards another that must be studied. Modeling of such film reactors is important, not only because it provides valuable fundamental insight into the effects of mass transfer in polymerization reactors, but also because such films are indeed generated by mechanical means in some special finishing reactors used industrially.

It may be pointed out that if a constant pressure is applied on the film for $t > 0$, Eq. (5.2.3b) predicts that the volume fraction v_W, of W, in the liquid just adjacent to the gas is independent of time. Since polymerization proceeds simultaneously, this implies that the molar concentration, $[W]^*$, in this surface "layer" changes with time (to keep v_W the same). However, incorporating this effect makes the computations considerably more involved, and as a first approximation, it is assumed that the molar concentration $[W]^*$ is constant.

The second geometry of interest is the situation depicted in Fig. 5.5b, wherein an inert gas bubbles through the reaction mass to remove W (or bubbles of W at a different concentration, generated by boiling or by some other means, are present), and thus gives a higher molecular weight product. One type of patented[4] nylon 6 reactor uses this method of removing W. A very simple model of this physical system can be made by assuming equal-sized bubbles of radii R_1, distributed uniformly as shown, with center-to-center distance between the bubbles as $2R_2$. Because of symmetry, one can associate a concentric sphere of reaction mass, extending over $R_1 \leq r \leq R_2$ with each gas bubble, and study the diffusion of W simultaneously with polymerization in this repeating spherical shell. Once again, it is assumed that for $t > 0$ the concentration in the liquid at $r = R_1$ is constant at $[W]^*$.

Several simplifying assumptions are made to model the effect of mass transfer in such diffusion-controlled reactors. The most important of these, made by Secor,[29] is that the polymer molecules are essentially immobilized. This is a fairly good approximation at high conversions, when the polymer molecules are long enough to be well entangled. A similar assumption has since been made in some of the early papers on free-radical polymerization.[32,33] The movement of low molecular weight material (W) through this immobile matrix can then be described by[34]

$$\frac{\partial[W]}{\partial t} = \mathscr{D}_W \nabla^2 [W] + \mathscr{R}_W \tag{5.3.1}$$

with the corresponding equation for the polymeric species given by

$$\frac{\partial[P_n]}{\partial t} = \mathscr{R}_{P_n}, \qquad n = 1, 2, \ldots \tag{5.3.2}$$

$\mathscr{D}_W$ is the diffusivity of W, assumed constant, and $\mathscr{R}_i$ is the molar rate of production of i by chemical reaction, per unit volume.

It must be emphasized that Eqs. (5.3.1) and (5.3.2) involve several assumptions, the effects of which possibly cancel out. There have been some recent studies[35] in free-radical polymerizations, in which the diffusion and swelling of polymer molecules in the pseudobinary reacting system have been accounted for more rigorously. These need to be extended to the modeling of step-growth polymerization reactors wherein mass transfer resistance is important. However, up till now, the assumptions used by Secor[29] have been applied quite extensively in the field of polymerization reaction engineering.[36-40]

For the film shown in Fig. 5.5a, and for the kinetic scheme of Eq. (5.1.1), Eqs. (5.3.1) and (5.3.2) can easily be shown to give Eqs. (a)-(d) in Table 5.1. Here, the moments λ_i and all the concentrations are functions of both the position, y, as well as time, t. The rate terms in these equations are identical *in form* to those in Section 2.9 and similar moment closure conditions are used. Equations (a)-(e) of Table 5.1 can be integrated using the finite difference technique[41] along with the initial and boundary conditions given in this table. Condition (g) of Table 5.1 represents the fact that at the impervious wall at $y = 0$, there can be no gradient or flux of W.

The integration of the set of equations in Table 5.1 gives the various moments, λ_i, and [W] as functions of position and time. Often, it is of interest to determine the spatial-average values of these quantities as a function of time. These are given by

$$\overline{[P_n]}(t) = \frac{\int_{y=0}^{y_0} [P_n](y, t)\, dy}{y_0}, \qquad \overline{\lambda_k}(t) = \frac{\int_{y=0}^{y_0} \lambda_k(y, t)\, dy}{y_0} \tag{5.3.3}$$

where the bars represent spatial-average values and the notation (t) and (y, t) represent the dependence on time, and on position and time, respectively.

TABLE 5.1. Equations for Film Reactor

$$\frac{\partial[W]}{\partial t} = \mathscr{D}_W \frac{\partial^2[W]}{\partial y^2} + k_p[P]^2 - k_p'[W](\lambda_1 - [P]) \tag{a}$$

$$\frac{\partial \lambda_0}{\partial t} = \frac{\partial[P]}{\partial t} = -k_p[P]^2 + k_p'[W](\lambda_1 - [P]) \tag{b}$$

$$\frac{\partial \lambda_1}{\partial t} = 0; \qquad \therefore \quad \lambda_1 = \lambda_{1,0} \tag{c}$$

$$\frac{\partial \lambda_2}{\partial t} = 2k_p\lambda_1^2 + k_p'[W]\frac{\lambda_1 - \lambda_3}{3} \tag{d}$$

$$\lambda_3 = \frac{\lambda_2(2\lambda_2\lambda_0 - \lambda_1^2)}{\lambda_1\lambda_0} \tag{e}$$

Initial conditions

$t = 0$, all y:

$[W] = [W]_0$;

$$\lambda_0 = \lambda_{0,0} = \sum_{n=1}^{\infty} [P_n]_0; \qquad \lambda_1 = \lambda_{1,0} = \sum_{n=1}^{\infty} n[P_n]_0; \qquad \lambda_2 = \lambda_{2,0} \tag{f}$$

Boundary conditions

$y = 0$, all t:

$$\frac{\partial[W]}{\partial y} = 0 \tag{g}$$

$y = y_0$, all t:

$[W] = [W]^* \tag{h}$

At times, the *mean* rate of desorption of the volatile component W over time t_f is of interest. This can be determined using Ficks' law,[34] giving

$$\bar{N}_{W,y}\big|_{y_0} = -\frac{1}{t_f}\int_{t=0}^{t_f} \mathscr{D}_W \frac{\partial[W]}{\partial y}\bigg|_{y=y_0} dt \tag{5.3.4}$$

The use of Eq. (5.3.4) leads to severe instabilities if $[W]$ (y, t) is differentiated numerically to obtain $\bar{N}_{W,y}\big|_{y_0}$. Amon and Denson[13] have suggested the following procedure to overcome this problem. Summing Eqs. (a) and (b) in Table 5.1, one gets

$$\frac{\partial}{\partial t}([W] + [P]) = \mathscr{D}_W\frac{\partial^2[W]}{\partial y^2} \tag{5.3.5}$$

which, on integrating with respect to y, gives

$$\int_{y=0}^{y_0} \frac{\partial}{\partial t}([\mathrm{W}] + [\mathrm{P}]) \, dy = \int_{y=0}^{y_0} \mathscr{D}_\mathrm{W} \frac{\partial^2[\mathrm{W}]}{\partial y^2} \, dy$$

$$= \mathscr{D}_\mathrm{W}\left\{ \left.\frac{\partial[\mathrm{W}]}{\partial y}\right|_{y_0} - \left.\frac{\partial[\mathrm{W}]}{\partial y}\right|_{y=0} \right\} = \mathscr{D}_\mathrm{W} \left.\frac{\partial[\mathrm{W}]}{\partial y}\right|_{y_0}$$

$$(5.3.6)$$

The boundary condition at $y = 0$ [Eq. (g), Table 5.1] has been used in this equation. Equations (5.3.4) and (5.3.6) give

$$\bar{N}_{\mathrm{W},y}|_{y_0} = \frac{1}{t_f} \int_{y=0}^{y_0} \{([\mathrm{W}] + [\mathrm{P}])_{t=0} - ([\mathrm{W}] + [\mathrm{P}])_{t_f}\} \, dy \qquad (5.3.7)$$

Use of the profiles of $[\mathrm{W}]$ and $[\mathrm{P}]$ at $t = t_f$ and integration over the thickness y gives the mean flux $\bar{N}_{\mathrm{W},y}|_{y_0}$.

Appropriate equations[31] for the concentric sphere geometry shown in Fig. 5.5b can be written similarly and are left as an exercise (the boundary condition at $r = R_2$ will be $\partial[\mathrm{W}]/\partial r = 0$ because of symmetry at that location).

The equations characterizing the film reactor of Fig. 5.5a can be made dimensionless in terms of the following variables:

$$C_\mathrm{W} = [\mathrm{W}]/\lambda_{1,0} \qquad\qquad\qquad (5.3.8a)$$

$$C = [\mathrm{P}]/\lambda_{1,0} \qquad\qquad\qquad (5.3.8b)$$

$$m_k = \lambda_k/\lambda_{1,0}, \qquad k = 0, 1, 2, 3, \dots (m_0 = C) \qquad (5.3.8c)$$

The following dimensionless independent variables and parameters emerge from this step:

$$\tau = k_p \lambda_{1,0} t \qquad\qquad\qquad (5.3.9a)$$

$$\xi = \left(\frac{k_p \lambda_{1,0}}{\mathscr{D}_\mathrm{W}}\right)^{1/2} y \qquad\qquad (5.3.9b)$$

$$\tau_f = k_p \lambda_{1,0} t_f \qquad\qquad\qquad (5.3.9c)$$

$$\xi_1 = \left(\frac{k_p \lambda_{1,0}}{\mathscr{D}_\mathrm{W}}\right)^{1/2} y_0 \qquad\qquad (5.3.9d)$$

$$C_\mathrm{W}^* = [\mathrm{W}]^*/\lambda_{1,0} \qquad\qquad\qquad (5.3.9e)$$

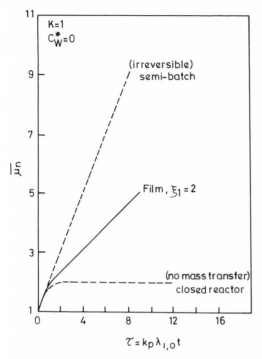

Figure 5.6. Variation[31] of the spatial average $\bar{\mu}_n$ with time for the film reactor of Fig. 5.5a. Feed of pure monomer, $K = 1$, $C_W^* = 0$, $\xi_1 = 2$. The case of no mass transfer ($\mathcal{D}_W = 0$) and that for infinitely rapid mass transfer (irreversible, semibatch) also shown for comparison.

The equations of Table 5.1 have been integrated numerically after nondimensionalization for several interesting situations[29-31] and some of the results[31] are shown in Fig. 5.6. In generating this figure, the feed has been assumed to be pure monomer, even though the model is *really* applicable for finishing stages where the feed to the reactor has a much higher μ_n. However, qualitative conclusions can still be drawn from these results. The results shown in Fig. 5.6 can be compared with those for two limiting cases: first, the infinitely fast mass transfer case (irreversible or semibatch conditions, with $C_W = C_W^* = 0$ throughout the reaction mass) and second, the "closed" reactor in which there is no mass transfer, i.e., the $\mathcal{D}_W \partial^2[W]/\partial y^2$ term in Eq. (a) of Table 5.1 is zero and the boundary conditions on [W] are not required. Results are found to lie between these two cases, depending on the value of ξ_1 chosen. It may be noted that ξ_1 is an important dimensionless parameter for this model, and incorporates the effect of the polymerization rate constant, the diffusivity, as well as the length, y_0, of the diffusion path. Figure 5.7 shows[29] numerically obtained results on the spatial variation of μ_n with time when the feed to the reactor is an *equilibrium* prepolymer (from a previous reactor) having a number average chain length, $\mu_{n,0}$, of 100 {and $C_{W,0} = K/[\mu_{n,0}(\mu_{n,0} - 1)]$, and $m_{2,0} = [2\mu_{n,0} - 1]$}. It is observed

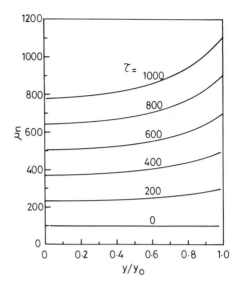

Figure 5.7. Variation of μ_n with position for a film reactor.[29] $1/m_{0,0} = \mu_{n,0} = 100$, $K = 1$, $C_W^* = 0$, $\xi_1 = 1$ (close to actual values for PET reactors). $C_{W,0} = K/\{\mu_{n,0}(\mu_{n,0} - 1)\}$. (Reprinted from Ref. 29 with permission from the American Institute of Chemical Engineers, New York.)

that near the vapor-liquid interface, the values of μ_n are significantly higher, as intuitively expected, since the value of [W] there is the lowest.

It may be emphasized that the equations in Table 5.1 are applicable over the *entire* range of operation of reactors wherein both mass transfer as well as polymerization are taking place, provided there are no mixing effects and provided the assumptions used are justified. At one end of the spectrum is the situation where $\mathscr{D}_W$ is infinitely high. Under these limiting conditions, $[W] = [W]^*$ throughout the film, with no spatial variations. The entire reaction mass would then behave similarly to the semibatch reactor discussed in Section 5.2, and the reactor is said to be kinetically controlled under these conditions. At the other end of the spectrum, mass transfer is nonexistent and the term $\mathscr{D}_W \partial^2[W]/\partial y^2$ in Eq. (a) of Table 5.1 becomes identically equal to zero, with the boundary conditions [Eqs. (g) and (h)] not applicable. The reactor then behaves like a "closed" reactor (in fact, in such a case, if one starts with an equilibrium situation, no further reaction would take place). Between these two limiting cases there lies a whole array of reactor behavior,[28] which may be simulated by the numerical integration of the equations in Table 5.1.

It may be mentioned here that for certain values of τ, there is a very sharp gradient in [W] existing in an extremely thin layer near the gas-liquid interface, while in the bulk of the liquid, the concentration of W is uniform. Chemical equilibrium is maintained in this main body of the liquid. Figure 5.8 shows a typical plot of [W] obtained numerically. One must be extremely careful in choosing small enough values of Δy or $\Delta \xi$ while integrating the equations *numerically* for low values of τ. As τ increases, the effect of the

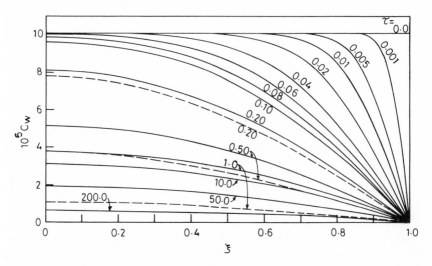

Figure 5.8. Spatial variation of the concentration of W for a film reactor: $K = 1$, $\mu_{n,0} = 100$, $C_W^* = 0$, $\xi_1 = 1$, $C_{W,0} = 1.0101 \times 10^{-4}$ (as for Fig. 5.7). Solid lines represent C_W when reaction is taking place, while dashed lines represent the case when no reaction takes place. Note that the affected depth $\xi_\delta \sim \tau^{1/2}$.

interfacial concentration [W]* *penetrates* farther into the film and gradually reaches the wall.

The *existence* of a sharp gradient in a thin region near the vapor interface when τ is small, or equivalently, when $\mathscr{D}_W$ is small, can also be predicted theoretically.[†] Equation (a) of Table 5.1 reduces to the following equation as $\mathscr{D}_W \to 0$:

$$\frac{\partial[W]}{\partial t} = k_p[P]^2 - k_p'[W](\lambda_1 - [P]) \tag{5.3.10}$$

which appears to imply that the boundary conditions on [W] cannot be satisfied. In order to find out the relevant differential equations for *low but nonzero* values of $\mathscr{D}_W$, which simultaneously satisfy the two boundary conditions [Eqs. (g) and (h), Table 5.1], Eqs. (a)–(e) of this table must be made dimensionless using a different set of independent variables than defined in Eq. (5.3.9). The time, t, and the length, y, must be made dimensionless using

$$\tau^* = t/t_f$$

$$\tag{5.3.11}$$

$$\xi^* = y/y_0$$

[†] This treatment is discussed in more detail in Ref. 42.

instead of using Eqs. (5.3.9a) and (5.3.9b). This leads to

$$\frac{\partial C_W}{\partial \tau^*} = \left(\frac{t_f \mathcal{D}_W}{y_0^2}\right)\frac{\partial^2 C_W}{\partial \xi^{*2}} + (k_p \lambda_{1,0} t_f)\left\{C^2 - \frac{C_W(m_1 - C)}{K}\right\} \quad (5.3.12a)$$

$$C_W = C_{W,0}; \quad \text{at } \tau^* = 0, \text{ all } \xi^* \quad (5.3.12b)$$

$$\frac{\partial C_W}{\partial \xi^*} = 0; \quad \xi^* = 0, \text{ all } \tau^* \quad (5.3.12c)$$

$$C_W = C_W^*; \quad \xi^* = 1, \text{ all } \tau^* \quad (5.3.12d)$$

It is possible to write similar equations for C and m_1. If $t_f \mathcal{D}_W / y_0^2 \to 0$, Eq. (5.3.12a) reduces to

$$\frac{\partial C_W}{\partial \tau^*} = (k_p \lambda_{1,0} t_f)\left\{C^2 - \frac{C_W(m_1 - C)}{K}\right\} \quad (5.3.13)$$

which implies that Eqs. (5.3.12c) and (5.3.12d) *cannot* be satisfied. It may be recalled that on omitting these conditions, one obtains the closed-reactor equations. Evidently there is something wrong in the above "scaling" of the equations using Eq. (5.3.11) to give results for low $\mathcal{D}_W$. The reason, on closer scrutiny, is the choice of a wrong characteristic length in the y direction. Instead of y_0, one must use a length δ ($\ll y_0$) to nondimensionalize y. Physically, δ would be the effective length (a function of time) where W is changing from the value at the interface to the equilibrium value in the bulk of the film as shown in Fig. 5.8. Hence, ξ^* and τ^* are redefined as

$$\xi^* = y/\delta$$
$$\tau^* = t/t_f \quad (5.3.14)$$

which gives

$$\frac{\partial C_W}{\partial \tau^*} = \left(\frac{y_0}{\delta}\right)^2\left(\frac{t_f \mathcal{D}_W}{y_0^2}\right)\frac{\partial^2 C_W}{\partial \xi^{*2}} + k_p \lambda_{1,0} t_f\left\{C^2 - \frac{C_W(m_1 - C)}{K}\right\} \quad (5.3.15a)$$

$$C_W = C_{W,0} \quad \text{at } \tau^* = 0, \text{ all } \xi^* \quad (5.3.15b)$$

$$\frac{\partial C_W}{\partial \xi^*} = 0 \quad \text{at } \xi^* = 0, \text{ all } \tau^* \quad (5.3.15c)$$

$$C_W = C_W^* \quad \text{at } \xi^* = y_0/\delta, \quad \text{all } \tau^* \quad (5.3.15d)$$

Now, as $t_f \mathscr{D}_w / y_0^2$ becomes small, it is observed from Eq. (5.3.15a) that the coefficient of $\partial^2 C_W / \partial \xi^{*2}$ does *not* approach zero (since $y_0 \gg \delta$) and does not drop out. In fact, an estimate of δ can be made by choosing the coefficient of $\partial^2 C_W / \partial \xi^{*2}$ to be unity, i.e.,

$$\frac{\delta}{y_0} \sim \left(\frac{t_f \mathscr{D}_w}{y_0^2} \right)^{1/2} \quad \text{or} \quad \delta \sim \mathscr{D}_w^{1/2} t_f^{1/2} \tag{5.3.16}$$

The above analysis is similar to that done for boundary layers in fluid mechanics.[42] Thus, sharp gradients in C_W exist at $(y_0 - \delta) \leq y \leq y_0$, while in the inner region called the "bulk," $0 \leq y \leq (y_0 - \delta)$, equilibrium conditions prevail with $\partial C_W / \partial \xi^* = 0$.

In the above discussion, attention was focused on the behavior of *stationary* films when the initial equilibrium is disturbed by application of a vacuum at the vapor–liquid interface. In several commercial reactors, there is a considerable amount of mixing present in the film simultaneously, because of which the above analysis for stationary films must be modified. A simple model to account for this effect is one where reaction with desorption (but no mixing) takes place for a time $\Delta \tau$ (with the equations of Table 5.1 applicable), and then there is an instantaneous mixing because of which the concentrations of $P_1, P_2, P_3, \ldots, P$, and W all become uniform across the entire film once again [given by Eq. (5.3.3)]. Obviously, chemical equilibrium no longer prevails in the bulk of the film, though $\partial C_W / \partial y$ in that zone is still zero. The use of extremely low values of Δy for simulating the reaction in such "mixed films" is extremely important, since the gradients in W remain confined to a very thin layer near the interface [of thickness approximately $\mathscr{D}_w^{1/2} (\Delta t)^{1/2}$, according to Eq. (5.3.16)], while "shifting" uniform concentrations slightly removed from the initial equilibrium conditions exist in the bulk of the film. Extreme numerical difficulties are encountered while solving this problem. It may be emphasized that the separation of the *simultaneous* reaction, desorption, and mixing into a series of *sequential* operations of reaction + desorption and mixing, is purely to keep the analysis simple. Rigorous modeling of these three steps would lead to a modification of the equations in Table 5.1.

The performance of the "mixed" film reactor can be best predicted analytically.[28] In this analytical approach, the film is modeled using a linearized perturbation theory.[28,36] The concentration [W] is written in terms of a perturbation, ε, about a pseudo-"equilibrium" value $[W]_{eq}$ at time t, existing in the "bulk" of the film, as

$$\varepsilon \equiv [W] - [W]_{eq} = [W] - \frac{K[P]_b^2}{\lambda_{1,0} - [P]_b} \tag{5.3.17}$$

where $[P]_b$ is the concentration of functional groups in the bulk at time t. $\lambda_{1,0}$ has been used for λ_1 since it does not vary with time.

If the film is now "exposed" for a *short* period of time $t \leq t \leq (t + t_f^*)$ (or alternatively, if a new time scale t^* is defined, so that this exposure extends over $0 \leq t^* \leq t_f^*$), some W is desorbed from the film. This can be obtained by substituting Eq. (5.3.17) in Eq. (a) of Table 5.1 to give

$$\frac{\partial \varepsilon}{\partial t^*} = \mathcal{D}_W \frac{\partial^2 \varepsilon}{\partial y^2} - \frac{k_p}{K}(\lambda_{1,0} - [P]_b)\varepsilon \qquad (5.3.18)$$

where $[P]_b$ has been assumed constant over this small time period. The boundary conditions on W are replaced by those on ε, given by

$$t^* = 0, \text{ all } y: \qquad \varepsilon = 0 \text{ (equilibrium)} \qquad (5.3.19a)$$

$$y = y_0, \text{ all } t: \qquad \varepsilon = \varepsilon^* = [W]^* - [W]_{eq} \qquad (5.3.19b)$$

$$y = -\infty, \text{ all } t: \qquad \varepsilon = 0 \text{ (equilibrium)} \qquad (5.3.19c)$$

It may be noted that in Eq. (5.3.19c), the impervious wall condition at $y = 0$ has been replaced by one at $y \rightarrow -\infty$ (a semi-infinite film) in order to reduce the mathematical problem to a standard one and obtain an analytical solution. Equations (5.3.18) and (5.3.19) lead to the following expression[36] for the time-averaged flux of W:

$$\overline{N_{W,y}}\Big|_{y_0} = -\frac{1}{t_f^*} \int_0^{t_f^*} \mathcal{D}_W \frac{\partial \varepsilon}{\partial y}\Big|_{y=y_0} dt^*$$

$$= -\varepsilon^*(\mathcal{D}_W k^*)^{1/2}\left\{\left[1 + \frac{1}{2k^*t_f^*}\right] \text{erf}(k^*t_f^*)^{1/2} + \frac{\exp(-k^*t_f^*)}{k^*t_f^*}\right\}$$

$$(5.3.20)$$

where t_f^* is the exposure time and

$$k^* = k_p\left(\frac{\lambda_{1,0} - [P]_b}{K}\right) \qquad (5.3.21)$$

For $k^*t_f^* > 10$, Eq. (5.3.20) simplifies to give

$$\bar{N}_{W,y}\Big|_{y_0} = -\varepsilon^*(\mathcal{D}_W k^*)^{1/2} = ([W]_{eq} - [W]^*)(\mathcal{D}_W k^*)^{1/2} \qquad (5.3.22)$$

which is observed to be independent of the exposure time t_f^*.

A pseudo-steady-state analysis can now be made as follows. If s is the interfacial area per unit volume of the reaction mass (it must be noted that since the gradients exist only in a very thin region near the interface, the results are expected to be independent of the geometry), the average rate of removal of W (per unit volume of the reaction mass) at time t is given as $(\mathscr{D}_W k^*)^{1/2}([W]_{eq} - [W]^*)s$. Since each mole of W formed is associated with the reaction of one mole of functional groups A and B, Eq. (5.3.22) leads to the following equation for the rate of change of $[P]_b$:

$$\frac{d[P]_b}{dt} = -s(k^* \mathscr{D}_W)^{1/2}([W]_{eq} - [W]^*) \tag{5.3.23}$$

with

$$[W]_{eq} = \frac{K[P]_b^2}{\lambda_{1,0} - [P]_b} \tag{5.3.24}$$

Equation (5.3.23) can be integrated after substitution of Eqs. (5.3.24) and (5.3.21) (or k^* may be assumed constant) to give $[P]_b$ as a function of time in the film reactor. This can then be used to give $\bar{\mu}_n$ $(=\lambda_{1,0}/[P]_b)$. It may be noted that in this analysis several approximations have been used, one of which is the use of the equilibrium concentration $[W]_{eq}$ at *every* time [Eq. (5.3.19a)]. Such approximations will not be required if an equivalent numerical solution is developed.

Hoftyzer and van Krevelen[28] have extended the above analysis to explain the behavior of PET and nylon 6 reactors wherein mass transfer is occurring from the bulk to gas bubbles generated by vaporization. The value of the surface area per unit volume, s, in such situations is difficult to estimate, particularly since it involves the nucleation and growth of gas bubbles. Nucleation is a very complex phenomenon and is usually a function of (exact functionality available in the form of experimental data) process variables like the local temperature, viscosity, degree of agitation, etc. In fact, s may vary with time. Hoftyzer and van Krevelen[28] have obtained $s(t)$ experimentally using *nonreacting* model systems and, using this in Eq. (5.3.23), they have generated $[P]_b$ and μ_n as functions of time. They report an excellent agreement between the results so generated and experimental data on nylon 6 and PET polymerizations. Unfortunately, these workers have presented only semiquantitative results because of proprietary reasons and do not give details of the experimental determination of $s(t)$. Ravindranath and Mashelkar[36] have integrated Eq. (5.3.23) for PET systems and have found that $\bar{\mu}_n(t)$ varies with time in a manner which is similar to that found experimentally by Yokoyama et al.[43] on batch reactors. These studies thus appear to justify the model.

Thus, if it is possible to estimate $\mathcal{D}_W$ by controlled experiments on a batch film reactor (with interfacial area well defined and measured) as done for nylon 6 by Nagasubramanian and Reimschuessel,[40] and then obtain $s(t)$ on nonreacting model systems,[28] one can integrate Eq. (5.3.23) numerically to predict $\bar{\mu}_n$ as a function of time. Alternatively, with $\mathcal{D}_W$ and $s(t)$ known, the equations in Table 5.1 can be integrated numerically after appropriate modifications to account for a time-varying interfacial area, since new bubbles nucleate and grow continuously. It may be added that since the gradient in W is limited to a very thin region near the vapor interface, the center-to-center distance between the bubbles will be unimportant. One can neglect curvature effects and solve the equations for the film for the same reason. At present, the development of a sound fundamental basis for predicting or estimating $s(t)$ seems to be a major problem in the reaction engineering of such polymerizations.

The dynamics of growth of gas bubbles (Fig. 5.5b) has been recently studied[44] to explain bubble growth in expanding foams. The effects of surface tension and other mechanical factors were investigated. The model is an improvement over the concentric sphere reactor of Fig. 5.5b, even though it has been applied to nonreactive systems, and can easily be used to account for mass transfer effects in step growth polymerizations.

5.4. FINISHING REACTORS

In this section, the modeling of some important finishing reactors is attempted. These are used industrially when the diffusivity of W becomes extremely low. As discussed in Section 5.1, there are several kinds of equipment that carry out polymerization with removal of the by-product by application of a vacuum. One of these is the wiped film reactor[13] (Fig. 5.1), in which some amount of liquid from a pool is applied as a thin film on an impervious wall and desorption of W takes place at the vapor–liquid interface. After some exposure time, t_f, in the film, depending on the number of blades and their rate of rotation, the film is scraped off and mixed with the bulk of the reaction mass again. Because of the high viscosity, it is reasonable to assume that there are little mixing effects in the axial direction in the pool of liquid. Another commonly used reactor is the intermeshing twin-screw extruder with deep channels. In these, the entire reaction mass exists as a thin film, subjected to vacuum, and at the end of every cycle, the entire mass is well mixed.[45]

The first model for finishing reactors was described by Ault and Mellichamp,[45] in which it was assumed that the volume of the liquid in the pool is negligible. The *entire* reaction mass is applied as a film on the walls at $t = 0$, and after time t_f, it is instantaneously well mixed and reapplied.

Thus, for ARB polymerization, the equations in Table 5.1 apply with just one modification. The values of the variables [W], [P] and the moments at $t = 0$, when the film is generated, are given by the average values in the film at the end of the previous cycle [obtained by using Eq. (5.3.3)], and then polymerization again proceeds[†] for a time t_f. A numerical integration of this modified set of equations can easily be carried out. The feed to the *reactor*, i.e., the initial conditions for the *first* cycle, is assumed to be a prepolymer having a number-average degree of polymerization of $\mu_{n,0}$ and a value of [P] which is in equilibrium with the concentration $[W]_0$ present in the *previous* reactor. That is, the feed to this film reactor is characterized by

$$\frac{[W]_0}{\lambda_{1,0}} = \frac{K}{\mu_{n,0}(\mu_{n,0} - 1)}$$

$$\frac{[P]_0}{\lambda_{1,0}} = \frac{1}{\mu_{n,0}} \tag{5.4.1}$$

$$\frac{\lambda_{2,0}}{\lambda_{1,0}} = 2\mu_{n,0} - 1$$

These initial conditions can easily be derived by setting the $\partial^2/\partial y^2$ and $\partial/\partial t$ terms in the equations of Table 5.1 as zero. Figure 5.9 (curve a) shows[46] typical results for this model for $K = 1$ and for N, the total number of cycles, as 100. The exit values of $\bar{\mu}_n$ are not found to be too high because of the high values of τ_f and ξ_1 [see Eq. (5.3.9)] used in generating these results. This model has recently been applied[37] to PET systems.

The second model, proposed by Amon and Denson,[13] accounts for the presence of a separate pool of liquid in the reactor but assumes that there is no *axial* mixing within the pool. Figure 5.10 shows the notation used. If the change in the density of the reaction mass is neglected, the volumetric (axial) rate of flow, $\mathring{Q}$, would be the same for all values of x. The mass balance equations for the differential control volume for reversible ARB polymerization at steady state can then be written as

$$\mathring{Q}[W]_b - \mathring{Q}\left\{[W]_b + \frac{d[W]_b}{dx}\delta x\right\} + A_b(\delta x)\{k_p[P]_b^2$$

$$- k_p'[W]_b(\lambda_{1,b} - [P]_b)\} - \bar{N}_W s(\delta x) = 0 \tag{5.4.2a}$$

[†] Note that this model differs from the "mixed" film model of the previous section since here mixing occurs after several $\Delta\tau$'s, instead of simultaneously.

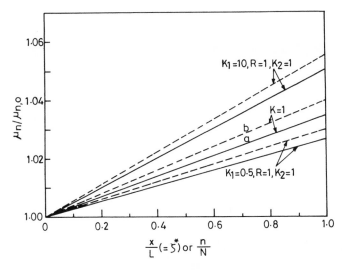

Figure 5.9. Performance of wiped film reactors.[46] $\mu_{n,0} = 50$, $C_W^* = 0$. Curves a and b are for the equal reactivity case with $K = 1$. Solid lines: Ault and Mellichamp $\{k_p\lambda_{1,0}t_f \equiv \tau_f = 2, (k_p\lambda_{1,0}/\mathcal{D}_w)^{1/2}y_0 \equiv \xi_1 = 2 \times 10^4, N = 100\}$. n is the number of cycles, going from zero to N. Dotted lines: Amon and Denson $\{k_p\lambda_{1,0}\bar{\theta} = 2000, k_p\lambda_{1,0}t_f = 2, t_f\mathcal{D}_w/y_0^2 = 10^{-4}, Sy_0/\mathring{Q}t_f = 100\}$.

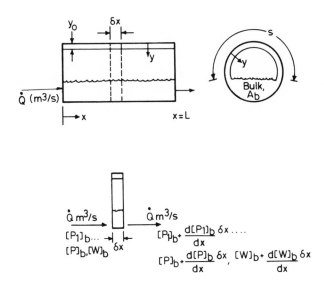

Figure 5.10. The control volume to be considered in modeling wiped film reactors.

$$\mathring{Q}[P_n]_b - \mathring{Q}\left\{[P_n]_b + \frac{d[P_n]_b}{dx}\delta x\right\} + A_b(\delta x)\mathscr{R}_{P_{n,b}} = 0, \qquad n = 1, 2, \ldots$$

$$(5.4.2b)$$

where $[W]_b, [P_n]_b, \lambda_{1,b}, \ldots$ etc., denote the values of $[W], [P_n], \lambda_1, \ldots$, etc., in the pool of liquid (called the bulk, hereafter) at axial position x, and A_b is the cross-sectional area of the bulk (so that $A_b\delta x$ is the volume of the

TABLE 5.2. Equations for the "Bulk" in Wiped-Film Reactors

Balance equations (bulk)

$$-\mathring{Q}\frac{d[W]_b}{dx} + A_b\{k_p[P]_b^2 - k_p'[W]_b(\lambda_{1,b} - [P]_b)\} - \bar{N}_W s = 0 \tag{a}$$

$$\mathring{Q}\frac{d\lambda_{0,b}}{dx} = \mathring{Q}\frac{d[P]_b}{dx} = A_b\{-k_p[P]_b^2 + k_p'[W]_b(\lambda_{1,b} - [P]_b)\} \tag{b}$$

$$\mathring{Q}\frac{d\lambda_{1,b}}{dx} = 0, \qquad \therefore \quad \lambda_{1,b} = \lambda_{1,0} \tag{c}$$

$$\mathring{Q}\frac{d\lambda_{2,b}}{dx} = A_b\left\{2k_p\lambda_{1,0}^2 + k_p'[W]_b\frac{\lambda_{1,0} - \lambda_{3,b}}{3}\right\} \tag{d}$$

Closure condition

$$\lambda_{3,b} = \frac{\lambda_{2,b}(2\lambda_{2,b}\lambda_{0,b} - \lambda_{1,0}^2)}{\lambda_{1,0}\lambda_{0,b}} \tag{e}$$

Equations for film
  Equations (a)–(e) of Table 5.1 (with subscript f)

Boundary and initial conditions
  $t = 0, 0 \leq y \leq y_0$

$$[W]_f = [W]_b(x) \tag{f}$$

$$[P]_f = [P]_b(x) \tag{g}$$

$$\lambda_{0,f} = \lambda_{0,b}(x) \tag{h}$$

$$\lambda_{1,f} = \lambda_{1,b}(x) = \lambda_{1,0} \tag{i}$$

$$\lambda_{2,f} = \lambda_{2,b}(x) \tag{j}$$

$$y = 0, 0 \leq t \leq t_f: \frac{\partial[W]_f}{\partial y} = 0 \tag{k}$$

$$y = y_0, 0 \leq t \leq t_f: [W]_f = [W]^* \tag{l}$$

Flux

$$\bar{N}_W = \frac{1}{t_f}\int_0^{y_0} \{([W]_b + [P]_b) - ([W]_f + [P]_f)_{t_f}\}\, dy \tag{m}$$

bulk in the differential control volume). In Eq. (5.4.2), s is the perimeter of the film exposed to the vacuum (so that $s\delta x$ is the differential interfacial area of the surface of desorption) and $\bar{N}_W$ is the time-averaged rate of desorption of W from the film at x. $\mathscr{R}_{P_{n,b}}$ is the net rate of production of species P_n by chemical reaction in the "bulk" and can be written in terms of the various concentrations using Eqs. (2.9.2) and (2.9.4). In writing Eq. (5.4.2), it is assumed that the term $s(\delta x)\mathscr{R}$ in the film is omitted in comparison with the term $A_b(\delta x)\mathscr{R}$ in the bulk, which is justified since the film is thin. In addition, it is assumed that desorption of W occurs solely from the film surface $s(\delta x)$ since this will be much larger than the desorption from the exposed surface of the pool. In view of the second assumption, no *radial* gradients are assumed to be present within the pool, and $[P_1]_b, \ldots$ are thus functions of x only. The contrast with the earlier model of Ault and Mellichamp is apparent.

Equation (5.4.2) can be simplified and appropriately summed up to give equations for the moments $\lambda_{i,b}$ in the bulk. The final equations are given in Table 5.2, including the moment closure condition.[46-48]

The only problem in integrating the set of equations for the bulk is the dependence of $\bar{N}_W$ on x. In order to obtain this, attention must be focused on the film at position x. A careful study reveals that the equations characterizing the film are identical to those given in Table 5.1 (with subscript f used on all the concentrations and moments to distinguish them from the concentrations in the bulk, denoted by subscript b). The boundary and initial conditions are given by Eqs. (f)-(l) in Table 5.2. It has been assumed that the initial film concentrations are identical to those in the bulk at the same axial position. The value of $[W]^*$ is obtained, as earlier, using thermodynamic relationships and is found to depend on the pressure in the gas phase. The exposure time, t_f, for the film depends upon the rate of rotation of the blades. The value of $\bar{N}_W(x)$ is given by Eq. (5.3.7) and is also included in Table 5.2. These equations can easily be nondimensionalized using variables similar to those given in Eq. (5.3.14). Such an exercise yields the following functional relationships for the dimensionless variables:

$$\frac{[P]_b}{\lambda_{1,0}}, \frac{[W]_b}{\lambda_{1,0}}, \frac{\mu_{n,b}}{\mu_{n,0}}, \frac{Q_b}{Q_0}$$

$$= fns\left\{\frac{x}{L}, \mu_{n,0}, \frac{[W]^*}{\lambda_{1,0}}, K, (k_p\lambda_{1,0}\bar{\theta}), \frac{Sy_0}{t_f\mathring{Q}}, \frac{t_f\mathscr{D}_W}{y_0^2}, (k_p\lambda_{1,0}t_f)\right\} \qquad (5.4.3)$$

where $\mu_{n,0}$ and Q_0 are the number average chain length and the polydispersity index of the feed to the reactor, $\bar{\theta}$ ($=LA_b/\mathring{Q}$) is the mean residence time in the reactor, and S ($\equiv Ls$) is the total desorption area in the reactor. It has been assumed that the feed to the reactor is an equilibrium prepolymer corresponding to conditions in a previous reactor [see Eq. (5.4.1)].

The various equations for the wiped film reactor may be integrated numerically using a Runge–Kutta subroutine for the bulk equations, a finite-difference subroutine (with several grid points since the gradient of W is very sharp near the vapor–liquid interface) for the film, and a Simpson's rule subroutine for the flux.†

Figure 5.9, curve b, shows the axial variation of $\mu_n/\mu_{n,0}$ for one set of parameter values[46] (which are in the same range as chosen for the previous model of Ault and Mellichamp). It is observed that the results of the two models of finishing reactors are not too different. In fact, a close scrutiny of these two models reveals that they can be interpreted as two limiting models of wiped film reactors, since in one, gradients are neglected in the bulk, while in the other, there is no pool. The behavior of actual reactors is expected to lie between these two. This is further justified by the fact that there is always some amount of axial mixing (convection) in the bulk, which is not accounted for in the model of Amon and Denson, and is overestimated in the model of Ault and Mellichamp.

Among the various parameters in Eq. (5.4.3), it has been found[13,48] that the values of $\mu_{n,b}$ are most sensitive to $Sy_0/t_f\mathring{Q}$. Figure 5.11 shows results[48] for a typical case. The variations of $k_p\lambda_{1,0}\bar{\theta}$, $t_f\mathcal{D}_W/y_0^2$ and $k_p\lambda_{1,0}t_f$ do not alter the results shown by more than about 20%. If the values of y_0, t_f, and $\mathring{Q}$ are taken as those encountered in practice (for common PET reactors,[13] these are 0.2 cm, 1 sec, and 500 cm^3/sec, respectively), it is found from Fig. 5.11 that values of S required to get polymer of $\mu_{n,b} \simeq 100$ at the end of the reactor are around 25×10^6 cm^2. A typical finishing reactor (25 cm diam, 3 m length) has a wall area of about 2.5×10^4 cm^2. Thus, mass transfer areas S required to attain a reasonable product are about 1000 times the "smooth" wall area. Since PET of $\mu_{n,b} \simeq 100$ are indeed obtained in such reactors, this simple calculation points out that S cannot be the wall area of the reactor and that several small gas bubbles must be forming inside the film (and possibly in the pool also) by vaporization, and the by-product W must be diffusing towards these also. A model of the type shown in Fig. 5.5b must necessarily be used to describe the behavior of such reactors, or else S must be treated as a curve-fit parameter. A similar discrepancy has also been referred to by Latinen[11] in his work on polymer devolatilization with no reaction taking place. In his work, Latinen curve-fitted his experimental results and found that the values of the diffusivity of the low molecular weight compound were almost an order of magnitude higher than values reported in the literature. One must therefore be extremely careful in using the various models of the finishing reactors, since all of

---

† Alternatively, the orthogonal collocation technique (see Chap. 8) may be used for the film equations.[48]

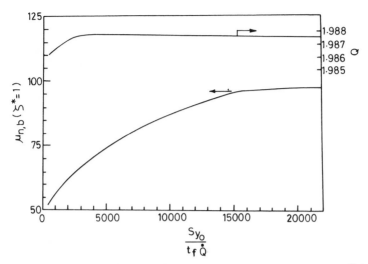

Figure 5.11. Variation of $\mu_{n,b}$ at the end ($\zeta^* \equiv x/L = 1$) of the wiped-film reactor[48] (model of Amon and Denson[13]). $\mu_{n,0} = 50$, $[W]^*/\lambda_{1,0} = 0$, $K = 0.5$, $k_p\lambda_{1,0}\bar{\theta} = 2000$, $t_f\mathcal{D}_W/y_0^2 = 10^{-5}$, $k_p\lambda_{1,0}t_f = 2$.

them parallel that of Latinen, and either use the diffusivity or (preferably) the surface area S as a curve-fit parameter with appropriate scale-up procedures. It may be added that if the gas bubbles formed are spaced far apart from each other [compared to $(\mathcal{D}_W t_f)^{1/2}$], and if curvature effects can be ignored, the above analysis is still applicable with S as the total mass-transfer area (to be determined experimentally on pilot-scale equipment). The results shown in Fig. 5.11 indicate the usefulness of generating bubbles of vapor to obtain high molecular weight product, but only up to a certain value, since beyond this, an asymptote is attained.

The effect of the vacuum applied (i.e., the effect of $[W]^*/\lambda_{1,0}$) is shown[46] as curve a in Fig. 5.12 for $K = 1$ (note the low value of $Sy_0/t_f\dot{Q}$ used to generate these results). It is found that application of a pressure below a certain value is useless since asymptotic behavior is observed below a certain value of $[W]^*/\lambda_{1,0}$. However, one must be careful in maintaining the reactor below a threshold pressure since the μ_n of the product falls very sharply above this value.

Figures 5.9 and 5.12 also show the effect of unequal reactivity in step growth polymerizations of $AR_1A + B'R_2B''$ monomers, where the rate and equilibrium constants are described by

$$-A + -B' \underset{k_1' = k_1/K_1}{\overset{k_1}{\rightleftharpoons}} -C'- + W$$

$$-A + -B'' \underset{k_2' = k_2/K_2}{\overset{k_2 = k_1R}{\rightleftharpoons}} -C''- + W$$

(5.4.4)

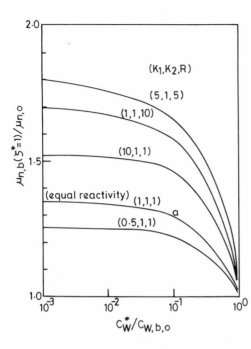

Figure 5.12. Effect of vacuum applied on the number average chain length of the polymer formed (at $\zeta^* \equiv x/L = 1$) in wiped film reactors.[46] $\mu_{n,0} = 50$, $k_p\lambda_{1,0}\bar{\theta} = 2000$, $Sy_0/t_f\bar{Q} = 10^3$, $t_f\mathscr{D}_W/y_0^2 = 10^{-4}$, $k_p\lambda_{1,0}t_f = 2$. $C_{W,b,0} = [W]_0/\lambda_{1,0}$ [see Eq. (5.4.1)]; $C_W^* = [W]^*/\lambda_{1,0}$.

Higher values of $\mu_{n,b}$ can be obtained under certain conditions. More detailed kinetic schemes, as for example for PET polymerization, have also been studied using these models, and some results are presented in Chapter 8. Some other types of finishing reactors, e.g., using interfacial surface generators or disk reactors (Fig. 5.1a), partially filled screw extruders, etc., have also been modeled to some extent. The disk reactor has been modeled[49] as a series of stirred tank reactors, each representing one disk, with some amount of interstage backmixing. A considerable amount of modeling has also been done on devolatilization in screw extruders,[11,18,50] much of which can be extended quite easily to incorporate polymerization taking place simultaneously. Polymerization in the solid state (i.e., above the glass transition temperature but below the melting point of the reaction mixture) has also been modeled.[51,52] In this, the solid polymer chips are heated in an inert atmosphere or under vacuum. The condensation by-product W diffuses out to the vapor phase. The conventional transient diffusion equation, with diffusivity $\mathscr{D}_W$ varying with time (obtained empirically), is solved after appropriate change of variables, and it is found that μ_n is a function of the size of the polymer chips, the initial value $\mu_{n,0}$, and of the integral $\int_0^t \mathscr{D}_W(t)\, dt$. More detailed theoretical models for $\mathscr{D}_W(t)$ are being developed, and a short discussion is presented in the following section.

5.5. STEP GROWTH POLYMERIZATIONS AT HIGH CONVERSIONS

In Sections 5.3 and 5.4, one of the assumptions made was that polymer molecules are "frozen" whereas the condensation by-product diffuses through the reaction mass. It may be recognized that this cannot be rigorously true for the following reason. In Chapter 2, it was pointed out that two polymer molecules can react only after they have diffused from the "bulk" to close proximity of each other, after which the chain segments having the functional groups rearrange through segmental diffusion. The entire process of bond formation for polymeric systems was represented by

$$
\text{www}A + \text{www}B \underset{\text{diffusion}}{\overset{\text{Bulk}}{\rightleftharpoons}} [\text{www}A \ B\text{www}]
$$

$$
\underset{\text{diffusion}}{\overset{\text{Segmental}}{\rightleftharpoons}} [\text{www}AB\text{www}] \xrightarrow{\underset{\text{reaction}}{\text{Chemical}}} [\text{www}A{-}B\text{www}]
$$

$$(5.5.1)$$

Thus, the assumption of polymer molecules being "frozen" while W diffuses through, really implies that sufficient bulk and segmental diffusion takes place so as to keep k_p constant. This assumption breaks down at still higher conversions. As the polymerization progresses to very high conversions, the polymer formed leads to an increase in the viscosity of the reaction mass to an extent that it slows down the bulk and segmental diffusion steps in Eq. (5.5.1), thus reducing k_p to a value below that encountered at low conversions.

The fall in the polymerization rate constant due to the fall in the rate of the bulk and segmental diffusion steps has been a subject widely studied in free-radical polymerization. These approaches can easily be adapted to explain step growth polymerization data at high conversions where k_p becomes diffusion-controlled. Work along these lines is currently being carried out, and only an introductory discussion is presented here, highlighting how some of the approaches used in free radical polymerization can be applied to step growth polymerizations. It is well recognized that at high concentrations of polymer in the reaction mass, the molecules get highly entangled. The movement of a polymer molecule cannot then be classified into bulk and segmental diffusion steps as done in Eq. (5.5.1). Instead, a polymer molecule can move only by a snake-like wriggling motion along its length, called reptation.[53-57]

One of the simplest *phenomenological* models to account for the effects of diffusion on the rate constants is based on the work of Chiu et al.[58] on free radical polymerization. This model[59] assumes that two functional groups react only when they lie within a minimum distance r_m (Fig. 5.13). If $k_{p,0}$ is the rate constant of the chemical reaction step alone in Eq. (5.5.1) (i.e.,

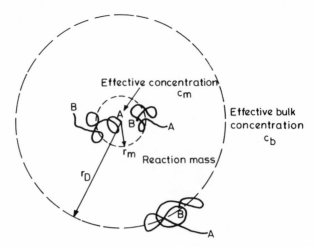

Figure 5.13. Illustration of the reaction of two polymer molecules represented by Eq. (5.5.1).

in the absence of diffusive limitations), this implies that whenever another polymer molecule sweeps within the spherical domain of radius r_m of a given molecule, the collision processes are described by $k_{p,0}$. It is further assumed that at a large distance, r_D, the concentration of polymer approaches the value in the bulk, c_b ($\equiv [P]_b$), whereas the effective concentration within the sphere of radius r_m is assumed to be c_m.

In Fig. 5.13, the chain carrying functional group B must migrate through the bulk, the B group must move by segmental diffusion towards A, and only then is there chemical reaction. If steady state is assumed and $\mathcal{D}$ is the overall diffusivity, then between r_m and r_D, one has

$$4\pi r^2 \mathcal{D}\frac{d[P]}{dr} = C_1 \tag{5.5.2a}$$

Boundary Conditions

$$\text{at } r = r_m; \quad [P] = c_m \tag{5.5.2b}$$

$$\text{at } r = r_D; \quad [P] = c_b \tag{5.5.2c}$$

where C_1 is a constant equal to the rate of consumption of functional groups B by chemical reaction at $r = r_m$. The rate of reaction of functional groups can be written in terms of the product of the probability of finding a central A group in the reaction mass (which is equal to c_b) and the probability of finding another B group within a distance r_m of the central A group (equal to c_m). The rate of reaction (moles/sec) of functional groups is, thus, equal to $k_{p,0} (\frac{4}{3}\pi r_m^3) c_m c_b$. On integrating Eq. (5.5.2) (with $r_D \to \infty$) and equating the diffusive flux at r_m to the rate of reaction, one has

$$4\pi \mathcal{D} r_m (c_b - c_m) = \frac{4}{3}\pi r_m^3 k_{p,0} c_b c_m = \frac{4}{3}\pi r_m^3 k_p c_b^2 \qquad (5.5.3)$$

since the overall polymerization rate is ordinarily written as $\frac{4}{3}\pi r_m^3 k_p c_b^2$, with k_p the *apparent* (overall) rate constant. Eliminating c_m and rearranging gives

$$\frac{1}{k_p} = \frac{1}{k_{p,0}} + \frac{r_m^2 c_b}{3\mathcal{D}} \equiv \frac{1}{k_{p,0}} + \left(\frac{r_m^2}{3\mathcal{D}_0}\right)\frac{c_b}{\mathcal{D}/\mathcal{D}_0} \qquad (5.5.4)$$

where $\mathcal{D}_0$ is the diffusivity at some reference state. The dependence of $\mathcal{D}/\mathcal{D}_0$ upon the polymer concentration and temperature can be approximated by the Fujita–Doolittle theory[56]

$$\log \frac{\mathcal{D}}{\mathcal{D}_0} = \frac{v_f}{b_1(T) + b_2(T) v_f} \qquad (5.5.5)$$

where v_f is the free volume fraction and $b_1(T)$ and $b_2(T)$ are empirically determined functions of the temperature, T. v_f is given by

$$v_f = 0.025 + (\alpha_l - \alpha_g)(T - T_g) \qquad (5.5.6)$$

where α_l and α_g are the volume-expansion coefficients of the liquid and glassy polymers and T_g is the glass transition temperature of the polymer. α_l, α_g, and T_g can be obtained, *a priori* through independent experiments. T_g is, in general, a function of the number average chain length or of c_b, and can be expressed as

$$\frac{1}{T_g} = \frac{1}{T_{g,\infty}} + \frac{\mathcal{K} c_b}{T_{g\infty}^2} \qquad (5.5.7)$$

Equations (5.5.4)–(5.5.7) can be used to predict the value of k_p at any time t and this may be used with appropriate mole-balance equations to predict the value of $[P]_b$ at time $t + \Delta t$. Thus, one can easily obtain the variation of μ_n, Q, etc., with time. $b_1(T)$, $b_2(T)$, and $\{r_m^2/3\mathcal{D}_0\}(T)$ can be treated as empirical functions to explain experimental data. Some preliminary results have been obtained[59] using $b_1(T)$, $b_2(T)$, and $r_m^2/3\mathcal{D}_0$ as parameters (assumed independent of T) and it is found that the $\mu_n(t)$ plot shows lower values of μ_n at high conversions than obtained in the absence of diffusive limitations.

There have been several other approaches taken in the literature to establish the diffusional effects on the rate constants of polymerization. All these are limited to chain polymerization. Marten and Hamielec[60,61] have taken a similar approach as that of Chiu $et\ al.$,[58] except that in Eq. (5.5.4), $k_{p,0}$ was assumed to be zero, i.e.,

$$k_p = k_1 \mathscr{D} \tag{5.5.8}$$

where k_1 is equivalent to $3/r_m^3 c_b$ and is taken to be a constant. The diffusion coefficient was subsequently related by the Bueche theory[62] as follows:

$$\mathscr{D} = \frac{\phi_0 \delta^2}{C_2 M} e^{-(C_3/v_f)} \tag{5.5.9}$$

where ϕ_0 is the jump frequency, δ is a jump distance, C_2 and C_3 are constants, and M is the molecular weight of the polymer (assumed monodisperse). O'Driscoll, Ito, and co-workers[63-67] have taken a slightly different phenomenological approach from the one described above. It was assumed that polymer molecules beyond a critical chain length, n_{c1}, exist in the entangled state, and because of their reduced mobility, they react with a lower rate constant, $k_{p,e}$. If the entangled and nonentangled molecules are denoted by $P_{n,e}$ and P_n, respectively, then the mechanism of polymerization can be modified (adapted for step growth polymerizations) to

$$\text{P}_{n,e} + \text{P}_{m,e} \xrightarrow{k_{p,e}} \text{P}_{(n+m),e} \tag{5.5.10a}$$

$$\text{P}_{n,e} + \text{P}_m \xrightarrow{(k_p k_{p,e})^{1/2}} \text{P}_{(n+m),e} \tag{5.5.10b}$$

$$\text{P}_n + \text{P}_m \xrightarrow{k_p} \text{P}_{(n+m)} \tag{5.5.10c}$$

where the molecule $P_{(n+m)}$ is entangled if $n + m > n_{c1}$. The rate constant, $k_{p,e}$, can be modeled as proportional to $1/v_P \mu_n$, where v_P is the volume fraction of the polymer ($\simeq 1$ at high conversions).

Tirrell $et\ al.$,[55,56,68,69] on the other hand, argue that there are two critical chain lengths, n_{c_1} and n_{c_2}. Above n_{c_1} but below n_{c_2}, polymer chains begin to entangle to the extent that segmental diffusion is the slowest step. However, for polymer chains beyond n_{c_2}, polymer chains are highly entangled and the only mechanism by which they can move from one place to another is by reptation. They have recently measured the self-diffusion coefficients of polymers using dynamic light scattering from refractive-index matched ternary solutions,[68] and have found that the diffusivity, $\mathscr{D}$, can be written empirically as

$$\mathscr{D} = M^{-\nu(c)} g(c) \tag{5.5.11}$$

where M is the molecular weight and c is the concentration of the polymer. The function $\nu(c)$ is given by

$$\nu(c) = \nu_0 + (2 - \nu_0)(1 - e^{-\xi c}) \qquad (5.5.12)$$

where ν_0 is the Flory exponent (0.5–0.6). A minimum amount of experimental data is necessary to determine ξ and $g(c)$. These correlations can be used to model the dependence of the rate constants on the properties of the reaction mass, and thus predict the increase of μ_n etc. with time.

It has been found that all these models, though differing fundamentally, are almost equally successful in explaining experimental data on PMMA and polystyrene systems. At the present state of knowledge there is no particular reason for recommending the use of any one model over the other. The studies of Tirrell *et al.* on chain-growth polymerizations appear promising, and are more appealing because of their fundamental nature. These can be easily extended to explain step-growth polymerization data at high conversions. A considerable amount of work, both experimental and theoretical, is necessary in this area of high-conversion polymerizations, some of which may be able to answer the unresolved problems of solid state step growth polymerizations.

5.6. CONCLUSIONS

In this chapter, an attempt has been made to model the effect of finite rates of mass transfer on step growth polymerizations. It is found that depending on the viscosity of the medium (which, in turn, is determined by the concentration and average chain length of the polymer), different simplified models are found to be applicable, using which one can simulate reactor behavior. The greatest drawback in modeling mass transfer phenomena is the lack of information on the effective mass transfer area. The reactor performance is sensitive to this variable, which must be obtained by model experiments at present. A fundamental understanding of the nucleation and growth of vapor bubbles and their correlation with process variables is essential for a significant advance in this area. The effect of diffusion on the rate constants at high conversions has also been discussed and it is seen to offer scope for considerable further work.

REFERENCES

1. S. K. Gupta and A. Kumar, Simulation of step-growth polymerizations, *Chem. Eng. Commun.* **20**, 1–52 (1983).

2. K. Ravindranath and R. A. Mashelkar, Modeling of poly (ethylene terephthalate) reactors: 5. A continuous prepolymerization process, *Polym. Eng. Sci.* **22**, 619-627 (1982).

3. H. K. Reimschuessel and K. Nagasubramanian, On the optimization of caprolactam polymerization, *Chem. Eng. Sci.* **27**, 1119-1130 (1972).

4. Vereinigte Glanzstoff Fabriken, German Patent, 1167021 (1962).

5. H. Gerrens, On selection of polymerization reactors, *Germ. Chem. Eng.* **4**, 1-13 (1981).

6. W. H. Li (Du Pont), US Patent, 3113843 (1959/63).

7. J. S. Vrentas and J. L. Duda, Molecular diffusion in polymer solutions, *AIChE J.* **25**, 1-21 (1979).

8. J. L. Duda, J. S. Vrentas, S. T. Ju, and H. T. Liu, Prediction of diffusion coefficients for polymer-solvent systems, *AIChE J.* **28**, 279-285 (1982).

9. R. A. Mashelkar, Gas-liquid contactors in non-Newtonian technology, *Chem. Indus. Dev.* **10**(9), 17-24 (1976).

10. F. Windmer, Behavior of viscous polymers during solvent stripping or reaction in an agitated thin film, *Adv. Chem. Ser.* **128**, 51-67 (1973).

11. G. A. Latinen, Devolatilization of viscous polymer systems, *Adv. Chem. Ser.* **34**, 235-246 (1962).

12. D. B. Todd and H. F. Irving, Axial mixing in a self-wiping reactor, *Chem. Eng. Prog.* **65**(9), 84-89 (1969).

13. M. Amon and C. D. Denson, Simplified analysis of the performance of wiped-film polycondensation reactors, *Ind. Eng. Chem., Fundam.* **19**, 415-420 (1980).

14. R. E. Emmert, US Patent, 3110547 (1963).

15. G. M. Turner, US Patent, 3161710 (1964).

16. Y. Murakami, K. Fujimoto, S. Kakimoto, and M. Sekino, On a high viscosity polymer finisher apparatus with two agitator axes having multidisks, *J. Chem. Eng. Jpn.* **5**, 257-263 (1972).

17. W. A. Mack and R. Herter, Extruder reactors for polymer production, *Chem. Eng. Prog.* **72**(1), 64-70 (1976).

18. R. W. Coughlin and G. P. Canevari, Drying polymers during screw extrusion, *AIChE J.* **15**, 560-564 (1969).

19. K. Ravindranath and R. A. Mashelkar, Modeling of poly (ethylene terephthalate) reactors. I. A semibatch ester interchange reactor, *J. Appl. Polym. Sci.* **26**, 3179-3204 (1981).

20. A. Ramagopal, A. Kumar, and S. K. Gupta, Computational scheme for the calculation of molecular weight distributions for nylon 6 polymerization in homogeneous, continuous-flow stirred-tank reactors with continuous removal of water, *Polym. Eng. Sci.* **22**, 849-856 (1982).

21. S. K. Gupta, D. Mohan, and A. Kumar, Simulation of ARB type reversible step growth polymerization in semibatch reactors, *J. Appl. Polym. Sci.* **30**, 445-460 (1985).

22. P. J. Flory, *Principles of Polymer Chemistry*, 1st ed., Cornell University Press, Ithaca, New York (1953).

23. A. Kumar and S. K. Gupta, *Fundamentals of Polymer Science and Engineering*, 1st ed., Tata McGraw-Hill, New Delhi, India (1978).

24. R. H. Perry and D. W. Green, *Chemical Engineers' Handbook*, 6th ed., McGraw-Hill, New York (1984).

25. E. J. Henley and E. M. Rosen, *Material and Energy Balance Computations*, 1st ed., Wiley, New York (1969).

26. D. A. Mellichamp, Reversible polycondensation in a semibatch reactor, *Chem. Eng. Sci.* **24**, 125-139 (1969).

27. J. W. Ault and D. A. Mellichamp, Complex linear polycondensation—I. Semi-batch reactor, *Chem. Eng. Sci.* **27**, 2219-2232 (1972).

28. P. J. Hoftyzer and D. W. van Krevelen, The rate of conversion in polycondensation processes as determined by combined mass transfer and chemical reaction, *Proc. 4th Eur.*

Sym. Chem. Rxn. Eng., Brussels, 9-11 September 1968, Pergamon, Oxford (1971), pp. 139-146.

29. R. M. Secor, The kinetics of condensation polymerization, *AIChE J.* **15**, 861-865 (1969).

30. J. W. Ault and D. A. Mellichamp, A diffusion and reaction model for simple polycondensation, *Chem. Eng. Sci.* **27**, 1441-1448 (1972).

31. S. K. Gupta, N. L. Agarwalla, and A. Kumar, Mass transfer effects in polycondensation reactors wherein functional groups are not equally reactive, *J. Appl. Polym. Sci.* **27**, 1217-1231 (1982).

32. J. P. A. Wallis, Ph.D. dissertation, University of Calgary, Alberta (1973).

33. A. Hussain and A. E. Hamielec, Bulk thermal polymerization of styrene in a tubular reactor—A computer study, *AIChE Symp. Ser.* **72**, 112 (1976).

34. R. B. Bird, W. E. Stewart, and E. N. Lightfoot, *Transport Phenomena,* 1st ed., Wiley, New York (1960).

35. J. W. Hamer, Ph.D. dissertation, University of Wisconsin, Madison (1983).

36. K. Ravindranath and R. A. Mashelkar, Modeling of poly (ethylene terephthalate) reactors: 6: A continuous process for final stages of polycondensation, *Polym. Eng. Sci.* **22**, 628-636 (1982).

37. K. Ravindranath and R. A. Mashelkar, Finishing stages of PET synthesis: A comprehensive model, *AIChE J.* **30**, 415-422 (1984).

38. S. K. Gupta, A. Kumar, and K. K. Agarwal, Simulation of AA + B'B" type reversible polymerizations with mass transfer of condensation product, *Polymer* **23**, 1367-1371 (1982).

39. S. K. Gupta, S. S. Rao, R. Agarwal, and A. Kumar, in *Recent Advances in the Engineering Analysis of Chemically Reacting Systems* (L. K. Doraiswamy, Ed.), 1st ed., pp. 480-496, Wiley Eastern, New Delhi, India (1984).

40. K. Nagasubramanian and H. K. Reimschuessel, Diffusion of water and caprolactam in nylon 6 melts, *J. Appl. Polym. Sci.* **17**, 1663-1677 (1973).

41. H. S. Mickley, T. K. Sherwood, and C. E. Reed, *Applied Mathematics in Chemical Engineering,* 2nd ed., McGraw-Hill, New York (1957).

42. V. Gupta and S. K. Gupta, *Fluid Mechanics and its Applications,* 1st ed., Wiley Eastern, New Delhi, India (1984).

43. H. Yokoyama, T. Sano, T. Chijiiwa, and R. Kajiya, Simulation of polyethylene terephthalate continuous polycondensation reaction, *Kagaku Kagaku Ronbunshu* **5**(3), 236-242 (1979).

44. M. Amon and C. D. Denson, A study of the dynamics of foam growth: Analysis of the growth of closely spaced spherical bubbles, *Polym. Eng. Sci.* **24**, 1026-1034 (1984).

45. J. W. Ault and D. A. Mellichamp, Complex linear polycondensation II. Polymerization rate enhancement in thick film reactors, *Chem. Eng. Sci.* **27**, 2233-2242 (1972).

46. S. K. Gupta, A. Kumar, and A. K. Ghosh, Simulation of reversible AA + B'B" polycondensations in wiped film reactors, *J. Appl. Polym. Sci.* **28**, 1063-1076 (1983).

47. A. Kumar, S. K. Gupta, S. Madan, N. G. Shah, and S. K. Gupta, Solution of final stages of polyethylene terephthalate reactors using orthogonal collocation technique, *Polym. Eng. Sci.* **24**, 194-204 (1984).

48. S. K. Gupta, A. K. Ghosh, S. K. Gupta, and A. Kumar, Analysis of wiped-film reactors using the orthogonal collocation technique, *J. Appl. Polym. Sci.* **29**, 3217-3230 (1984).

49. H. Yokoyama, T. Sano, T. Chijiiwa, and R. Kajiya, An effect of reduced pressure on the formation of poly (ethylene terephthalate), *J. Jpn Petrol Inst.* **21**, 77-79 (1978).

50. K. G. Powell and C. D. Denson, Presented at 75th Annual AIChE Meeting, Washington, D.C., 1 November, 1983.

51. T. M. Chang, Kinetics of thermally induced solid state polycondensation of poly (ethylene terephthalate), *Polym. Eng. Sci.* **10**, 364-368 (1970).

52. F. C. Chen, R. G. Griskey, and G. H. Beyer, Thermally induced solid state polycondensation of nylon 6-6, nylon 6-10 and polyethylene terephthalate, *AIChE J.* **15**, 680-685 (1969).

53. P. G. de Gennes, Reptation of a polymer chain in the presence of fixed obstacles, *J. Chem. Phys.* **55**, 572-579 (1971).

54. T. J. Tulig and M. Tirrell, On the onset of the Trommsdorff effect, *Macromolecules* **15**, 459-463 (1982).

55. T. J. Tulig and M. Tirrell, Toward a molecular theory of the Trommsdorff effect, *Macromolecules* **14**, 1501-1511 (1981).

56. D. T. Turner, Autoacceleration of free-radical polymerization. 1. The critical concentration, *Macromolecules* **10**, 221-226 (1977).

57. H. B. Lee and D. T. Turner, Autoacceleration of free-radical polymerization 2. Methyl methacrylate; 3. Methyl methacrylate plus diluents, *Macromolecules* **10**, 226-230; 231-235 (1977).

58. W. Y. Chiu, G. M. Carratt, and D. S. Soong, A computer model for the gel effect in free-radical polymerization, *Macromolecules* **16**, 348-357 (1983).

59. A. Kumar, K. Saksena, J. P. Foryt, and S. K. Gupta, Effect of segmental diffusion on irreversible, step growth polymerizations of ARB monomers, *Polym. Eng. Sci.*, in press.

60. F. L. Marten and A. E. Hamielec, in *Polymerization Reactors and Processes* (J. N. Henderson and T. C. Bouton, Eds.), 1st ed., pp. 43-70, American Chemical Society, Washington, D.C. (1979).

61. F. L. Marten and A. E. Hamielec, High conversion diffusion-controlled polymerization of styrene. I, *J. Appl. Polym. Sci.* **27**, 489-505 (1982).

62. F. Bueche, *Physical Properties of Polymers*, 1st ed., Wiley, New York (1962).

63. J. N. Cardenas and K. F. O'Driscoll, High conversion polymerization I. Theory and Application to methyl methacrylate; II. Influence of chain transfer on the gel effect; III. Kinetic behavior of ethyl methacrylate, *J. Polym. Sci., Polym. Chem. Ed.* **14**, 883-897 (1976); **15**, 1883-1888; 2097-2108 (1977).

64. H. K. Mahabadi and K. F. O'Driscol, Concentration dependence of the termination rate constant during the initial stages of free-radical polymerization, *Macromolecules* **10**, 55-58 (1977).

65. K. Ito and K. F. O'Driscoll, The termination reaction in free-radical copolymerization. I. Methyl methacrylate and butyl- or dedecyl methacrylate, *J. Polym. Sci., Polym. Chem. Ed.* **17**, 3913-3921 (1979).

66. K. Ito, Estimation of molecular weight in terms of the gel effect in radical polymerization, *Polym. J.* **12**, 499-506 (1980).

67. K. F. O'Driscoll, J. M. Dionisio, and H. K. Mahabadi, in *Polymerization Reactors and Processes* (J. N. Henderson and T. C. Bouton, Eds.), 1st ed., pp. 361-374, American Chemical Society, Washington, D.C. (1979).

68. B. Hanley, M. Tirrell, and T. Lodge, The behavior of the tracer diffusion coefficient of polystyrene in isorefractive "solvents" composed of polyvinylmethyl ether and *o*-fluorotoluene, *Polym. Bull.* **14**, 137-142 (1985).

69. B. Hanley and M. Tirrell, An empirical correlation for polymer self-diffusion data in the dilute and semidilute concentration regimes, *Polym. Eng. Sci.* **25**, 947-950 (1985).

EXERCISES

1. Model the semibatch reactor and develop the numerical technique required to solve the set of equations using
   a. Raoult's law (note that this is a poor representation of the physical situation, but is simpler to use and so is good for classroom exercise);

b. The Flory-Huggins equation, but without assuming that $\mu_n \to \infty$ and $v_{P_2 \to P_\infty}$ is unity. Compare with results shown in Figs. 5.3 and 5.4.

2. Model the semibatch reactor where the monomer is not returned to the reactor after condensation. Develop the numerical algorithm from the set of equations, assuming Raoult's law behavior as an approximation.

3. In a well-stirred semibatch ARB polymerization reactor, the vaporization of monomer and W (because of heating) builds up the total pressure in the vapor space till a certain value $P_{T,set}$, before a valve opens. Write down the relevant equations for this system and develop an algorithm to solve them. Assume the heating jacket temperature to be constant, and the overall heat transfer coefficient, U, to be known. Also, assume Raoult's law behavior to simplify the analysis. Such a situation is analogous to what actually takes place in some industrial batch nylon 6 reactors.

4. Nondimensionalize the equations for the film reactor of Fig. 5.5a in terms of the variables indicated in Eqs. (5.3.8) and (5.3.9).

5. Write the finite difference[41] form of the equations in Table 5.1 and the corresponding flow chart for their numerical integration. Pay particular attention to the boundary conditions, particularly to Eq. (g) of the table.

6. Write the complete set of equations[31] including those for mean moments and mean rate of desorption for the concentric sphere model of a polymerization reactor, shown in Fig. 5.5b. Again, write the finite difference algorithm, paying special attention to the symmetry boundary condition at $r = R_2$.

7. Show that the equations relating $C_{W,0}$ to $\mu_{n,0}$ and $m_{2,0}$ to $\mu_{n,0}$ for the feed to the film reactor (for which results are shown in Fig. 5.7) are indeed correct.

8. Under the limiting situation of infinitely rapid diffusion of W through the reaction mass, see how you would model a (semibatch) reactor starting with an equilibrium prepolymer with degree of polymerization $\mu_{n,0}$ and a corresponding water concentration of $[W]_0$. The vapor-liquid equilibrium relation fixes $[W]$ at a constant value of $[W]^*$ at the interface for $t \geq 0$. Discuss the differences with respect to the case discussed in Sec. 5.2 where one starts with pure monomer, with no W present initially.

9. Nondimensionalize the equations for the wiped film reactor for both the bulk and the film. Thus verify Eq. (5.4.3).

10. Model the unsteady diffusion of W out of an infinite slab of the solid reaction mass with $\mathcal{D}_W$ an unknown function of time. Define new variables, as in Ref. 52, and obtain how the spatial average μ_n varies with time.

11. Try to extend the model proposed in Eq. (5.5.10) and develop an algorithm to obtain $\mu_n(t)$ and $Q(t)$.

6

OPTIMAL CONTROL OF STEP GROWTH POLYMERIZATIONS

6.1. INTRODUCTION

In the earlier chapters, the simulation of step growth polymerization in ideal (batch, plug flow or homogeneous, continuous flow stirred tank) reactors has been discussed. In order to simulate real reactors, the effects of several factors like residence time distribution in a continuous flow stirred tank reactor (for the case of perfect macromixing, Section 2.8), recycle in a plug flow reactor (Section 2.7), mass transfer limitations (Chapter 5), etc., were also investigated. It was shown that these variables have considerable effect on the performance of the reactor. In this chapter, attention is focused on the optimal control and operation of *ideal* reactors (PFRs and HCSTRs). For a given residence time of a batch reactor or an HCSTR, the variables that can be independently changed (degrees of freedom) are (a) the time history of the reactor temperature and (b) feed composition. Optimal conditions for these independent (or control) variables are determined in this chapter for simple ARB systems in order to establish the principles used. Optimization of industrially important systems like nylon 6 or PET polymerization is discussed in later chapters.

6.2. THE BASIC ALGORITHM

In this section, some simple concepts required for the optimization of polymerization reactors are presented without proof or derivation. The interested reader may consult Refs. 1 and 2 for more details.

Most of the optimization studies on polymerization reactors have been carried out either for steady state operation in plug flow reactors or for unsteady operations (during start-up) in HCSTRs. In either case, one has several dependent or *state* variables whose time dependence can be described by equations of the following form:

$$\frac{dx_i}{dt} \equiv \dot{x}_i = f_i(x_1, x_2, \ldots, x_n; u_1, u_2, \ldots, u_m), \qquad i = 1, 2, \ldots, n \quad (6.2.1)$$

where $u_1(t)$, $u_2(t), \ldots, u_m(t)$, are the independent or *control* variables. In the case of polymerization reactors, the state variables could be the concentration of unreacted monomer, various moments of the MWD, concentration of unwanted side products, etc., whereas the control variables could be the temperature and pressure histories or profiles. The initial conditions on the state variables are usually known (values for the feed) and can be written as

$$x_i(t = 0) = x_{i,0}, \qquad i = 1, 2, \ldots, n \qquad (6.2.2)$$

Equations (6.2.1) and (6.2.2) may be written in a more compact form in matrix notation and are given as Eq. (c) in Table 6.1, with the various matrices defined in Eq. (h) in that table.

A design engineer is usually interested in choosing the control variables $u_1(t)$, $u_2(t), \ldots, u_m(t)$ such that some function of the state variables, called the *objective function*, *I*, is minimized. A very general form of this objective function is

$$I \equiv G(x_1, x_2, \ldots, x_n)_{t_f} + \int_0^{t_f} F(x_1, x_2, \ldots, x_n; u_1, u_2, \ldots, u_m) \, dt \quad (6.2.3)$$

with the equivalent matrix form given by Eq. (a) of Table 6.1. In this equation, t_f is the residence time (presently assumed fixed). A typical example of an objective function relevant to polymerization reactors is

$$I = \alpha_1 [P_1]_{t_f}^2 + \alpha_2 (\mu_{w,t_f} - \mu_{w,d})^2 \qquad (6.2.4)$$

which would minimize the final value (at $t = t_f$) of the concentration, $[P_1]_{t_f}$, of the unreacted monomer (thus reducing postreactor separation costs) and simultaneously, bring the final weight-average chain length, μ_{w,t_f}, as *close as possible* to a desired value, $\mu_{w,d}$. These two requirements are given relative weightages of α_1 and α_2, respectively. It may be mentioned that for ARB polymerization, the concentration of the monomer, $[P_1]$, and the weight average chain length, μ_w, are not independent, as discussed in Chapter 2, and so cannot be chosen as the state variables simultaneously. However, for more complex step growth polymerization schemes, as for example for

TABLE 6.1. Optimization Equations in a Compact Form for Specified t_f

$$\min I \equiv G(\mathbf{x})_{t_f} + \int_0^{t_f} F(\mathbf{x}, \mathbf{u})\, dt \tag{a}$$

subject to $\mathbf{u}_* \leq \mathbf{u} \leq \mathbf{u}^*$; t_f fixed

Hamiltonian

$$H \equiv F(\mathbf{x}, \mathbf{u}) + \mathbf{z}^T \mathbf{f}(\mathbf{x}, \mathbf{u}) \equiv F(\mathbf{x}, \mathbf{u}) + \mathbf{z}^T \dot{\mathbf{x}}(\mathbf{x}, \mathbf{u}) \tag{b}$$

State variables

$$\frac{d\mathbf{x}}{dt} = \mathbf{f}(\mathbf{x}, \mathbf{u}) = \dot{\mathbf{x}}(\mathbf{x}, \mathbf{u}); \; \mathbf{x}(t = 0) = \mathbf{x}_0 \tag{c}$$

Adjoint variables

$$\left(\frac{d\mathbf{z}}{dt}\right)^T = -\frac{\partial H}{\partial \mathbf{x}} = -\frac{\partial F(\mathbf{x}, \mathbf{u})}{\partial \mathbf{x}} - \mathbf{z}^T \frac{\partial \mathbf{f}}{\partial \mathbf{x}} = -\frac{\partial F}{\partial \mathbf{x}} - \mathbf{z}^T \frac{\partial \dot{\mathbf{x}}}{\partial \mathbf{x}} \tag{d}$$

$$\mathbf{z}^T(t = t_f) = \frac{\partial G(\mathbf{x})}{\partial \mathbf{x}}\bigg|_{t = t_f} \tag{e}$$

Necessary conditions (for analytical solutions)

$$\frac{\partial H}{\partial \mathbf{u}} = \frac{\partial F}{\partial \mathbf{u}} + \mathbf{z}^T \frac{\partial \mathbf{f}}{\partial \mathbf{u}} = \frac{\partial F}{\partial \mathbf{u}} + \mathbf{z}^T \frac{\partial \dot{\mathbf{x}}}{\partial \mathbf{u}} = \mathbf{0} \qquad \text{if } \mathbf{u}_* \leq \mathbf{u} \leq \mathbf{u}^* \tag{f}$$

H minimum on constrained portions of control trajectory.
Conditions for numerical solutions

$$\mathbf{u}^{\text{new}} = \mathbf{u}^{\text{old}} - \varepsilon (\partial H / \partial \mathbf{u})^T \tag{g}$$

Definitions

$$\mathbf{x} = [x_1, x_2, \ldots, x_n]^T$$

$$\mathbf{u} = [u_1, u_2, \ldots, u_m]^T$$

$$\mathbf{z} = [z_1, z_2, \ldots, z_n]^T$$

$$\mathbf{f}(\mathbf{x}, \mathbf{u}) = [f_1(\mathbf{x}, \mathbf{u}), f_2(\mathbf{x}, \mathbf{u}), \ldots, f_n(\mathbf{x}, \mathbf{u})]^T = [\dot{x}_1(\mathbf{x}, \mathbf{u}), \ldots, \dot{x}_n(\mathbf{x}, \mathbf{u})]^T$$

$$\frac{\partial F}{\partial \mathbf{x}} = \left[\frac{\partial F}{\partial x_1}, \frac{\partial F}{\partial x_2}, \ldots, \frac{\partial F}{\partial x_n}\right], \text{ etc.} \tag{h}$$

$$\partial \mathbf{f} / \partial \mathbf{x} = \begin{bmatrix} \partial f_1(\mathbf{x}, \mathbf{u}) / \partial x_1, \ldots, \partial f_1(\mathbf{x}, \mathbf{u}) / \partial x_n \\ \partial f_2(\mathbf{x}, \mathbf{u}) / \partial x_1, \ldots, \partial f_2(\mathbf{x}, \mathbf{u}) / \partial x_n \\ \vdots \qquad\qquad \vdots \\ \partial f_n(\mathbf{x}, \mathbf{u}) / \partial x_1, \ldots, \partial f_n(\mathbf{x}, \mathbf{u}) / \partial x_n \end{bmatrix}, \text{ etc.}$$

nylon 6 or PET, this restriction is not present, and Eq. (6.2.4) is then a perfectly valid objective function.

Another objective function relevant for polymerization reactor optimization could be

$$I \equiv \int_0^{t_f} (\mu_w - \mu_{w,d})^2 \, dt \tag{6.2.5}$$

Such a function reduces the difference between the actual μ_w at any time and the desired value, $\mu_{w,d}$, at the end of the reactor *at every value of t*. In other words, the minimization of I would lead to as rapid an increase in μ_w as possible, and so gives an idea of what is necessary to reduce t_f. Use of objective functions of the form given in Eq. (6.2.5) is very common, even though more complex algorithms to minimize t_f simultaneously are also available. Several other forms of I are discussed in this chapter, and some more, relevant to specific systems, are considered in Chapters 7 and 8. It may be added that a maximization of I is mathematically equivalent to minimizing $(-I)$, and, therefore, any maximization problem can be easily solved using concepts of minimization presented here.

The solution of the optimization problem is facilitated by defining a *Hamiltonian*, H, given in Eq. (b) of Table 6.1. In this equation, $z_i(t)$, $i = 1, 2, \ldots, n$, are referred to as adjoint variables and are given by Eq. (d) of Table 6.1. It may be noted that conditions on $z_i(t)$ are known, through Eq. (e) of this table, at $t = t_f$, rather than at $t = 0$. A variational analysis of this problem can be carried out to obtain the necessary conditions for minimization of I. The results are given in Eq. (f) (Table 6.1).

In several problems of engineering interest, the control variables are constrained to lie between bounds:

$$u_{i*} \leq u_i(t) \leq u_i^*, \qquad i = 1, 2, \ldots, m \tag{6.2.6}$$

where u_{i*} and u_i^* are the lower and upper limiting values, respectively, which $u_i(t)$ can take. A good example is the one encountered in nylon 6 polymerization, where $220°C \leq T(t) \leq 280°C$. The lower limit represents the melting point of nylon 6 (and is used to ensure single-phase polymerization) and the upper value represents the boiling point of the monomer, ε-caprolactam, at atmospheric pressures (and is used to reduce the costs of condensing and recycling the vaporized monomer). In such *constrained* optimization problems, the necessary conditions for the minimization of the objective function are slightly different and are given by

$$\frac{\partial H}{\partial u_i} = 0 \qquad \text{if } u_{i*} < u_i < u_i^* \qquad \text{(unconstrained portion)} \tag{6.2.7a}$$

$$\frac{\partial H}{\partial u_i} < 0 \qquad \text{if } u_i(t) = u_i^* \tag{6.2.7b}$$

$$\frac{\partial H}{\partial u_i} > 0 \qquad \text{if } u_i(t) = u_{i*} \qquad i = 1, 2, \ldots, m \tag{6.2.7c}$$

Equation (6.2.7) represents the fact that for optimality, it is necessary that $\partial H/\partial \mathbf{u}$ be zero (i.e., H is constant) whenever the constraints on $\mathbf{u}$ are not violated and that H be a minimum along the constrained portions of the control trajectory [e.g., when u_i is at its upper limit, any small change, δu_i in u_i must be negative, and Eq. (6.2.7b) predicts that δH must be positive, i.e., H must increase as one goes away from the boundary, $u_i = u_i^*$]. This is also referred to as the *weak minimum principle*. The equations for the adjoint variables remain unchanged. Table 6.1 gives the entire set of equations in one place and in a very compact manner. Necessary conditions for other optimization problems, as for example, when t_f is not fixed but must be chosen optimally, are available in more advanced textbooks.[3,4]

Usually it is difficult to obtain analytical solutions of the optimization problem posed above, particularly for problems of engineering interest, and one must resort to numerical techniques. The set of differential equations for $\mathbf{x}$ and $\mathbf{z}$ along with their boundary conditions constitute a two-point boundary value problem since the conditions on $\mathbf{x}$ and $\mathbf{z}$ are not specified at the same value of time. In this particular case, the conditions on $\mathbf{x}$ are specified at $t = 0$, while those on $\mathbf{z}$ are specified at $t = t_f$. Fortunately, since the $\mathbf{x}$ equations are independent of $\mathbf{z}$, it is possible to integrate these equations for an *assumed* set of profiles $u_1(t)$, $u_2(t)$, ..., $u_m(t)$, in the forward direction, using the Runge–Kutta technique (integration of the $\mathbf{x}$ equations from $t = 0$ to $t = t_f$ is numerically stable). After reaching $t = t_f$, the equations for $\mathbf{z}$ are integrated in the "backward" direction, using the same technique. It must be emphasized that the assumed profiles $u_1(t), \ldots, u_m(t)$ for the control variables must satisfy the necessary conditions, Eq. (6.2.7). A numerical technique to change the assumed profiles so as to approach the optimal conditions is thus required. Several search techniques[1,2] are available for this purpose. Alternatively, an adaptation of the method of steepest descent (called the control vector iteration procedure[1]) may be used in which the new profiles are obtained from the previous ones using

$$u_i^{\text{new}}(t) = u_i^{\text{old}}(t) - \varepsilon \left(\frac{\partial H}{\partial u_i} \right), \qquad i = 1, 2, \ldots, m \tag{6.2.8}$$

where ε is an arbitrarily chosen *positive* constant.[2,5] The constraints on $u_i(t)$

are accounted for easily in a computer program by using IF statements, whenever any of these bounds are violated.

The control vector iteration method suggested in this section suffers from the problem that convergence to the optimal profiles $u_1(t)$, $u_2(t), \ldots, u_m(t)$ is extremely sluggish, particularly after the first few cycles of computation. Better (second-order) techniques have also been developed[1-4] in which Eq. (6.2.8) (for a single control variable, i.e., $m = 1$) is replaced by

$$u_1^{\text{new}} = u_1^{\text{old}} - \frac{\partial H/\partial u_1}{\partial^2 H/\partial u_1^2} \qquad (6.2.9)$$

The profiles obtained using the first-order techniques, however, give reasonable results and are used in this text. In fact, the objective function is relatively insensitive to the changes in the profiles after the first few iterations. This is really a blessing in disguise for an engineer since minor changes in $\mathbf{u}(t)$ during operation of the reactor are found not to affect I very much.

6.3. OPTIMIZATION OF ARB POLYMERIZATION IN BATCH REACTORS

Very few optimization studies have been reported on simple step growth polymerizations of ARB monomers. This is not surprising in view of the fact that such systems give almost trivial results, as shown in this section. The optimization of irreversible ARB polymerization in batch reactors using the temperature history, $T(t)$, as the only control variable[6] is discussed. For the kinetic scheme

$$P_n + P_m \xrightarrow{\ k_p\ } P_{m+n} + W \qquad (6.3.1)$$

the mole balance and the moment equations have already been derived [Eqs. (2.4.5), (2.4.7), (2.4.8), (2.9.10)]. A design engineer may be interested in the following state variables:

   a. The monomer concentration $[P_1]$, since the unconverted monomer may have to be removed;
   b. The degree of polymerization $= \lambda_1/\lambda_0 = \lambda_{1,0}/[P]$; and
   c. The polydispersity index $Q = (\lambda_2/\lambda_1)/(\lambda_1/\lambda_0) = \lambda_2[P]/\lambda_{1,0}^2$.

Instead of these three variables, one may equivalently consider $[P_1]$, $[P]$, and λ_2 as an alternate set of state variables. However, these three variables are not all independent, even when the temperature is a function of time, as discussed in Section 2.6 for the case of pure monomer feed. This conclusion is valid for the more general situation also, where the feed is a mixture of various oligomers, as shown below.

If the equation for [P] is divided by that for λ_2, one obtains

$$\frac{d[P]}{d\lambda_2} = -\frac{[P]^2}{2\lambda_{1,0}^2} \qquad (6.3.2)$$

which, on integration and use of the initial condition $[P] = [P]_0$ at $\lambda_2 = \lambda_{2,0}$, gives

$$[P] = \frac{2[P]_0}{2 + ([P]_0/\lambda_{1,0}^2)(\lambda_2 - \lambda_{2,0})} \qquad (6.3.3)$$

Similarly, division of the equation for $[P_1]$ by that for λ_2, and use of Eq. (6.3.3) gives

$$\frac{d[P_1]}{[P_1]} = -\frac{2[P]_0 d\lambda_2}{2\lambda_{1,0}^2 + [P]_0(\lambda_2 - \lambda_{2,0})} \qquad (6.3.4)$$

Integration of this equation along with the initial condition $[P_1] = [P_1]_0$ when $\lambda_2 = \lambda_{2,0}$, leads to

$$\frac{[P_1]}{[P_1]_0} = \frac{4}{\left\{ 2 + \dfrac{[P]_0(\lambda_2 - \lambda_{2,0})}{\lambda_{1,0}^2} \right\}^2} \qquad (6.3.5)$$

Equations (6.3.3) and (6.3.5) show that for the most general case of irreversible ARB polymerization in a batch reactor, with *any* feed and *any* temperature history, [P] and $[P_1]$ are uniquely determined, once λ_2 is specified. In other words, there is only one degree of freedom in the system. A design engineer can therefore choose λ_2 (or equivalently, $[P_1]$ *or* [P]) as the state variable. The optimization problem is, thus, to choose a temperature history $T(t)$, subject to $T_* \leq T(t) \leq T^*$, such that for a given feed, an appropriate objective function, $I(\lambda_2)_{t_f}$, is minimized.

A common optimization problem is to obtain polymer having a specified (desired) degree of polymerization, $\mu_{n,d}$ in the shortest possible time,† t_f. Since there is only one degree of freedom in the system, this is equivalent to finding the optimal $T(t)$ so as to produce polymer having a specified value $\lambda_{2,d}$ (uniquely determined by $\mu_{n,d}$ through the equation $\lambda_{2,d} = \lambda_{2,0} + 2\lambda_{1,0}\{\mu_{n,d} - \lambda_{1,0}/[P]_0\}$) of the second moment, in the minimum t_f. In order

---

† It may be observed that this problem is one having final state variable constraints, with t_f not fixed, and is different from that discussed in Section 6.2.

to obtain the optimal $T(t)$, one observes from Eq. (2.9.10) (with $k'_p = 0$) that λ_2 is a monotonically increasing function of t, with a positive slope of $2k_p\lambda^2_{1,0}$, and that k_p increases as T increases. Hence to attain the desired polymer in the shortest time, the reactor must be operated at the highest permissible temperature, T^*, with t_f given by

$$\lambda_{2,\text{desired}} = \lambda_{2,0} + 2\lambda^2_{1,0}t_f k_p(T^*) \qquad (6.3.6)$$

6.4. OPTIMIZATION OF IRREVERSIBLE ARB POLYMERIZATIONS IN HCSTRs

It is possible to formulate several types of optimization problems for irreversible ARB polymerization in a cascade of N HCSTRs. One of these is to find the temperature of each of these reactors operating at steady state, such that a polymer having desired characteristics is obtained from a given feed to the first reactor in the minimum *total* residence time. Another problem could be to find out how to reach the steady state operation (i.e., obtain optimal temperatures during start-up) in as short a time as possible so as to minimize "off-spec" products. In order to analyze the first problem, one must write the appropriate mole balance equations for a single HCSTR in the cascade. Equation (2.8.1) (with associated moment equations) can easily be extended for the case where the feed to the Jth reactor in the cascade is a mixture of oligomers from the $(J-1)$th reactor (Fig. 6.1) to give, in general,

$$\frac{d[P_1]_J}{dt} = -2k_{p,J}[P_1]_J[P]_J + \frac{[P_1]_{J-1} - [P_1]_J}{\bar{\theta}_J} \qquad (6.4.1a)$$

$$\frac{d\lambda_{0,J}}{dt} = \frac{d[P]_J}{dt} = -k_{p,J}[P]^2_J + \frac{[P]_{J-1} - [P]_J}{\bar{\theta}_J} \qquad (6.4.1b)$$

$$\frac{d\lambda_{1,J}}{dt} = 0 + \frac{\lambda_{1,J-1} - \lambda_{1,J}}{\bar{\theta}_J} \qquad (6.4.1c)$$

$$\frac{d\lambda_{2,J}}{dt} = 2k_{p,J}\lambda^2_{1,J} + \frac{\lambda_{2,J-1} - \lambda_{2,J}}{\bar{\theta}_J} \qquad (6.4.1d)$$

where $\lambda_{i,J}$, etc., represent the values of the ith moment in the Jth reactor (and so, in the corresponding output stream), etc., at time t.

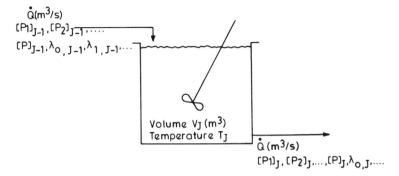

Figure 6.1. Notation used for the Jth HCSTR in a cascade.

Under steady state operation denoted by subscript s, the left-hand sides of Eq. (6.4.1) drop out to give

$$\lambda_{1,J,s} = \lambda_{1,J-1,s} \tag{6.4.2a}$$

$$[P_1]_{J,s} = -2k_{p,J}[P_1]_{J,s}[P]_{J,s}\bar{\theta}_J + [P_1]_{J-1,s} \tag{6.4.2b}$$

$$[P]_{J,s} = -k_{p,J}[P]^2_{J,s}\bar{\theta}_J + [P]_{J-1,s} \tag{6.4.2c}$$

$$\lambda_{2,J,s} = 2k_{p,J}\lambda^2_{1,J,s}\bar{\theta}_J + \lambda_{2,J-1,s} \tag{6.4.2d}$$

As in Section 6.3, it can be seen that for a given feed to the Jth reactor, $\lambda_{2,J,s}$ is a function of the temperature T_J (through $k_{p,J}$) and of the mean residence time, $\bar{\theta}_J$, and algebraic manipulations similar to those performed in Section 6.3 lead to

$$\frac{[P]_{J,s} - [P]_{J-1,s}}{\lambda_{2,J,s} - \lambda_{2,J-1,s}} = -\frac{[P]^2_{J,s}}{2\lambda^2_{1,J-1,s}} \tag{6.4.3a}$$

$$\frac{[P_1]_{J,s} - [P_1]_{J-1,s}}{\lambda_{2,J,s} - \lambda_{2,J-1,s}} = -\frac{[P_1]_{J,s}[P]_{J,s}}{\lambda^2_{1,J-1,s}} \tag{6.4.3b}$$

This indicates that for a given feed to the Jth reactor, both $[P_1]_{J,s}$ and $[P]_{J,s}$ can be written purely in terms of $\lambda_{2,J,s}$. In other words, there is only one independent state variable characterizing the system. Again, using arguments similar to those presented in Section 6.3, it can easily be observed from Eq. (6.4.2d) that one must select the highest permissible temperature, T_J^*, in the Jth reactor to minimize $\bar{\theta}_J$ to give a desired value of $\lambda_{2,J,s}$. The required mean residence time, $\bar{\theta}_J$, is then

$$\bar{\theta}_J = \frac{\lambda_{2,J,s,\text{desired}} - \lambda_{2,J-1,s}}{2k_p(T_J^*)\lambda_{1,J-1,s}^2} \tag{6.4.4}$$

The optimal policy is, therefore, observed to be very similar to that for batch reactors, i.e., one must hold every HCSTR at the maximum permissible temperature. The total mean residence time $\sum_{J=1}^{N} \bar{\theta}_J$ is fixed by the desired value of λ_2 (or, alternatively, the degree of polymerization). The monomer conversion and the polydispersity index are automatically determined and can be computed using Eqs. (6.4.2).

After establishing the optimal policy under steady operating conditions, one may talk about optimizing an HCSTR under unsteady conditions starting from specified initial conditions. This is referred to as the "start-up" problem. A simple situation is a *single* HCSTR filled with pure monomer initially, in which *pure monomer feed* starts to flow from time $t = 0$. The control variable to be manipulated is the temperature history, $T(t)$, of the reactor (through k_p). In this situation, a simplification of the type given in Eqs. (6.4.2) and (6.4.3) is not possible and one may select $[P_1]$, $[P]$, and λ_2 as three independent state variables. The governing equations are written in terms of the following dimensionless state variables:

$$x_1 \equiv [P_1]/[P_1]_0$$

$$x_2 \equiv [P]/[P_1]_0 = \lambda_0/[P_1]_0$$

$$x_3 \equiv (\lambda_2 - \lambda_1)/[P_1]_0 \tag{6.4.5}$$

where the subscript J has been omitted since there is only one reactor and $[P_1]_0$ is the molar concentration of the pure monomer in the feed. Equations (6.4.1a), (6.4.1b), and (6.4.1d) are rewritten as

$$\frac{dx_1}{d\tau} = -ux_1x_2 + (1 - x_1) \tag{6.4.6a}$$

$$\frac{dx_2}{d\tau} = -\frac{u}{2}x_2^2 + (1 - x_2) \tag{6.4.6b}$$

$$\frac{dx_3}{d\tau} = u - x_3 \tag{6.4.6c}$$

where the control variable, u, is $2k_p[P_1]_0\bar{\theta}$ [defined in Eq. (2.8.3c) as τ^*] and the dimensionless time τ [see Eq. (2.8.3b)] is defined as $t/\bar{\theta}$. Equations

(6.4.6a) and (6.4.6b) are identical to Eqs. (2.8.2a) and (2.8.6), respectively. In obtaining Eq. (6.4.6c), the integrated form of Eq. (6.4.1c), $\lambda_1/[P_1]_0 = 1$, has been used. The initial conditions for the three state variables are

$$x_1(\tau = 0) = 1 \tag{6.4.7a}$$

$$x_2(0) = 1 \tag{6.4.7b}$$

$$x_3(0) = 0 \tag{6.4.7c}$$

A typical objective function for this start-up problem is

$$I = \int_0^{\tau_f} \left\{ \alpha_1 \left(\frac{\mu_{n1}}{\mu_{n1,s}} - 1 \right)^2 + \alpha_2 \left(\frac{\mu_{w1}/\mu_{n1}}{\mu_{w1,s}/\mu_{n1,s}} - 1 \right)^2 + \alpha_3 \left(\frac{x_1}{x_{1s}} - 1 \right)^2 \right\} \tau^2 \, d\tau \tag{6.4.8}$$

where μ_{n1} and μ_{w1} are the number and weight average chain lengths, with monomer P_1 excluded from the summations, at time τ and are given by

$$\mu_{n1} = \frac{1 - x_1}{x_2 - x_1}$$

$$\mu_{w1} = \frac{(x_3 + 1) - x_1}{1 - x_1} = 1 + \frac{x_3}{1 - x_1} \tag{6.4.9}$$

The subscript s in Eq. (6.4.8) represents the steady state values obtained from Eqs. (6.4.2) and (6.4.9) or, alternatively, from Eqs. (6.4.6) (with $d/d\tau = 0$) and (6.4.9). The objective function is of the same form as Eq. (6.2.5) and tries to bring the values of μ_{n1} as close to $\mu_{n1,s}$, Q_1 as close to $Q_{1,s}$, and x_1 as close to the steady state value $x_{1,s}$, with weightage factors α_1, α_2, and α_3 respectively, as rapidly as possible. The equations for the adjoint variables and $\partial H/\partial u$ are given in Table 6.2. Results are shown[6] in Fig. 6.2 for $\alpha_1 = \alpha_2 = \alpha_3 = 1$. It is observed that the use of the optimal temperature profile leads to a much more rapid attainment of the steady state values of the state variables and the related parameters of the MWD, particularly Q_1, than when the control variable is set at the steady-state value of 300 throughout. Operationally, an easier control scheme would be a bang-bang type (which means that the control variable is changed like a step function, $u = 500 = u^*$, for $0 \leq \tau \leq \tau_1$, and $u = 300 = u_s$ for $\tau \leq \tau$), and the value of the objective function is found to be almost unaffected when this simpler control variable policy is imposed.

TABLE 6.2. Equations for the Adjoint Variables for the Optimal Start-up Policy of a Single HCSTR

$$\frac{dz_1}{d\tau} = -\frac{\partial F}{\partial x_1} + z_1(1 + ux_2); \qquad z_1(\tau_f) = 0$$

$$\frac{dz_2}{d\tau} = -\frac{\partial F}{\partial x_2} + z_1 ux_1 + z_2(1 + ux_2); \qquad z_2(\tau_f) = 0$$

$$\frac{dz_3}{d\tau} = -\frac{\partial F}{\partial x_3} + z_3; \; z_3(\tau_f) = 0$$

where

$$\frac{F}{\tau^2} = \alpha_1 \left[\frac{1 - x_1}{(x_2 - x_1)\mu_{n1,s}} - 1 \right]^2 + \alpha_2 \left[\frac{(1 - x_1 + x_3)(x_2 - x_1)}{(1 - x_1)^2 Q_{1,s}} - 1 \right]^2 + \alpha_3 \left(\frac{x_1}{x_{1,s}} - 1 \right)^2$$

and

$$u^{(\text{new})} = u^{(\text{old})} - \varepsilon \left(-z_1 x_1 x_2 - \frac{z_2 x_2^2}{2} + z_3 \right)$$

Hicks *et al.*[6] have also studied the start-up problem using the dimensionless flow rate $u_2 \equiv \mathring{Q}(\tau)/\mathring{Q}_s$ of the feed as the control variable instead of temperature. The equations are slightly different [$\bar{\theta}_J$ in the last term of Eq. (6.4.1) is replaced by $V_J/\mathring{Q}_J(t)$, leading to an additional factor of u_2 in the last term in Eq. (6.4.7)], but this method is found to be less effective than the use of temperature as the control variable.

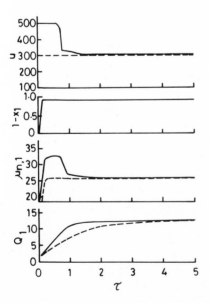

Figure 6.2. Optimal (solid line) and steady state (dashed line) profiles[6] of $u(\tau)$ for the start-up problem in a single HCSTR. $\alpha_1 = \alpha_2 = \alpha_3 = 1$. $u_s = 300$, $u^* = 500$, $u_* = 100$. $x_1(\tau)$, $\mu_{n1}(\tau)$, and $Q_1(\tau)$ shown for optimal $u(\tau)$ (solid lines) and for u_s (dashed lines). (Reprinted from Ref. 6 with permission of the Chemical Institute of Canada, Ottawa.)

6.5. PERIODIC OPERATION OF HCSTRs

In the previous section, it was observed that only one degree of freedom exists in the design of HCSTRs operating at steady state, even when the temperature varies with time. Thus, a design engineer does not have any control over the MWD of the polymer formed in such a reactor, once he chooses the degree of polymerization of the product (or the monomer conversion). It is shown in this section that when the monomer concentration in the feed of an HCSTR is varied periodically (with a time period $t^* = 2\pi/\omega$), the time-average polydispersity index of the product from such a reactor can be altered over some range of values keeping μ_n the same, by a proper choice of the period of oscillation and the amplitude of the feed concentration.[7] This provides an added degree of flexibility to a design engineer for tailor-making polymers for specific end uses.

It is assumed that the total feed flow rate $\overset{\circ}{Q}$ is constant but that the monomer (with inert) concentration varies sinusoidally with time as

$$[P_1]_0(t) = \overline{[P_1]_0}\{1 + a \sin(\omega t)\} \tag{6.5.1}$$

In Eq. (6.5.1), the bar on $[P_1]_0$ denotes the average over one time period and is given by

$$\overline{[P_1]_0} = \frac{1}{t^*} \int_t^{t+t^*} [P_1]_0(t)\, dt \tag{6.5.2}$$

The balance equations for an HCSTR of volume V can be easily written for this case as

$$\frac{d[P_1]}{dt} = -2k_p[P_1][P] + \frac{[P_1]_0(t) - [P_1]}{\bar{\theta}} \tag{6.5.3a}$$

$$\frac{d\lambda_0}{dt} = \frac{d[P]}{dt} = -k_p[P]^2 + \frac{[P_1]_0(t) - [P]}{\bar{\theta}} \tag{6.5.3b}$$

$$\frac{d\lambda_1}{dt} = \frac{[P_1]_0(t) - \lambda_1}{\bar{\theta}} \tag{6.5.3c}$$

$$\frac{d\lambda_2}{dt} = 2k_p\lambda_1^2 + \frac{[P_1]_0(t) - \lambda_2}{\bar{\theta}} \tag{6.5.3d}$$

where $\lambda_{i,0} = [P_1]_0(t)$ since the feed is assumed to contain only monomer and an inert solvent. This equation differs from Eq. (6.4.1) in that the time-varying feed concentration $[P_1]_0(t)$ is used. Equation (6.5.3) is non-dimensionalized using the following variables:

$$\tau = t/\bar{\theta}, \qquad x_2 = [P]/\overline{[P_1]_0} = \lambda_0/\overline{[P_1]_0}$$

$$u = 2k_p\overline{[P_1]_0}\bar{\theta}, \qquad x_3 = (\lambda_2 - \lambda_1)/\overline{[P_1]_0}$$

$$x_1 = [P_1]/\overline{[P_1]_0}, \qquad x_4 = \lambda_1/\overline{[P_1]_0} \tag{6.5.4}$$

$$x_{1,0} = [P_1]_0(t)/\overline{[P_1]_0} = 1 + a\sin(\omega\bar{\theta}\tau)$$

to give

$$\frac{dx_1}{d\tau} = -ux_1x_2 + x_{1,0} - x_1$$

$$\frac{dx_2}{d\tau} = -\frac{u}{2}x_2^2 + x_{1,0} - x_2$$

$$\frac{dx_4}{d\tau} = x_{1,0} - x_4 \tag{6.5.5}$$

$$\frac{dx_3}{d\tau} = ux_4^2 - x_3$$

The similarity between Eqs. (6.5.5) and (6.4.6) may again be observed.

The concentrations in the output from the reactor will also vary sinusoidally with the time period t^* and the time-average concentration of P_n in this stream can be written in terms of an equation similar to Eq. (6.5.2). The time-average value of the dimensionless variables x_1 to x_4 can thus be written as

$$\overline{x_i} = \frac{1}{t^*}\int_t^{t+t^*} x_i(t)\,dt$$

$$= \frac{\omega\bar{\theta}}{2\pi}\int_\tau^{(\tau+2\pi/\omega\bar{\theta})} x_i(\tau)\,d\tau \tag{6.5.6}$$

The time-average monomer conversion and the number and weight average chain lengths, $\overline{\mu_{n1}}$ and $\overline{\mu_{w1}}$ (with monomer excluded from the summations), are given by [see Eq. (6.4.9)]

$$\overline{\text{Monomer conversion}} = \frac{\overline{[P_1]_0} - \overline{[P_1]}}{\overline{[P_1]_0}} = 1 - \bar{x}_1 \tag{6.5.7a}$$

$$\overline{\mu_{n1}} = \frac{\bar{x}_4 - \bar{x}_1}{\bar{x}_2 - \bar{x}_1} \tag{6.5.7b}$$

$$\overline{\mu_{w1}} = 1 + \frac{\bar{x}_3}{\bar{x}_4 - \bar{x}_1} \tag{6.5.7c}$$

$$\overline{Q_1} = \overline{\mu_{w1}}/\overline{\mu_{n1}} \tag{6.5.7d}$$

Equation (6.5.5) has been integrated[7] numerically for various values of a and ω and Eqs. (6.5.6) and (6.5.7) used to give the time-average product characteristics. Some results are shown in Figs. 6.3 and 6.4. The steady state values denoted by subscript s are those for a feed having a time-invariant monomer concentration equal to $\overline{[P_1]_0}$, and are obtained from Eqs. (6.5.6) and (6.5.7) by putting $d/d\tau = 0$, as

$$x_{4,s} = x_{1,0,s} = 1 \tag{6.5.8a}$$

$$x_{3,s} = u \tag{6.5.8b}$$

$$x_{2,s} = \frac{(2u + 1)^{1/2} - 1}{u} \tag{6.5.8c}$$

$$x_{1,s} = \frac{1}{(2u + 1)^{1/2}} \tag{6.5.8d}$$

$$\mu_{n1,s} = 1 + (2u + 1)^{1/2} \tag{6.5.8e}$$

$$\mu_{w1,s} = \frac{3 + 2u + (2u + 1)^{1/2}}{2} \tag{6.5.8f}$$

Figure 6.3 shows[7] significant differences in the product properties because of periodic operation, specially when the frequency, ω, is small. At high frequencies, the reactor acts as a filter and steady-state values are obtained. Figure 6.4 brings out the effect of periodic operation more dramatically.[7] It shows how the time-averaged polydispersity index, $\overline{Q_1}$, deviates from the steady-state value when the number average chain length, $\overline{\mu_{n1}}$, is fixed. This figure shows that the polydispersity index can be made to increase

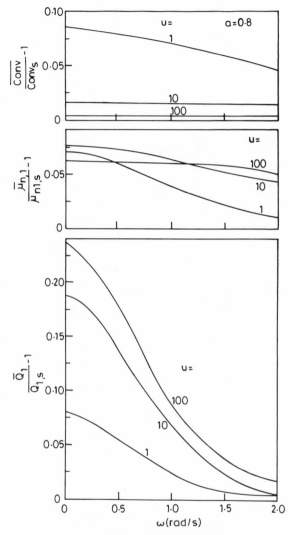

Figure 6.3. Effect of varying the frequency of the monomer concentration in the feed on product properties. $u = $ const, $a = 0.8$. (Reprinted with permission from Ref. 7. Copyright 1968, American Chemical Society, Washington, DC.)

by as much as 16% and decrease by as much as 6% by the use of periodic operation under certain conditions.

Lee and Bailey[8] have studied the bang-bang periodicity of the monomer concentration (instead of sinusoidal oscillation), which would mean easier operability, and have also considered the effect of imperfect micromixing in the HCSTR (modeled in terms of two interconnected, perfectly mixed

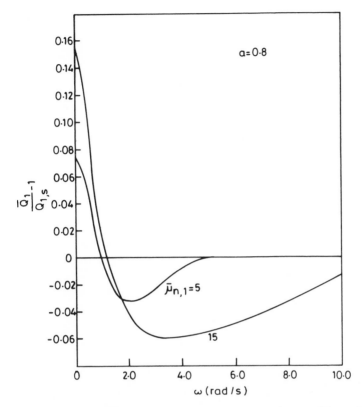

Figure 6.4. Effect of frequency of the feed monomer concentration. $a = 0.8$. u varied to achieve fixed $\overline{\mu_{n,1}}$. (Reprinted with permission from Ref. 7. Copyright 1968, American Chemical Society, Washington, DC.)

reactors). The mole balance equations for a single perfectly mixed reactor are the same as in Eq. (6.5.5), though now more algebraic manipulations are required to obtain appropriate product properties. These workers find that the effect of mixing is relatively small. The effect of periodic variation in the feed concentration for the case of multifunctional R_1A_f polymerization has been found to be even more interesting since the onset of gelation adds on an added degree of complexity. More extensive work on the periodic operation of reactors for step growth polymerizations needs to be done, before the final word is said in this area.[10]

6.6. CONCLUSIONS

In this chapter, typical optimization and control problems have been discussed for step growth polymerization reactors. It is possible to formulate many other similar problems, and one of them is to maintain the control variables such that a plant operates as close as possible to the design point, in spite of fluctuations in the feed (regulation problem).[7,11] Very little work along these lines has been reported in the open literature for step growth polymerizations, though some work on free-radical polymerizations exists.[12] Another problem of great interest is the optimization and control of polymerization reactors with energy and mass transfer limitations (such that the temperature and the condensation by-product concentration become state variables and the coolant flow rate and pressure become the control variables). Such problems are important particularly at high conversions and have not received much attention, at least for step growth polymerizations. (Some work on the stability of *addition* polymerization reactors with coolant flow has been reported by Amundson and coworkers[13-15] and others,[16] and the use of various types of controllers to maintain isothermal conditions in HCSTRs manufacturing PVC using suspension polymerization[11] has been investigated.) The next decade will probably witness a spate of published information in the areas of the dynamics, stability, and control of step growth polymerization reactors.

REFERENCES

1. W. H. Ray and J. Szekely, *Process Optimization*, 1st ed., Wiley, New York (1973).
2. M. M. Denn, *Optimization by Variational Methods*, 1st ed., McGraw-Hill, New York (1969).
3. A. E. Bryson and Y. C. Ho, *Applied Optimal Control*, 1st ed., Blaisdell, Waltham, Massachusetts (1969).
4. L. Lapidus and R. Luus, *Optimal Control of Engineering Processes*, 1st ed., Blaisdell, Waltham, Massachusetts (1967).
5. A. Ramagopal, A. Kumar, and S. K. Gupta, Computational scheme for the calculation of molecular weight distributions for nylon-6 polymerization in homogeneous, continuous-flow stirred-tank reactors with continuous removal of water, *Polym. Eng. Sci.* **22**, 849–856 (1982).
6. J. Hicks, A. Mohan, and W. H. Ray, The optimal control of polymerization reactors, *Can. J. Chem. Eng.* **47**, 590–597 (1969).
7. W. H. Ray, Periodic operation of polymerization reactors, *Ind. Eng. Chem., Proc. Des. Dev.* **7**, 422–426 (1968).
8. C. K. Lee and J. E. Bailey, Influence of mixing on the performance of periodic chemical reactors, *AIChE J.* **20**, 74–81 (1974).
9. S. K. Gupta, S. Nath, and A. Kumar, Forced oscillations in CFSTRs with nonlinear step growth polymerization, *J. Appl. Polym. Sci.* **30**, 557–569 (1985).
10. G. R. Meira, Forced oscillations in continuous polymerization reactors and molecular weight distribution control. A survey, *J. Macromol. Sci. Rev. Macromol. Chem. C* **20**(2), 207–241 (1981).
11. C. Kiparissides and S. R. Ponnuswamy, Hierarchial control of a train of continuous polymerization reactors, *Can. J. Chem. Eng.* **59**, 752–759 (1981).
12. T. A. Kenat, R. I. Kermode, and S. L. Rosen, Dynamics of a continuous stirred-tank polymerization reactor, *Ind. Eng. Chem., Proc. Des. Dev.* **6**, 363–370 (1967).

13. R. B. Warden and N. R. Amundson, Stability and control of addition polymerization reactions. A theoretical study, *Chem. Eng. Sci.* **17**, 725–734 (1962).
14. S. L. Liu and N. R. Amundson, Polymerization reactor stability, *Z. Elektrochem.* **65**, 276–282 (1961).
15. R. P. Goldstein and N. R. Amundson, Analysis of chemical reactor stability and control. Xa. Polymerization models in two immiscible phases in physical equilibrium; Xb. Polymerization models in two immiscible phases with interphase heat and mass transfer resistances; XI. Further considerations with polymerization models; XII. Special problems in polymerization models, *Chem. Eng. Sci.* **20**, 195–236, 449–476, 477–499, 501–527 (1965).
16. P. J. Hoftyzer and T. N. Zwietering, The characteristics of a homogenized reactor for the polymerization of ethylene, *Chem. Eng. Sci.* **14**, 241–251 (1961).

EXERCISES

1. Rewrite the matrix equations of Table 6.1 in long-hand notation [as in Eq. (6.2.1)].

2. From the discussion in the text, set up the flow chart for an optimization algorithm using the first-order, control vector iteration method. Keep in mind the high memory-storage requirements ordinarily required.

3. Derive the equations in Table 6.2.

4. Write the appropriate balance equations for the following kinetic scheme relevant for nylon 6 polymerization:

$$M + W \underset{k_1'}{\overset{k_1}{\rightleftharpoons}} P_1$$

$$P_n + P_m \underset{k_2'}{\overset{k_2}{\rightleftharpoons}} P_{n+m} + W$$

$$P_n + M \underset{k_3'}{\overset{k_3}{\rightleftharpoons}} P_{n+1}$$

where M is the monomer and W, the condensation by-product (added to the feed to start the formation of P_1). Then, for the following objective function[5]

$$\min I = \alpha_1 [M]_{t_f}^2 + \int_0^{t_f} \left(\frac{\mu_n - \mu_{n,d}}{\mu_{n,d}} \right)^2 dt$$

and with $T_* \leq T \leq T^*$, set up the necessary equations to obtain the optimal temperature profile $T(t)$ for a given feed having concentrations $[W]_0$ and $[M]_0$.

5. Set up the optimization problem referred to in Section 6.4 for the optimal start-up policy in a single HCSTR using the feed flow rate as the control variable.[6]

6. Set up the equations for periodic operation (of feed concentration) of an HCSTR in which multifunctional $R_1 A_f$ step growth polymerization is being carried out.[9] Note the possibility of gelation.

7. Set up the energy balance equation for the semibatch stirred reactor of Chapter 5. Then see if a reasonable optimization problem can be formulated for this system along the lines discussed in Section 6.3. Note that the system equations are more complex for this case than for the reactor without vaporization.

NYLON REACTORS

7.1. INTRODUCTION

Nylons are polymers having amide

$$
\begin{array}{cc}
O & H \\
\| & | \\
(C - N &)
\end{array}
$$

linkages, because of which they are also called polyamides. Since proteins are polyamides of various amino acids, they, too, fall into this category, but the discussion in this chapter is limited to industrially important synthetic polyamides. There are two classes of synthetic nylons. One of these is formed from cyclic monomers (or amino acids), as for example, nylon 6 $\{H[HN-(CH_2)_5-CO]_n-OH\}$, which has six carbon atoms per repeat unit and is made from ε-caprolactam, nylon 12 $\{H[NH-(CH_2)_{11}-CO]_n-OH\}$, having 12 carbon atoms in the repeat unit, and made from the lactam of 12-amino dodecanoic acid, etc. The single index used in describing these nylons indicates the number of carbon atoms in the repeat unit. The second class of synthetic nylons is formed from diamines and diacids, and is represented in terms of two indices indicating the number of carbon atoms in the diamine and diacid units, respectively, e.g., nylon-6,6

$$
\{H[NH-(CH_2)_6-NH-CO-(CH_2)_4-CO]_n-OH\}
$$

made from hexamethylene diamine and adipic acid.

Of the various nylons available, nylon 6 and nylon-6,6 are commercially the most important and form the subject of this chapter. In fact, the importance of nylon 6 in particular, and of nylons in general, can be gauged from the fact that several review articles[1-5] have been written on these polymers in the last few years.

7.2. POLYMERIZATION OF ε-CAPROLACTAM

Two routes are used for the polymerization of ε-caprolactam {C_1 or CL: $HN-(CH_2)_5-CO$}. The most commonly used process is the hydrolytic polymerization of CL, in which water (W) is used[6,7] to open the ε-caprolactam ring to give a linear molecule, amino caproic acid {P_1 or ACA: $H_2N(CH_2)_5-COOH$}. Polymerization then proceeds by the step growth mechanism of this bifunctional compound giving linear polymers, P_n{$H-[HN(CH_2)_5-CO]_n-OH$}, with water as the condensation by-product. In addition to these reactions, the caprolactam ring can also be directly opened by the amino end group[8] of any linear polymer molecule, P_n. This reaction leads to the growth of the molecule by one monomeric unit at a time, and so has characteristics of chain growth polymerization. Thus, the major reactions in the hydrolytic process can be represented schematically by[1-3,5-8]

$$C_1 + W \rightleftharpoons P_1 \tag{7.2.1a}$$

$$P_n + P_m \rightleftharpoons P_{n+m} + W, \qquad n, m = 1, 2, \ldots \tag{7.2.1b}$$

$$C_1 + P_n \rightleftharpoons P_{n+1}, \qquad n = 1, 2, \ldots \tag{7.2.1c}$$

The reversible nature of these reactions had been recognized by Carothers[6] very early in his work on polyamides. The equilibrium product formed at conditions usually encountered in industrial reactors, contains about 8%–10% (by weight) of unreacted monomer. This is removed by various techniques, e.g., evaporation in falling film evaporators under vacuum, drying of solid chips in fluidized beds or in tumble driers, leaching of the solid polymer chips in hot water, etc. In these finishing operations, further polymerization also occurs as the CL is removed and concepts of mass transfer with chemical reaction are required for modeling them.

The second route of ε-caprolactam polymerization is by the ionic chain growth mechanism.[4] The advantage of this technique is that it can be carried out below the melting point of nylon 6 ($\approx 220°C$) and can be used for the production of large cast articles. The molecular weight of the polymer formed is usually much higher than in the hydrolytic polymerization route. In this chapter, the discussion is limited to the hydrolytic polymerization of nylon 6 since ionic polymerization falls outside the scope of this book.

7.3. HYDROLYTIC POLYMERIZATION OF ε-CAPROLACTAM IN BATCH REACTORS

In this section the polymerization of nylon 6 in sealed tubes is simulated. This can also be used to predict the performance of plug flow and batch reactors in which there is no vaporization of water. In later sections, the polymerization of nylon 6 in stirred tank flow reactors, and in cascades of different types of reactors used industrially, are discussed.

The kinetic scheme for nylon 6 polymerization includes the three major reactions: ring opening, polycondensation, and polyaddition, given in Eq. (7.2.1). In addition, the formation of higher cyclic oligomers[7]

$$C_2, C_3, \ldots, \left\{ \begin{matrix} H & & O & H & & O \\ | & & \| & | & & \| \\ N-(CH_2)_5 \underbrace{[C-N-(CH_2)_5]}_{n-1} C \end{matrix} \right\}$$

is an important side reaction. Even though the total amount of these compounds formed is small (below about 2%–3% by weight), it is known that they cause problems in the spinning and molding of the final polymer. As a result, hot water extraction is used before these processing operations to remove the unreacted monomer, as well as the higher cyclic compounds. Since this is an expensive, energy-intensive process, a design engineer endeavors to keep the formation of these compounds below a critical value to minimize costs. The incorporation of these side reactions is thus necessary in any realistic model. The cyclic oligomers C_n can also undergo ring opening and polyaddition reactions as follows[9]:

$$C_n + W \rightleftharpoons P_n, \qquad n = 2, 3, \ldots \qquad\qquad (7.3.1a)$$

$$C_n + P_m \rightleftharpoons P_{n+m}, \qquad n = 2, 3, \ldots, \qquad m = 1, 2, \ldots \qquad (7.3.1b)$$

Usually, the two reactions with C_2 only are incorporated in the kinetic scheme. This is because it has been found experimentally that the formation of the cyclic dimer predominates and because incorporating the other cyclic oligomers makes the analysis more difficult.

Sometimes, monofunctional acids, e.g., acetic acid, are added to the reaction mass to control the molecular weight of the polymer formed or to increase the rate of polymerization. The acid end groups of the monofunctional compounds,

$$P_{mx} \left\{ \begin{matrix} O & H \\ \| & | \\ X[C-N-(CH_2)_5]_{\overline{m-1}} COOH, \text{ with X unreactive} \end{matrix} \right\}$$

react with the amino groups of P_n, but do not participate in ring opening or polyaddition reactions. The complete kinetic scheme of nylon 6 polymerization including reactions with C_2 and P_{mx} is given in Table 7.1. At times, monofunctional amines like n-butyl amine are used for the same

TABLE 7.1. Kinetic Scheme for Nylon 6 Polymerization

---

1. Ring opening

$$C_1 + W \underset{k_1' = k_1/K_1}{\overset{k_1}{\rightleftharpoons}} P_1$$

2. Polycondensation

$$P_n + P_m \underset{k_2' = k_2/K_2}{\overset{k_2}{\rightleftharpoons}} P_{n+m} + W, \qquad n, m = 1, 2, 3 \ldots$$

3. Polyaddition

$$P_n + C_1 \underset{k_3' = k_3/K_3}{\overset{k_3}{\rightleftharpoons}} P_{n+1}, \qquad n = 1, 2, \ldots$$

4. Ring opening of cyclic dimer

$$C_2 + W \underset{k_4' = k_4/K_4}{\overset{k_4}{\rightleftharpoons}} P_2$$

5. Polyaddition of cyclic dimer

$$P_n + C_2 \underset{k_5' = k_5/K_5}{\overset{k_5}{\rightleftharpoons}} P_{n+2}, \qquad n = 1, 2, \ldots$$

6. Reaction with monofunctional acid

$$P_n + P_{mx} \underset{k_2' = k_2/K_2}{\overset{k_2}{\rightleftharpoons}} P_{m+n,x} + W, \qquad n, m = 1, 2, \ldots$$

---

purpose, and appropriate modifications to the kinetic scheme of Table 7.1 must be made since these compounds can also react with $C_1, C_2, \ldots$, etc. Other side reactions[1,3] like decarboxylation,[10] desamination,[10] peroxidation of caprolactam,[11] etc. are not included in Table 7.1 in order to keep the analysis simple.

The various reactions of Table 7.1 have been found to be catalyzed by the carboxyl groups and the forward rate constants can be represented empirically (accounting for all the mechanistic steps) by Eq. (a) of Table 7.2, where [—COOH] is equal to $\sum_{n=1}^{\infty} ([P_n] + [P_{nx}])$. The equilibrium constants, K_i, are given by the standard thermodynamic equation [Eq. (b) of Table 7.2]. This table gives the various rate and equilibrium constants as determined recently by Tai *et al.*[3,12] These workers performed an extensive series of experiments[13,14] at 230–280°C with initial water concentrations varying in the range of 0.42–1.18 gmole/kg mixture. They measured concentrations of P_1 by high-pressure liquid chromatography (HPLC), of —NH$_2$ and —COOH groups by titration, and of C_2 by gas chromatography.[12] The rate constants of Table 7.2 were obtained using a nonlinear regression technique, and since they are more precise than the ones reported earlier,[7,8,15-17] they are now used in all simulation and optimization studies on nylon 6 reactors.

TABLE 7.2. Rate and Equilibrium Constants for Nylon 6 Polymerization[a]

$$k_i = k_i^0 + k_i^c[-\text{COOH}] = A_i^0 \exp(-E_i^0/RT) + A_i^c \exp(-E_i^c/RT)[-\text{COOH}] \qquad \text{(a)}$$

$$K_i = \exp\left(\frac{\Delta S_i}{R} - \frac{\Delta H_i}{RT}\right), \qquad i = 1, 2, \ldots, 5 \qquad \text{(b)}^b$$

i	A_i^0 (kg/mol hr)	E_i^0 (cal/mol)	A_i^c (kg/mol^2 hr)	E_i^c (cal/mol)	ΔH_i (cal/mol)	ΔS_i (eu)
1	5.9874×10^5	1.9880×10^4	4.3075×10^7	1.8806×10^4	1.9180×10^3	-7.8846×10^0
2	1.8942×10^{10}	2.3271×10^4	1.2114×10^{10}	2.0670×10^4	-5.9458×10^3	9.4374×10^{-1}
3	2.8558×10^9	2.2845×10^4	1.6377×10^{10}	2.0107×10^4	-4.0438×10^3	-6.9457×10^0
4	8.5778×10^{11}	4.2000×10^4	2.3307×10^{12}	3.7400×10^4	-9.6000×10^3	-1.4520×10^1
5	2.5701×10^8	2.1300×10^4	3.0110×10^9	2.0400×10^4	-3.1691×10^3	5.8265×10^{-1}

[a] References 3 and 12. (Reprinted with permission from Ref. 3. Copyright 1983, American Chemical Society, Washington, DC.)
[b] ΔH_i and ΔS_i assumed to be independent of temperature.

It may be worthwhile to observe the slightly unconventional units used for the rate constants in Table 7.2. Instead of a volumetric basis, a mass basis has been used to express the k_i's. Another point to observe is the fact that k_i and K_i are the rate and equilibrium constants associated with functional groups, and therefore, appropriate multiplicative factors are required in the mole balance equations of molecular species. It can be assumed that the rate constants associated with the acid end groups of P_{nx} and P_m are equal.

Mole balance equations can now be written for an isothermal batch reactor. These equations, when integrated numerically using the Runge-Kutta technique, give the concentrations of the individual molecular species $C_1, C_2, W, P_1, P_2, \ldots, P_{1x}, P_{2x}, \ldots$ as a function of time. This has recently been done and the molecular weight distributions (called "exact" MWDs) for some common feed and reactor conditions[18,19] are shown in Fig. 7.1.

The numerical integration of the equations for the individual molecular species is extremely cumbersome and time consuming. To simplify the analysis, it is possible to study the MWD in terms of the moments:

$$\lambda_i \equiv \sum_{n=1}^{\infty} n^i[P_n], \qquad \lambda_{ix} \equiv \sum_{n=1}^{\infty} n^i[P_{nx}], \qquad i = 0, 1, 2, 3, \ldots \qquad (7.3.2)$$

The equations for the molecular species can be appropriately summed up to give equations for these moments. The final set of equations is given[20] in Table 7.3. $\mathcal{R}_i$ represents the rate of "generation" of species or moment i by chemical reaction.

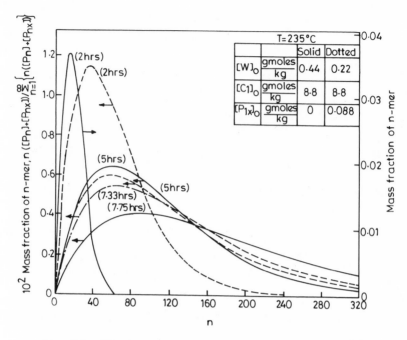

Figure 7.1. MWDs for nylon 6 formed in isothermal batch reactors.[18,19]

Several subsets of equations can be chosen from Table 7.3 and solved simultaneously, using appropriate closure equations. A close scrutiny of Table 7.3 reveals that in the absence of monofunctional species, the equations for $[C_1]$, $[P_1]$, λ_0, λ_1, $[C_2]$, and $[W]$ do not involve any other variable (like λ_2, λ_3) except themselves and $[P_2]$ and $[P_3]$. If the equations for $[P_2]$ and $[P_3]$ are incorporated into this subset, additional variables $[P_4]$ and $[P_5]$ arise on the right-hand side. In order to solve for these variables more efficiently, one breaks this hierarchy of equations by *arbitrarily* assuming[1] both $[P_2]$ and $[P_3]$ equal to $[P_1]$. With this closure approximation substituted in the equations of Table 7.3, it is found that the subset of equations for $[C_1]$, $[P_1]$, λ_0, λ_1, $[C_2]$, and $[W]$ becomes a complete set of coupled ordinary differential equations which may be solved. Fortunately, the results are insensitive[13] to this closure assumption. The solution of this subset of equations, however, does not give the polydispersity index.

In order to obtain the polydispersity index also, the complete set of equations in Table 7.3 (for $[C_1]$, $[P_1]$, λ_0, λ_1, λ_2, $[C_2]$, $[W]$, λ_{1x}, λ_{2x}, and $[P_{1x}]$) must be solved. These equations incorporate the extra variables $[P_2]$, $[P_3]$, λ_3, and λ_{3x}. Since moment relations for λ_3 and λ_{3x} incorporate still higher moments, moment closure approximations are also required to break the hierarchy of equations. The closure equations m in Table 7.3, based on

TABLE 7.3. Mass Balance Equations[a] for a Batch Reactor
with No Water Removal

$$\frac{d[C_1]}{dt} = \mathcal{R}_{C_1} = -k_1[C_1][W] + k_1'[P_1] - k_3[C_1]\lambda_0 + k_3'(\lambda_0 - [P_1]) \tag{a}$$

$$\frac{d[P_1]}{dt} = \mathcal{R}_{P_1} = k_1[C_1][W] - k_1'[P_1] - 2k_2[P_1]\lambda_0 + 2k_2'[W](\lambda_0 - [P_1]) - k_3[P_1][C_1]$$
$$+ k_3'[P_2] - k_2'\lambda_{0x}[P_1] + k_2'[W](\lambda_{0x} - [P_{1x}]) - k_5[P_1][C_2] + k_5'[P_3] \tag{b}$$

$$\frac{d\lambda_0}{dt} = \mathcal{R}_{\lambda_0} = k_1[C_1][W] - k_1'[P_1] - k_2\lambda_0^2 + k_2'[W](\lambda_1 - \lambda_0)$$
$$- k_2\lambda_0\lambda_{0x} + k_2'[W](\lambda_{1x} - \lambda_{0x}) + k_4[W][C_2] - k_4'[P_2] \tag{c}$$

$$\frac{d\lambda_1}{dt} = \mathcal{R}_{\lambda_1} = k_1[C_1][W] - k_1'[P_1] + k_3[C_1]\lambda_0 - k_3'(\lambda_0 - [P_1]) - k_2\lambda_{0x}\lambda_1 - k_2'[W]\left(\frac{\lambda_{1x} - \lambda_{2x}}{2}\right)$$
$$+ 2k_5[C_2]\lambda_0 - 2k_5'(\lambda_0 - [P_1] - [P_2]) + 2k_4[W][C_2] - 2k_4'[P_2] \tag{d}$$

$$\frac{d\lambda_2}{dt} = \mathcal{R}_{\lambda_2} = k_1[C_1][W] - k_1'[P_1] + 2k_2\lambda_1^2 + \frac{k_2'}{3}[W](\lambda_1 - \lambda_3) + k_3[C_1](\lambda_0 + 2\lambda_1)$$
$$+ k_3'(\lambda_0 - 2\lambda_1 + [P_1]) - k_2\lambda_2\lambda_{0x} + \frac{k_2'}{6}[W](2\lambda_{3x} - 3\lambda_{2x} + \lambda_{1x})$$
$$+ 4k_5[C_2](\lambda_1 + \lambda_0) + 4k_5'(\lambda_0 - \lambda_1 + [P_2]) + 4k_4[W][C_2] - 4k_4'[P_2] \tag{e}$$

$$\frac{d[C_2]}{dt} = \mathcal{R}_{C_2} = -k_4[C_2][W] + k_4'[P_2] - k_5[C_2]\lambda_0 + k_5'(\lambda_0 - [P_1] - [P_2]) \tag{f}$$

$$\frac{d[W]}{dt} = \mathcal{R}_W = -k_1[C_1][W] + k_1'[P_1] + k_2(\lambda_0)^2 - k_2'[W](\lambda_1 - \lambda_0)$$
$$+ k_2\lambda_0\lambda_{0x} - k_2'[W](\lambda_{1x} - \lambda_{0x}) - k_4[C_2][W] + k_4'[P_2] = \frac{d\lambda_0}{dt} \tag{g}$$

$$\frac{d\lambda_{1x}}{dt} = \mathcal{R}_{\lambda_{1x}} = k_2\lambda_1\lambda_{0x} - \frac{k_2'[W]}{2}(\lambda_{2x} - \lambda_{1x}) \tag{h}$$

$$\frac{d\lambda_{2x}}{dt} = \mathcal{R}_{\lambda_{2x}} = k_2(2\lambda_1\lambda_{1x} + \lambda_2\lambda_{0x}) - \frac{k_2'[W]}{6}(4\lambda_{3x} - 3\lambda_{2x} - \lambda_{1x}) \tag{i}$$

$$\frac{d[P_{1x}]}{dt} = \mathcal{R}_{P_{1x}} = -k_2[P_{1x}]\lambda_0 + k_2'[W](\lambda_{0x} - [P_{1x}]) \tag{j}$$

$$\frac{d\lambda_{0x}}{dt} = 0 \quad \text{or} \quad \lambda_{0x} = [P_{1x}]_0 \tag{k}$$

Closure conditions

$$[P_2] = [P_3] = [P_1] \tag{l}$$

$$\lambda_3 = \frac{\lambda_2(2\lambda_2\lambda_0 - \lambda_1^2)}{\lambda_1\lambda_0}, \quad \lambda_{3x} = \frac{\lambda_{2x}(2\lambda_{2x}\lambda_{0x} - \lambda_{1x}^2)}{\lambda_{1x}\lambda_{0x}} \tag{m}$$

Initial conditions: at $t = 0$, use feed conditions

[a] Reference 20.

TABLE 7.4. Moments[20] Obtained from the Exact
MWDs and from the Equations of Table 7.3[a]

Variable	Exact MWDs	Moment equations (Table 7.3)
λ_1/λ_0	64.275	64.394
λ_2/λ_1	131.12	133.05
λ_{0x}	0.08796	0.088
λ_{1x}	5.538	5.414
λ_{2x}	705.35	703.47

[a] $[C_1]_0 = 8.8$ moles/kg, $[W]_0 = 0.14$ moles/kg, $[P_{1x}]_0 = 0.088$ moles/kg,
$T = 235°C$, $t = 8$ hr, no cyclic dimer formation ($k_4 = k_5 = 0$).

the Schulz–Zimm distribution, have been found to be successful[13,20,21] (see Section 2.9) for this purpose, and must be used along with Eq. (1) of the table. The insensitivity of the final results to the choice of the moment closure conditions was first demonstrated by Cave[21] and later by Tai et al.[13] However, a more rigorous test of its success was made later[20] by comparing results obtained by integrating the equations in Table 7.3 with corresponding moments computed from the "exact" MWD. An excellent agreement between these two for isothermal batch reactors is found, as seen in Table 7.4 for one typical set of conditions.[20] The computer time (DEC 1090) required to solve the moment equations is about 45 sec for a polymerization time of 16 hr as compared to an exhorbitant 4 hr for integrating the equations for the individual molecular species.[18,19]

Instead of using the moment closure equation based on the Schulz–Zimm distribution, one can use a higher-order method[22-24] in which the unknown MWD is written in terms of an expansion in Laguerre polynomials about a standard distribution for which the relations among the moments can be calculated exactly. It has been shown,[21,25] however, that results obtained from higher-order methods are almost identical to those obtained from the conceptually simpler set of equations in Table 7.3, and so the latter is recommended for simulation and optimization studies. In fact Eq. (m) of Table 7.3 has been successfully used for simulating not only nylon 6 systems, for which it was first established, but also for the simpler ARB polymerizations as well as for the more complex PET system.

Yet another subset of equations can be selected from Table 7.3 for the special case when there is no cyclization and no monofunctional species present. The corresponding equations are

$$\frac{d[C_1]}{dt} = -k_1[C_1][W] + k_1'[P_1] - k_3[C_1]\lambda_0 + k_3'(\lambda_0 - [P_1]) \qquad (7.3.3a)$$

$$\frac{d\lambda_0}{dt} = k_1[C_1][W] - k_1'[P_1] - k_2\lambda_0^2 + k_2'[W](\lambda_1 - \lambda_0) = \frac{-d[W]}{dt} \quad (7.3.3b)$$

$$\frac{d[P_1]}{dt} = k_1[C_1][W] - k_1'[P_1] - 2k_2[P_1]\lambda_0 + 2k_2'[W](\lambda_0 - [P_1])$$

$$- k_3[P_1][C_1] + k_3'[P_2] \quad (7.3.3c)$$

with the closure condition $[P_2] = [P_1]$. From Eq. (7.3.3b), $\lambda_0 + [W]$ is observed to be independent of time, and if the feed is assumed to be C_1 and water, the initial values of λ_0 and $[W]$ are zero and $[W]_0$, respectively. This implies that at any time,

$$[W] = [W]_0 - \lambda_0 \quad (7.3.4)$$

Also, since $\lambda_1(\equiv \sum_{n=1}^{\infty} n[P_n])$ now represents the total concentration of $-HN-(CH_2)_5-CO-$ repeat units in the linear chains, it is equal to the concentration of reacted caprolactam molecules, i.e.,

$$\lambda_1 = [C_1]_0 - [C_1] \quad (7.3.5)$$

Earlier workers[1,7,8,15,17] used this set of equations to curve-fit their experimentally measured conversions and evaluate the rate and equilibrium constants.

At equilibrium (denoted by subscript e), the rates of the *individual* reactions in Table 7.1 are zero and Eqs. (7.3.3)–(7.3.5) lead to

$$K_1 = \frac{k_1}{k_1'} = \frac{[P_1]_e}{[C_1]_e[W]_e} \quad (7.3.6a)$$

$$K_3 = \frac{k_3}{k_3'} = \frac{\lambda_{0,e} - [P_1]_e}{[C_1]_e\lambda_{0,e}} \approx \frac{1}{[C_1]_e} \quad (7.3.6b)$$

$$K_2 = \frac{k_2}{k_2'} = \left(\frac{\lambda_{1,e}}{\lambda_{0,e}} - 1\right)\left(\frac{[W]_0}{\lambda_{0,e}} - 1\right) \quad (7.3.6c)$$

where $[P_1]_e$ has been neglected in comparison with $\lambda_{0,e} = \sum_{n=1}^{\infty}[P_n]_e$ to obtain Eq. (7.3.6b). Since the number-average chain length at equilibrium, $\mu_{n,e}$, is equal to $\lambda_{1,e}/\lambda_{0,e}$, Eq. (7.3.5) can be used to give

$$\lambda_{0,e} = \frac{\lambda_{1,e}}{\mu_{n,e}} = \frac{[C_1]_0 - [C_1]_e}{\mu_{n,e}} \approx \frac{[C_1]_0}{\mu_{n,e}} \quad (7.3.7)$$

In this equation, $[C_1]_e$ has been neglected because the monomer conversion at equilibrium is usually above 90%. Substituting Eq. (7.3.7) in Eq. (7.3.6c), one obtains

$$K_2 \simeq (\mu_{n,e} - 1)\left(\frac{[W]_0 \mu_{n,e}}{[C_1]_0} - 1\right) \qquad (7.3.8)$$

This equation indicates that for a feed of CL and water only, and with no cyclization, the equilibrium degree of polymerization is controlled both by the temperature (through K_2) and the initial water concentration, while Eq. (7.3.6b) indicates that the final value of the monomer conversion is determined by temperature (through K_3) alone. Results of detailed simulations[1,17] are found to agree with these general conclusions, as shown[26] in Figs. 7.2 and 7.3. The sensitivity of the final value of μ_n to $[W]_0$ and temperature should be observed in Fig. 7.3. It indicates that in order to get a high *final* value of μ_n, the water concentration should be kept as low as possible. However, this would simultaneously slow down the reaction, as evident from Fig. 7.2. Thus, instead of using a single isothermal reactor to manufacture nylon 6, a better strategy appears to be a sequence of two reactors, the first one having a higher water concentration so as to achieve a rapid increase in the caprolactam conversion, and the second having a low water concentration in order to drive the polycondensation step in the forward direction

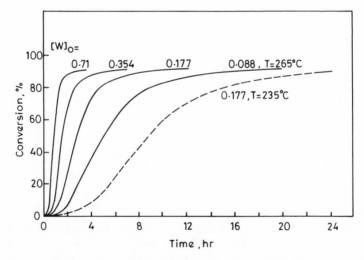

Figure 7.2. Monomer conversion[1,17,26] as a function of time for isothermal polymerization of C_1 with water in batch reactors. No monofunctional species, no cyclization. Solid lines: $T = 265°C$; dashed line: $T = 235°C$. $[W]_0$ in moles/kg mixture.

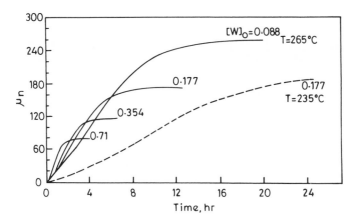

Figure 7.3. Number-average chain length[1,17,26] as a function of time. Conditions same as for Fig. 7.2.

and get higher molecular weight products. Simulations of such reactor sequences are discussed later in this chapter.

The effect of various design variables, e.g., T, $[W]_0$, $[P_{1x}]_0$, $[P_1]_0$, etc., on the product composition[20] is further illustrated in Figs. 7.4–7.7. Table 7.5 shows the conditions used for the simulation of the isothermal batch reactors. It is possible to deduce the strategies of reactor operation from these figures as follows. On comparing curves A, B, and C of Fig. 7.4, it is observed that the addition of either monofunctional acid or aminocaproic acid, P_1, to the feed leads to a very rapid conversion of caprolactam. In the first case, this is because the reaction of P_n with the monofunctional species produces more water, which speeds up the ring opening reaction. In the second case, this is because the addition of P_1 to the feed speeds up the polyaddition reaction. This effect is further illustrated by computing the *initial* rate constants for the relevant reaction for cases A and C: for case A, $k_{1,0}$ is $0.00156 + 0.327 \sum [P_n]_0 = 0.00156$ kg/mol hr while for case C, $k_{3,0} = 0.391 + 34.09 \sum [P_n]_0 = 15.39$ kg/mol hr.

The alternative strategy of using P_1 in the feed to speed up the reaction is not as attractive as it appears because P_1 is not available easily. However, P_1 can possibly be manufactured by depolymerizing scrap nylon 6 available in a plant by heating it with an excess of hot water at high pressures, provided the overall economics justifies such a step. Alternatively, since P_1 gets produced in the reactor itself when water is used in the feed, it may be worthwhile to recycle a fraction of the reaction mass to the feed point, as in Section 2.7. One can also use an HCSTR wherein, because of backmixing, the reaction occurs in the presence of P_1.

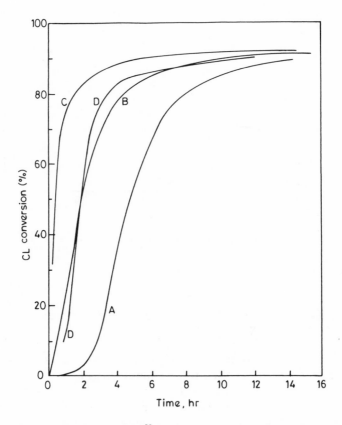

Figure 7.4. Caprolactam conversion[20] for various conditions given in Table 7.5.

TABLE 7.5. Conditions[20] Used for Generating Figs. 7.4–7.7

Curve	$[W]_0$ (gmole/kg mixture)	$[P_{1x}]_0$ (gmole/kg)	$[P_1]_0{}^a$ (gmole/kg)	T (°C)
A	0.44	0	0	235
B	0.44	0.088	0	235
C	0	0	0.44	235
D	0.44	0	0	265

^{a} Aminocaproic acid (P_1) can be used, if available, with CL in the feed.

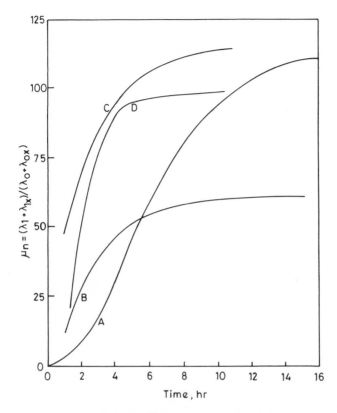

Figure 7.5. Degree of polymerization[20] for various conditions shown in Table 7.5.

The effect of varying the feed conditions on the degree of polymerization of the product is shown in Fig. 7.5. It is found that the addition of P_1 simultaneously leads to a faster increase in the value of μ_n, though the equilibrium value is almost the same as that obtained using water in the feed. The use of monofunctional acid, on the other hand, cuts down the equilibrium value of μ_n substantially. This suggests that a design engineer can use both P_1 and P_{1x} in the feed to the reactor to speed up the rate of reaction and still be able to obtain products having the desired molecular weight.

The effect of the various conditions given in Table 7.5 on the polydispersity index and the amount of cyclic dimer formed are depicted in Figs. 7.6 and 7.7, respectively. The equilibrium value of Q is found to be close to the most probable value of 2, irrespective of the feed conditions. The use of monofunctional compounds is found to lead to values of Q above the

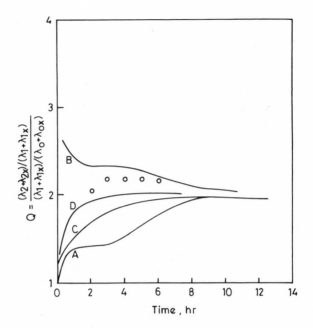

Figure 7.6. Polydispersity index[20] for various conditions shown in Table 7.5. Circles represent experimental values of Tai *et al.*[27] for $[W]_0 = 0.8$ moles/kg, $T = 240°C$.

theoretical limit of 2 only at small residence times in batch reactors. The values[27] of Q obtained by GPC and shown in Fig. 7.6 have been obtained in the absence of monofunctional compounds and are above 2.0, possibly because of experimental errors.

The use of P_1 in the feed also leads to a more rapid increase in the cyclic dimer concentration. However, a comparison of Figs. 7.5 and 7.7 shows that under conditions of equal μ_n, the concentration of C_2 does not differ considerably. Thus, the use of P_1 to reduce the residence time of the reactor offers attractive advantages. The tremendous increase in the concentration of C_2 at higher temperatures should be noted and leads to interesting ramifications in optimization studies. Another interesting point to be noted is that even though values of conversion, Q, and μ_n attain asymptotic values at reaction times of about 14–16 hr, the same is not true for $[C_2]$.

7.4. NYLON 6 POLYMERIZATION IN ISOTHERMAL RECYCLE REACTORS

The discussion in the previous section focused on the polymerization of nylon 6 in batch reactors. It was deduced that the use of a recycle reactor,

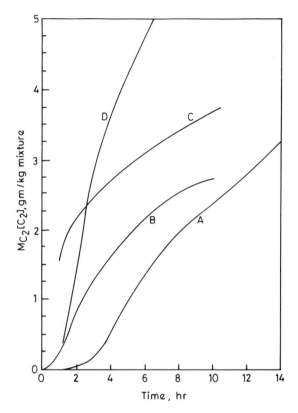

Figure 7.7. Grams of cyclic dimer[20] per kilogram of mixture for various conditions shown in Table 7.5. M_{C_2} is the molecular weight of C_2.

CSTR, or a cascade of CSTRs is a better strategy for nylon 6 polymerization. In this section, the nonvaporizing isothermal plug flow reactor with recycle (Fig. 7.8) is considered. The equations to be solved for the *reactor* are the same as in Table 7.3, but the feed to the *reactor* (point 2) is related to the

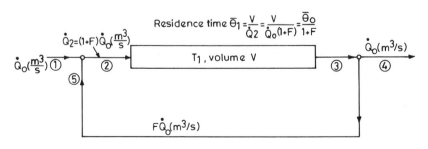

Figure 7.8. A recycle reactor.

feed to the system (point 1) by a "mixing" relationship. For P_n the mixing relationship takes the form

$$\overset{\circ}{Q}_0(1 + F)[P_n]_2 = \overset{\circ}{Q}_0[P_n]_1 + F\overset{\circ}{Q}_0[P_n]_5$$

or

$$[P_n]_2 = \frac{1}{1 + F}\{[P_n]_1 + F[P_n]_3\}, \qquad n = 1, 2, \dots \qquad (7.4.1)$$

where it has been assumed that the density of the reaction mass is the same at points 1, 5, and 2, so that the volumetric flow rate at 2 is the sum of those at 1 and 5. In this equation, $[P_n]_1$, $[P_n]_2$, $[P_n]_3$, and $[P_n]_5$ represent the concentrations of P_n at points 1, 2, 3, and 5, respectively, as shown in Fig. 7.8. "Mixing" equations for λ_0, λ_{0x}, λ_1, λ_{1x}, λ_2, λ_{2x}, $[C_1]$, $[P_{nx}]$, $[C_2]$, or $[W]$ can be written by replacing the $[P_n]$'s by any of these variables in Eq. (7.4.1).

The balance equations for the recycle reactor can best be solved numerically, since the procedure of Section 2.7 becomes extremely cumbersome to use. The computational method used[28] is that of successive substitutions. Results can be obtained for various values of the residence time $\bar{\theta}_0$ ($\equiv V/\overset{\circ}{Q}_0$), recycle ratio F, temperature T_1 of the reactor, and the feed composition (at point 1).

The effect of recycle on the *overall* monomer conversion $\{=1 - [C_1]_4/[C_1]_1\}$ and the μ_n of the product stream (3) is shown[28] in Fig. 7.9. It may be mentioned that $F = 0$ represents a plug flow reactor without recycle whereas (at constant $\bar{\theta}_0$) $F \to \infty$ signifies an infinite amount of backmixing and, therefore, represents an HCSTR. It is observed that for a fixed value of $\bar{\theta}_0$ of 1.5 hr (i.e., for a fixed reactor volume), the monomer conversion shows a maximum at an F of approximately 0.6. It is thus concluded that under certain values of $\bar{\theta}_0$, a recycle reactor gives higher conversions than either a PFR or an HCSTR. This effect is attributed to the presence of P_1 in the feed *to the plug flow reactor*, and leads to savings in operational costs.

7.5. NYLON 6 POLYMERIZATION IN HCSTRs AND IN CASCADES OF REACTORS

Figure 7.9 also shows that an HCSTR (for $F \to \infty$) is better than a PFR under some conditions (e.g., at $\bar{\theta}_0 = 1$). This is because in an HCSTR, the entire reaction mass is perfectly mixed on a molecular level (micromixed) and its concentration is the same as at the exit, with a finite concentration of P_1. This speeds up the overall polymerization as discussed in Section 7.3. It is possible that a sequence of an HCSTR and a PFR, or sequences of several HCSTRs, may be better for nylon 6 manufacture than a PFR alone. Indeed, several industrial processes with cascades of stirred tank

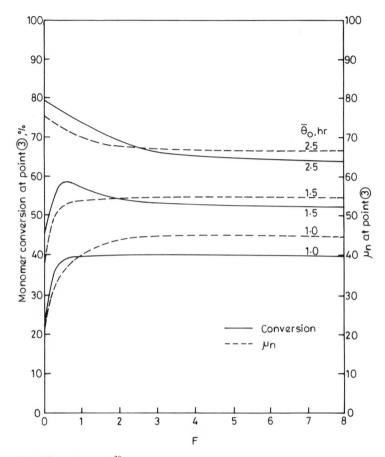

Figure 7.9. Effect of recycle[28] on overall monomer conversion and μ_n. Feed to *system* is monomer + 0.44 (mole/kg) of water. $T_1 = 265°C$.

reactors are currently being used.[29,30] Such reactors have the added advantage of good heat and mass transfer rates.

Instead of obtaining the performance of HCSTRs as a special case ($F \rightarrow \infty$) of recycle reactors discussed in Section 7.4, it is more convenient to solve for it *directly* using relevant mole balance equations. The appropriate equations for steady state operation of HCSTRs are [see Eq. (2.8.1) with d/dt terms zero]

$$\frac{[C_n] - [C_n]_0}{\bar{\theta}} = \mathscr{R}_{C_n}, \qquad n = 1, 2, \ldots \qquad (7.5.1a)$$

$$\frac{[P_n] - [P_n]_0}{\bar{\theta}} = \mathscr{R}_{P_n}, \qquad n = 1, 2, \ldots \qquad (7.5.1b)$$

where $\bar{\theta}$ is the mean residence time (=volume/flow rate), subscript 0 indicates feed concentrations, the concentrations $[C_1]$, $[P_1]$, etc., are those present in the reactor as well as in the output stream, and $\mathcal{R}_i$ is the rate of "generation" of species i by chemical reaction, similar to the right-hand side of the equations in Table 7.3. The equations for $[W]$ and $[P_{nx}]$ have the same form as Eq. (7.5.1). The equations for the various *molecular species* form a set of nonlinear, coupled, algebraic equations which have been solved[25] using Brown's technique[31] to generate the "exact" MWDs. This procedure takes a prohibitive amount of computer time and it is more efficient, and almost equally useful, to solve the equations for the moments

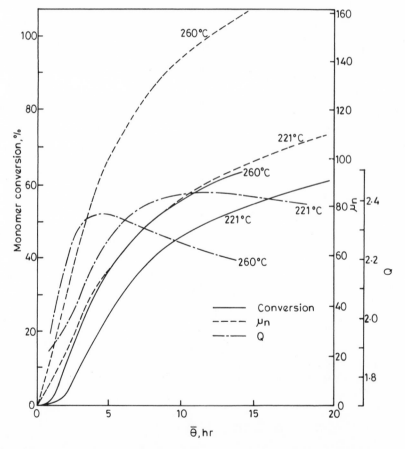

Figure 7.10. Nylon 6 polymerization in a single (nonvaporizing) isothermal HCSTR, with feed containing caprolactam ($[C_1]_0 = 8.8$ moles/kg) and water only. $[W]_0 = 0.2411$ moles/kg at 221°C and 0.111 moles/kg at 260°C. $[P_{1x}]_0 = 0$.

using appropriate closure conditions. The set of equations to be solved is similar to those in Table 7.3, with the LHS replaced by flow terms {e.g., $([C_1] - [C_1]_0)/\bar{\theta}$, $([P_1] - [P_1]_0)/\bar{\theta}$, $(\lambda_0 - \lambda_{0,0})/\bar{\theta}$, $(\lambda_1 - \lambda_{1,0})/\bar{\theta}$, etc.}. These coupled equations, with closure conditions on λ_3, λ_{3x}, $[P_2]$, and $[P_3]$, comprise a set of much fewer equations which have been solved recently[25] by Brown's technique and earlier (without λ_2 equations)[32,33] by the Newton-Raphson technique. It has been found[25] that the use of the closure relations [Eqs. (l) and (m) in Table 7.3] give results for the moments which are extremely close to those obtained by summing up the concentrations of the individual species, thus justifying their use in HCSTR simulations *also*. Figure 7.10 shows the variation of C_1 conversion, μ_n, and Q with $\bar{\theta}$ for two typical conditions. The decrease in the value of Q after a maximum is in sharp contrast to the monotonic increase of Q with $\bar{\theta}$ observed for ARB polymerizations, and is a result of the complex kinetic scheme for nylon 6 polymerization.

It was conjectured in Section 7.3 that the use of cascades of reactors may represent a better strategy for nylon 6 polymerization than single reactors. Some results shown in Fig. 7.11 illustrate[32,33] how significant reductions in the total residence time are possible, particularly if W is removed at an intermediate stage. This figure clearly establishes the need for carrying out optimization studies for nylon 6 polymerization, a subject discussed in Section 7.8.

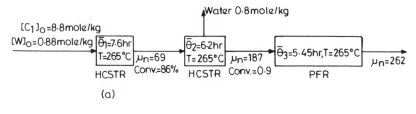

(a)

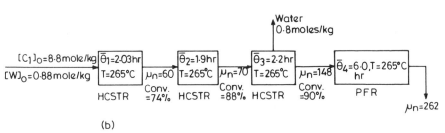

(b)

Figure 7.11. Polymerization of nylon 6 in two typical trains of reactors,[32] leading to the same final value of μ_n. Total residence times are (a) 19.25 hr and (b) 12.13 hr. (Reprinted from Ref. 32 with permission of John Wiley & Sons, New York.)

7.6. EFFECT OF NONIDEALITIES

Even though most of the studies on nylon 6 polymerization have been carried out for a single or a cascade of ideal reactors, it is now well established that several nonidealities exist in real tubular or continuous flow, stirred tank reactors. The effects of some of these nonidealities are studied in this section.

The effect of imperfect micromixing in continuous flow, stirred tank reactors is first studied. As discussed in Section 2.8, this is usually done by studying the other extreme—the continuous flow, stirred tank reactor with complete segregation at the microscopic level (SCSTR). The performance of real CSTRs lies between that of an HCSTR and an SCSTR.

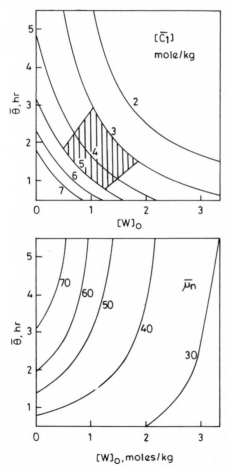

Figure 7.12. Results[34] for an SCSTR at 260°C. Shaded region represents industrially important values. $[P_{1x}]_0 = 0$. ($\bar{\mu}_n = \bar{\lambda}_1/\bar{\lambda}_0$ and $[\bar{C}_1]$ represent product characteristics.) (Reprinted from Ref. 34 with permission from John Wiley & Sons, New York.)

In an SCSTR, the distribution of the residence time, t, of any fluid element is given by Eq. (2.8.14). Since the fluid elements in an SCSTR are completely segregated, they react independently, and so each of them can be described by equations for a batch reactor. The average values of the concentrations of the various species or the moments in the product stream can then be obtained from Eqs. (2.8.15) and (2.8.16). Tai *et al.*[34] have simulated the performance of SCSTRs and have presented their results in the form of contour maps. Figure 7.12 shows some of these results[34] in which the shaded region represents the range of industrially encountered values of the variables. Trains of SCSTRs have been studied by Nagasubramanian and Reimschuessel,[32] and Fig. 7.13 shows results for the same system as shown in Fig. 7.11b, with HCSTRs replaced by SCSTRs. It is found that a lower total residence time is required in a cascade of SCSTRs than in a cascade of HCSTRs, to produce nylon 6 of the same number average chain length.

Another nonideality which can easily be accounted for is the presence of a nonuniform (i.e., non-plug-flow) velocity profile in real tubular reactors.[35] This is expected to improve the agreement between the reactor model and the actual reactor performance. In order to account for the velocity profile, it is assumed that each fluid element is completely segregated. Batch reactor equations can then be solved for each fluid element, as was done for an SCSTR, and the mean values of the concentrations of the various molecular species at the exit may be computed in a similar manner, using an equation similar to Eq. (2.8.15). If the flow is laminar, the velocity

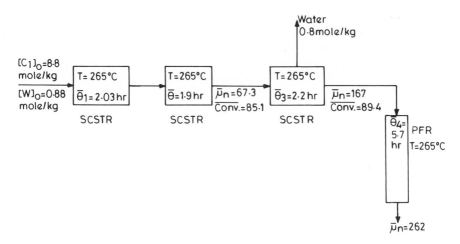

Figure 7.13. Conditions at different stages[32] when SCSTRs are used instead of HCSTRs in the system of Fig. 7.11b. (Reprinted from Ref. 32 with permission from   John Wiley & Sons, New York.)

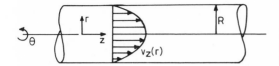

Figure 7.14. Velocity profile in a tubular reactor under laminar flow conditions.

distribution in a tubular reactor is parabolic (as shown in Fig. 7.14), and is given by

$$v_z(r) = 2\bar{v}(1 - r^2/R^2) \tag{7.6.1}$$

where $\bar{v}$ is the mean velocity. An element of fluid at radius r travels the length L of the reactor in a time $t(r)$ given by

$$t(r) = \frac{L}{v_z(r)} = \frac{L}{2\bar{v}(1 - r^2/R^2)} \equiv \frac{\bar{\theta}}{2(1 - r^2/R^2)} \tag{7.6.2}$$

where $\bar{\theta}$ is the mean residence time $(= L/\bar{v})$. Thus, the mean value of the concentration of species P_n at the end of the reactor is given by

$$\overline{[P_n]} = \frac{\int_{r=0}^{R} v_z(r) 2\pi r \, dr [P_n]_b \{t = \bar{\theta}/2(1 - r^2/R^2)\}}{\int_{r=0}^{R} v_1(r) 2\pi r \, dr}, \qquad n = 1, 2, \ldots, \tag{7.6.3}$$

with similar equations for the other concentrations and moments. In the integral of Eq. (7.6.3), $[P_n]_b\{t\}$ represents the value of $[P_n]$ for a batch reactor at time t (which is the same as that in the fluid element exiting at time t). In the above equation, the numerator represents the flow (moles/sec) of the desired component over the entire cross section while the denominator represents the total volumetric flow (m³/sec). Equation (7.6.3) can be simplified [since t and r are related by Eq. (7.6.2)] to give

$$\overline{[P_n]} = \frac{\int_{t=\bar{\theta}/2}^{\infty} [P_n]_b\{t\}(1/t^3) \, dt}{\int_{t=\bar{\theta}/2}^{\infty} (1/t^3) \, dt}$$

$$= \int_{t=\bar{\theta}/2}^{\infty} [P_n]_b\{t\} \frac{\bar{\theta}^2}{2t^3} \, dt \tag{7.6.4}$$

A comparison of Eqs. (7.6.4) and (2.8.15) indicates that $\bar{\theta}^2/2t^3$ can be interpreted as a residence time distribution {with $f(t)$ in Eq. (2.8.14), equal to zero for $0 \leq t \leq \bar{\theta}/2$} of a tubular reactor under laminar flow conditions. Tai *et al.*[34] have obtained product conditions ($\overline{[C_1]}$ and $\bar{\mu}_n = \bar{\lambda}_1/\bar{\lambda}_0$) for various values of $\bar{\theta}$ and T and have once again presented their results in

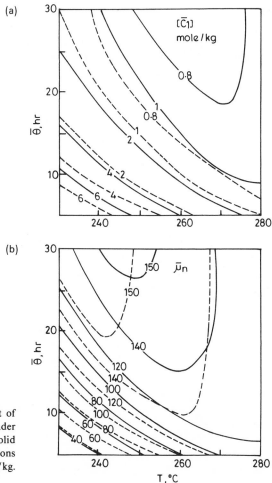

Figure 7.15. Conditions at outlet of an isothermal tubular reactor under (a) laminar flow conditions[34] (solid lines) and (b) plug flow conditions (dotted lines). $[W]_0 = 0.22$ mole/kg. $[P_{1x}]_0 = 0$.

the form of contour plots. Figure 7.15 shows some typical results. This figure also shows results for a plug reactor {with $f(t) = 1$ when $t = \bar{\theta}$ and $f(t) = 0$ otherwise}. The results on the laminar flow reactor have been found to match[34] with some data available in patents.

In this section, the effect of two important nonidealities, namely, micromixing and residence time distribution, have been studied. The models used represent real reactors far more closely than do ideal models like the PFR, HCSTR, etc. In fact, tracer studies can be carried out on industrial reactors or pilot plants to obtain the *actual* residence time distribution function, $f(t)$, and these may be used in Eq. (2.8.15) to obtain the mean values of the various concentrations and moments in the product stream.

Alternatively, if it is possible to carry out experimental studies on nonreacting systems in industrial reactors, one can approximate the latter as sequences of PFRs and HCSTRs or as a PFR with a recycle.[36] Appropriate mole balance equations for these models can then be solved to predict the performance of real reactors much more closely than possible with a single ideal reactor. In the next section, one such model for an industrial reactor is presented.

7.7. SIMULATION OF INDUSTRIAL REACTORS

The reactor models (including nonidealities) discussed in the previous sections explain the behavior of several industrial reactors and reactor cascades fairly well, *provided there is no evaporation of water*. As mentioned in Section 7.3, high concentrations of W are normally used in the initial stages so as to speed up the slow ring opening step, and thereafter W is removed to yield a product having high μ_n. As a result, industrial reactors for nylon 6 polymerization incorporate at least one stage (or have some zone) wherein W is removed from the reaction mass. Figure 7.16 shows[20] how the use of 0.577 moles/kg of water in the feed leads to an equilibrium value of μ_n of about 80 if no W is removed. But if all the water present in the reaction mass is removed instantaneously (so that [W] becomes zero before the reaction starts again) at $t = 4.67$ hr, much higher values of μ_n are obtained. The instantaneous removal of all W is an oversimplification of the physical process and it is necessary to model the mass transfer in nylon 6 polymerization in a manner similar to that discussed in Chapter 5, depending on the relative resistances of the diffusion and chemical reaction steps.

In addition to accounting for effects of mass transfer, heat transfer aspects of industrial reactors must also be studied, particularly at high conversions where the viscosity of the reaction mass becomes high and the thermal conductivity, low. In fact, adiabatic operation of tubular reactors is a better description of the reactor under such conditions than is the isothermal model. In this section, both mass and energy transfer effects are considered.

A commonly used industrial reactor for nylon 6 polymerization is the VK column (Vereinfacht Kontinuierliches Rohr).[1-3] This consists of a vertical tube operating at atmospheric pressure. The feed enters the top of the column and is heated up to about 220–270°C using heat exchangers in the form of internal gratings. In this top region, water and ε-caprolactam evaporate continuously. The bubbles of vapor formed in the reaction mass cause intense agitation as they rise up to a reflux condenser, which condenses and returns the caprolactam to the column. The hydrostatic head some

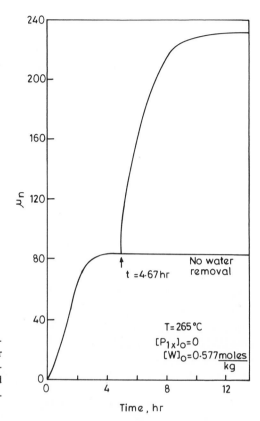

Figure 7.16. μ_n vs. time[20] for a two-stage PFR, both at 265°C with *all* water present at t = 4.67 hr removed instantaneously. No water is evaporated inside either of the two reactors. $[P_{1x}]_0 = 0$. $[W]_0 = 0.577$ moles/kg.

distance downstream prevents further vaporization, and so the remainder of the column is a nonvaporizing tubular reactor. Gratings are also used in this section both to facilitate heat removal and to ensure almost plug flow velocity profiles.

Another patented reactor[36,37] is shown in Fig. 7.17 in which vaporization is avoided both at the top of the column (stage I) and in the last section (stage III) by use of pressures above atmospheric. Water is removed by bubbling inert gas in a part (stage II) of the reactor between stages I and III. This reactor has the simplicity of the VK column, yet avoids the early vaporization of W and, therefore, gives the desired degree of polymerization in a shorter time[36] than the VK column. These two reactors are now modeled to illustrate the role of mass transfer in nylon 6 polymerization.

Jacobs and Schweigman[36] have carried out extensive pilot-plant tests and have also collected data on industrial VK columns. Based on temperature measurements and tracer studies, they have shown that the top zone of the column is well mixed owing to the agitation caused by the vapor

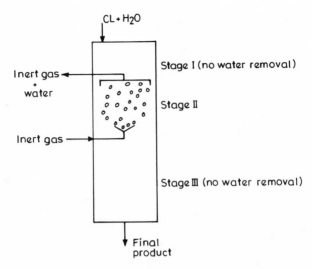

Figure 7.17. A three-stage reactor with no water removal in stages I and III (due to high pressures) but water removal by inert gas sparging in stage II.

bubbles. They find that the concentration of W in this zone is a function of temperature alone and is given by the following empirical relation:

$$[W] \, (\text{moles/kg}) = \{1.76 - 0.0060 \, T(^\circ C)\}/1.8 \qquad (7.7.1)$$

$[W]$ is found to be independent both of the residence time in this zone and of the feed conditions. This indicates that in the top zone of the VK column, the water concentration is determined solely by vapor–liquid equilibrium considerations, and that there is negligible mass transfer resistance. The modeling of this zone of the reactor, thus, is similar to that of semibatch reactors discussed in Section 5.2, with flow terms incorporated.

Based on residence time distribution measurements, Jacobs and Schweigman[36] suggest that the top zone of the VK column be modeled either as a single HCSTR (with appropriate vaporization terms) or as a series of two HCSTRs with backmixing and vaporization. They find a small difference between the performances predicted by these two models, but because of its simplicity, the first of these two is discussed in detail. Figure 7.18 shows the notation used. The appropriate set of algebraic equations to be solved for the top zone is given in Table 7.6. For a given residence time, $\bar{\theta}$, of the HCSTR, temperature T, and feed concentrations $[W]_0$ and $[C_1]_0$, these equations (with the closure conditions) yield a complete set for the variables $[C_1]$, $[P_1]$, λ_0, λ_1, λ_2, $[C_2]$, λ_{1x}, λ_{2x}, $[P_{1x}]$, and $F_0/(F_0 - F_v)$. It is to be noted that $[W]$ can be determined by the empirical Jacobs–Schweigman equilibrium relationship and the mole balance equation for

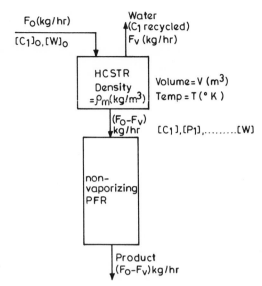

Figure 7.18. Simple model of VK column.

TABLE 7.6. Equations for the Top Zone of the VK Column Operating at Steady State[a]

$$\frac{(F_0 - F_v)[C_1] - F_0[C_1]_0}{\rho_m V} = \mathscr{R}_{C_1}$$

$$\frac{(F_0 - F_v)[P_n] - F_0[P_n]_0}{\rho_m V} = \mathscr{R}_{P_n}, \qquad n = 1, 2, \dots$$

$$\frac{(F_0 - F_v)\lambda_i - F_0\lambda_{i,0}}{\rho_m V} = \mathscr{R}_{\lambda_i}, \qquad i = 0, 1, 2$$

$$\frac{(F_0 - F_v)[W] - F_0[W]_0 + F_v(1000/18)}{\rho_m V} = \mathscr{R}_W$$

$$[W] = \frac{1}{1.8}\{1.76 - 0.0060\,(T - 273)\}, \qquad \text{Jacobs–Schweigman}$$

Closure conditions: $[P_2] = [P_3] = [P_1]$

$$\lambda_3 = \frac{\lambda_2(2\lambda_2\lambda_0 - \lambda_1^2)}{\lambda_1\lambda_0}, \qquad \lambda_{3x} = \frac{\lambda_{2x}(2\lambda_{2x}\lambda_{0x} - \lambda_{1x}^2)}{\lambda_{1x}\lambda_{0x}}$$

[a] $\mathscr{R}_i$ is the rate of generation of i by chemical reaction, given in Table 7.3. Similar equations can be written for $[P_{nx}]$, $[C_2]$, and λ_{ix}. $\rho_m V = (F_0 - F_v)\bar{\theta}$; T in K.

W can be solved for the variable $F_0/(F_0 - F_v)$. These nonlinear equations have been solved using Brown's technique,[31] and Fig. 7.19 presents some typical results.[25] Similar results obtained earlier by Jacobs and Schweigman[36] have been found to agree with some industrial data, details of which, however, are not presented. The variation of the polydispersity index with $\bar{\theta}$ for the HCSTR *with vaporization of W* differs from that in the absence of vaporization, as shown in Fig. 7.10.

The above model for the top of the VK column can be improved by using either better or more fundamental equations for vapor–liquid equilibrium instead of the empirical Eq. (7.7.1). Fukumoto[38] suggests the use of the following equation based on experimental thermodynamic equilibrium data on nylon $6\text{-}C_1\text{-}W$ systems:

$$[W]\ (\text{moles/kg}) = P_T\ (\text{mm Hg}) \exp\left\{\frac{8220}{T(\text{K})} - 24.0734\right\} \qquad (7.7.2)$$

where P_T is the total pressure. Table 7.7 compares the predictions of $[W]$ at equilibrium when P_T is 1 atm. It is observed that these two equations

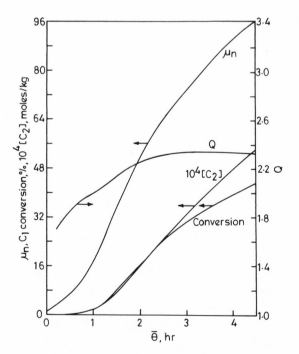

Figure 7.19. Results[25] for the top zone of the VK column as modeled in Fig. 7.18. Temperature = 260°C, $[C_1]_0 = 8.407$ moles/kg, $[W] = 0.11$ moles/kg (equilibrium). $[P_{1x}]_0 = 0$.

TABLE 7.7. Comparison of Equilibrium
Water Content at 1 atm

T (°C)	[W] (moles/kg)	
	Eq. (7.7.1)	Eq. (7.7.2)
220	0.244	0.465
240	0.178	0.242
260	0.111	0.133

yield similar values of [W], particularly at higher temperatures. Since the number of average chain length, μ_n, is sensitive to the water content in the reaction mass, there is a need to develop more precise correlations.

Recently, some research[34,39,40] has been directed towards developing a better theoretical analysis of vapor–liquid equilibrium in nylon 6 reactors. Tai et al.[34] have used the following relationships:

$$\log_{10} P_W^o = -\frac{2080}{T(K)} + 8.55 \qquad (7.7.3a)$$

$$\log_{10} P_{C_1}^o = -\frac{3280}{T(K)} + 9.03 \qquad (7.7.3b)$$

$$\log_{10}\{x_W/P_W\} = \frac{3570}{T(K)} - 11.41 \qquad (7.7.3c)$$

$$\log_{10} P_{C_1} = -\frac{4100}{T(K)} + 9.6 \qquad (7.7.3d)$$

$$x_W = \frac{[W]}{[W] + [C_1] + [C_2] + \lambda_0 + \lambda_{0x}} \qquad (7.7.3e)$$

$$P_T = P_W + P_{C_1} \qquad (7.7.3f)$$

where P_W^o and $P_{C_1}^o$ (in mm Hg) are the vapor pressures of pure water and ε-caprolactam, P_W and P_{C_1} (mm Hg) are the partial pressures of caprolactam and water over a nylon 6-C_1-W mixture [Eq. (7.7.3c) represents some average over several practical concentration ranges of the individual components], and x_W is the mole fraction of water in the liquid phase. Equations (7.7.3c), (7.7.3d), and (7.7.3f) can be solved to obtain x_W or [W] at a desired total pressure, P_T, and temperature. For $P_T = 1$ atm and $T = 260°C$, this gives $x_W = 0.01318$ or [W] $\simeq 0.105$ mole/kg (for a caprolactam conversion of about 8%–10% usually present at the top of the VK column). This may be compared with the equilibrium value of 0.111 moles/kg using Eq. (7.7.1). The need for more detailed studies on vapor–liquid equilibrium is apparent in view of the extreme sensitivity of μ_n to [W]. Also models discussed in Chapter 5 may be extended to apply to nylon 6 reactors with vaporization present.

Gupta and Gandhi[39,40] suggest the use of

$$\ln\{P^{\circ}_{C_1}\,(\text{atm})\} = 13.006 - \frac{7024.023}{T(\text{K})} \qquad (7.7.4a)$$

$$\ln\{P^{\circ}_{W}\,(\text{mm Hg})\} = 18.3036 - \frac{3816.44}{T(\text{K}) - 46.8} \qquad (7.7.4b)$$

to represent experimental vapor-pressure data.[41,42] These are much better than Eqs. (7.7.3a) and (7.7.3b). These may be used along with Eq. (7.7.3f) and the following, more detailed equations:

$$P_{W} = x_{W} P^{\circ}_{W}(T)\gamma_{W}(T, x_{W}, x_{C_1}, \lambda_0, \lambda_{0x}) \qquad (7.7.5a)$$

$$P_{C_1} = x_{C_1} P^{\circ}_{C_1}(T)\gamma_{C_1}(T, x_{W}, x_{C_1}, \lambda_0, \lambda_{0_x}) \qquad (7.7.5b)$$

where x_{C_1} is the mole fraction of caprolactam and γ_W and γ_{C_1} represent the activity coefficients of W and C_1 in the liquid phase. The activity coefficients are functions of the temperature, mole fraction of W and C_1, and the concentration of the polymer, $\lambda_0 + \lambda_{0x}$, in the liquid phase, and are much lower[43,44] than the value of unity given by Raoult's law. Wilson's equation for the γ's has been used, with the constants determined by curve fitting some experimental vapor–liquid equilibrium data of Giori and Hayes.[43] Unfortunately, there is quite a large amount of scatter in the data so plotted, and much more work needs to be done to get good expressions for γ_W and γ_{C_1}.

Another possible improvement in the reactor model of the top of the VK tube is the use of a sequence of two or more HCSTRs with backmixing as well as vapor flow. Figure 7.20 shows the notation used. The compositions of the liquid in the two HCSTRs differ, and consequently, there would be a difference in the compositions of the equilibrium vapors. An amount αF_1 kg/hr of liquid having the composition of the liquid in HCSTR 1 (in addition to F_1) flows into the second HCSTR, while an equal amount of liquid having the composition of the liquid in HCSTR 2 flows back to reactor 1. These two streams represent the backmixing present in the top of the VK column. Mole balance equations can be written for this system[40] and solved using appropriate equilibrium relationships. Figure 7.21

shows[39,40] μ_n for $\alpha = 0$ and $\alpha = 0.5$ for the two HCSTRs, as well as for the isothermal PFR following it. It is observed that even though the results for the first HCSTR differ significantly depending on the value of α, the μ_n for the final product is almost the same for the two cases studied. There is, however, a significant difference in the values of Q obtained for the two values of α. Thus, if one is interested in the MWD of the product, the second model is recommended.

In contrast to the modeling of the top of the VK column, wherein the mass transfer resistance is negligible, a more detailed analysis of the mass transfer of W is necessary for modeling reactors of the type shown in Fig. 7.17. The exact modeling of stage II of such a reactor is extremely cumbersome. It involves concepts of reactive distillation[45,46] and no study along these lines is available for polymerization systems yet. In one of the studies,[20] however, co-current passage of inert gas (instead of the more exact countercurrent flow) has been assumed, as shown in Fig. 7.22. As discussed in

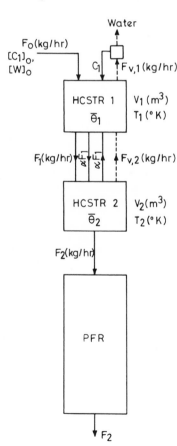

Figure 7.20. Improved model for top of VK column. Solid lines represent liquid flow, dashed lines represent vapor flow.

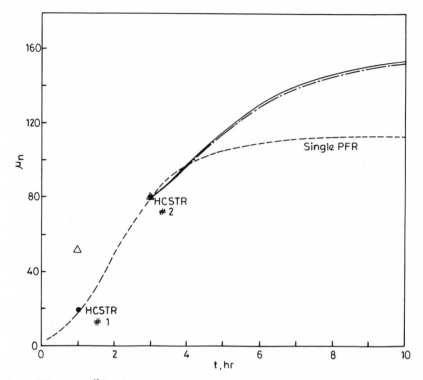

Figure 7.21. Results[40] for the reactor configuration shown in Fig. 7.20. Solid lines and circles: $\alpha = 0$; — · — and triangles: $\alpha = 0.5$. $T_1 = 230°C$, $T_2 = 253.5°C$, PFR isothermal at 253.5°C for both these cases. Feed to first reactor is C_1 and 0.44 (moles/kg) of W. $F_0 = 1000$ kg/hr. Dashed line represents behavior of a single, nonvaporizing PFR at 253.5°C starting from same feed as that to HCSTR1. (Reprinted from Ref. 40 with permission of Professor K. S. Gandhi.)

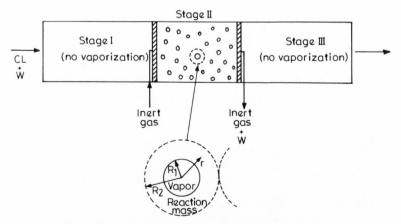

Figure 7.22. Model for co-current water removal in Stage II.

Section 5.3, a region of the reaction mass can be associated with each vapor bubble and water can be assumed to diffuse towards the vapor–liquid interface, $r = R_1$. Real systems would have some amount of mixing because of which the results of this simplified model would not hold exactly. The corresponding equations can be derived easily from those given in Table 5.1 (for spherical polar coordinates). Some typical results are shown in Fig. 7.23. It is assumed that stage II ends when the spatial-average water concentration becomes some fraction (say f) of the water concentration at the beginning of this stage. The values of the radii R_1 and R_2 have been assumed somewhat arbitrarily but cover the range of likely values. The

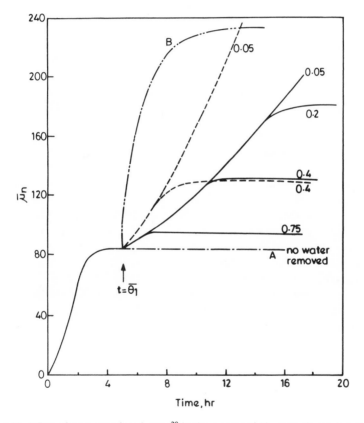

Figure 7.23. Effect of mass transfer of water[20] in the reactor of Fig. 7.22. Curve A represents no water removal while curve B represents *instantaneous* removal of *all* water present *at* $t = \bar{\theta}_1$ before sending directly to stage III. $T = 265°C$ in all three stages. Stage I: $[W]_0 = 0.58$ moles/kg, $[P_{1x}]_0 = 0$, $\bar{\theta}_1 = 4.67$ hr, $\mu_n(\bar{\theta}_1) = 83.8$. Stage II: (a) $R_1 = 3.79$ mm, $R_2 = 22.76$ mm (solid lines); (b) $R_1 = 2.68$ mm, $R_2 = 16.1$ mm (dashed lines); $\mathcal{D}_W = 0.9 \times 10^{-4}$ m²/hr, $[W]^* = 0$ or 0.001 moles/mole CL (same results in either case). Stage III: Feed is well mixed having uniform concentrations equal to the spatial average concentrations at the end of stage II.

value of the interfacial water concentration, $[W]^*$, which actually varies with time and depends on thermodynamic equilibrium between the liquid at $r = R_1$ and the vapor, has been taken as a constant as a first approximation, and has been treated as a parameter. The value of the diffusivity of water $\mathcal{D}_W$ in the liquid medium is that obtained experimentally (by curve-fitting some experimental results[47] with theoretically predicted values). It is found that the increase in μ_n in stage III is not very significant and that it is advantageous to conduct polymerization with a continuous removal of water till the desired value of μ_n is attained. It may be added that some controversy exists in the literature[44,47] on the value of $\mathcal{D}_W$ to be used (with values differing by several orders of magnitude) and some more experimental data needs to be taken before this is resolved.

While modeling the lower part of the VK tube as a PFR, isothermal conditions have been assumed till now. In order to improve the model, the following energy balance equation must also be solved simultaneously (see Fig. 7.14 for notation):

$$v_z(r)\frac{\partial T}{\partial z} = \alpha_m\left(\frac{\partial^2 T}{\partial r^2} + \frac{1}{r}\frac{\partial T}{\partial r}\right) + \frac{1}{c_{p,m}}\sum_{i=1}^{5}(-\Delta H_i)\mathcal{R}_i^* \qquad (7.7.6)$$

$$\underset{\text{convection)}}{\text{(axial}} \qquad\qquad \underset{\text{conduction)}}{\text{(radial}} \qquad\qquad \underset{\text{generation)}}{\text{(heat}}$$

where the axial conduction and radial convection terms have been neglected. In Eq. (7.7.6), $\mathcal{R}_i^*$ is the rate (in moles/kg hr) of the ith reaction in Table 7.1, ΔH_i is the corresponding heat of reaction (Table 7.2), $v_z(r)$ (in m/hr) is the velocity at radial position r, and α_m and $c_{p,m}$ are the thermal diffusivity (m^2/hr) and the specific heat of the reaction medium, respectively. The appropriate boundary condition at $r = R$ accounts for the heat supplied (or removed) from the jacket. Equation (7.7.6) must be solved with appropriate mole balance equations similar to those in Section 7.6, since the concentrations will also be functions of both z and r. For a plug flow reactor wherein there are no radial variations in the temperature or velocity across a section, a simpler semi-integral energy balance equation must be used instead of Eq. (7.7.6):

$$c_{p,m}v(dT/dz) = \sum_{i=1}^{5}(-\Delta H_i)\mathcal{R}_i^* + 2U(T_j - T)/(\rho_m R) \qquad (7.7.7)$$

where T_j is the jacket temperature, U is the overall heat transfer coefficient, and ρ_m is the density of the reaction medium. In Eq. (7.7.6) or (7.7.7), the following correlations may be used:[34,36]

$$k_m \text{ (thermal conductivity)} = 0.21\frac{W}{m\,K} \qquad (7.7.8a)$$

$$\rho_m \, (\text{kg/m}^3) = 1000\{1.0065 + 0.0123[\text{C}_1]$$

$$+ [T(\text{K}) - 495](0.00035 + 0.00007[\text{C}_1])\}^{-1} \qquad (7.7.8\text{b})$$

$$c_{p,m}\left(\frac{\text{cal}}{\text{kg K}}\right) = 659.3\frac{[\text{C}_1]}{[\text{C}_1]_0} + \left(1 - \frac{[\text{C}_1]}{[\text{C}_1]_0}\right)\{486.1 + 0.337\,T(\text{K})\} \qquad (7.7.8\text{c})$$

$$\alpha_m = k_m/(\rho_m c_{p,m}) \qquad (7.7.8\text{d})$$

Equation (7.7.6) has been solved with appropriate mole balance equations to simulate nonisothermal, nonvaporizing tubular reactors for some simplified cases,[34] but more work needs to be done in this area. Particularly important is the need to develop good correlations for the viscosity of the reaction medium as a function of temperature and conversion, since this influences the heat transfer coefficient quite critically. Most simulation studies[34,36] have approximated $U = 0$, i.e., the reactor operates adiabatically.

In this section an attempt has been made to model a few industrial nylon 6 reactors. Several of the concepts developed in earlier chapters have been made use of, and several gaps in the current state of information available have been pointed out, both in terms of concepts and in terms of estimation of physicochemical data. Several other types of reactors are also used industrially and a few are given as modeling exercises.

7.8. OPTIMIZATION OF NYLON 6 REACTORS

In the preceding sections, it has been shown that high water concentrations must be used at the beginning of the nylon 6 polymerization to speed up the conversion, followed by low concentrations of water to obtain a high value ($\sim$150–200) of μ_n. Some typical results on cascades of several reactors accomplishing this have already been presented. In these studies,[17,32,33] several reactor sequences were considered and the corresponding total reaction time required to give polymer of a desired μ_n was computed, thus giving an idea of the optimal set up.

In this section, the optimization of tubular reactors is considered, since these reactors are more common now. Most studies[48–50] which have been carried out in the past on such reactors have determined optimal profiles for the temperature and the water concentration to minimize a chosen objective function. These studies are preliminary in nature and mostly report semiquantitative results because of proprietary reasons. In addition, the kinetic schemes used in these early optimization studies incorporate several assumptions, particularly for the rate of formation of cyclic oligomers. With more complete and precise kinetic information now available,[3,12] a more systematic optimization study[51-53] can be carried out.

In a recent study,[51] a nonvaporizing plug flow reactor is considered and the optimal temperature profile $T(t)$ is obtained which minimizes the objective function

$$I\{T(t)\} = \alpha_3[C_1]_{t_f} + \int_0^{t_f} \left\{ \frac{\alpha_2}{\mu_{n,d}^2} \left(\frac{\lambda_1 + \lambda_{1x}}{\lambda_0 + \lambda_{0x}} - \mu_{n,d} \right)^2 + \alpha_1[C_2]^2 \right\} dt \quad (7.8.1)$$

subject to the constraint

$$220°C \leq T \leq 270°C \quad (7.8.2)$$

where t is z/v (see Fig. 7.14), v is the plug-flow velocity, t_f is the residence time (assumed fixed at L/v) in the PFR, $\mu_{n,d}$ is the desired value of the number average chain length, and α_1, α_2, and α_3 are the weightage factors representing the relative importance of the various terms in Eq. (7.8.1). This objective function is similar to that discussed in Section 6.2 and is a good choice for the design of a *new* reactor. The lower limiting value of the temperature in Eq. (7.8.2) is the melting point of nylon 6 (which ensures one-phase polymerization), while the upper limiting value is the boiling point of caprolactam at 1 atm pressure, and is chosen to avoid too much vaporization.

Equations for the Hamiltonian, the adjoint variables, and the changes in $T(t)$ can be easily written using Table 6.1, keeping in mind that the rate constants depend both on the acid end group concentration $\sum_{n=1}^{\infty} ([P_n] + [P_{nx}])$ or $(\lambda_0 + [P_{1x}]_0)$ as well as the temperature.

The following are chosen as "reference" conditions: $\alpha_1 = 10^6$, $\alpha_2 = 10^3$, $\alpha_3 = 50$, $\mu_{n,d} = 140$, $t_f = 10$ hr, $[C_1]_0 = 8.8$ mole/kg, $[W]_0 = 0.16$ mole/kg, $[P_{1x}]_0 = 0$; and the effect of varying these design variables on the optimal temperature profile[51] is obtained. Figure 7.24 shows the optimal temperature profile for the reference run (curve 1). It is observed that the temperature must be maintained at the highest permissible value in the initial stages of the reactor, after which it must be lowered and then, increased again in the last stages. The study shows that the high initial temperatures in the beginning lead to a rapid increase in the μ_n of the polymer [and are required by the first term in the integral in Eq. (7.8.1)]. The drop in the temperature is necessary to prevent the cyclic dimer concentration from becoming too high [see Fig. 7.7; this is required by the last term in the integral in Eq. (7.8.1)] and the final increase in the temperature is to increase the monomer conversion to reasonably high values (determined by the $\alpha_3[C_1]_{t_f}$ in I). The choice of the three parameters α_1, α_2, and α_3 in the objective function

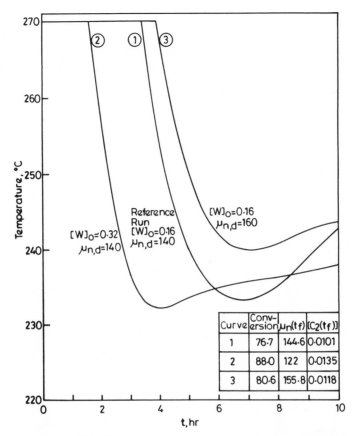

Figure 7.24. Optimal temperature profiles[51] for nylon 6 polymerization. Parameter values as in text.

determines the extent of the drop or increase in the temperature and its location.

Figure 7.24 also shows the effect of increasing the feed water concentration from 0.16 to 0.32 moles/kg (curve 2). Lower overall temperatures are now required though the objective function is minimized with a lower value of $\mu_n(t_f)$ of 122 and a higher monomer conversion of 88%. The effect of increasing $\mu_{n,d}$ is to increase the overall temperatures as shown by curve 3 in Fig. 7.24. The effect of changing the residence time,[51] t_f, is shown in Fig. 7.25. An increase in t_f, keeping $\mu_{n,d}$ at 140, leads to lower temperatures in the reactor, though the final conversion, $\mu_n(t_f)$, and cyclic dimer concentration remain unaffected. Clearly, the existence of a global optimum for a chosen set of $\mu_{n,d}$, α_1, α_2, α_3 and $[W]_0$ is indicated.

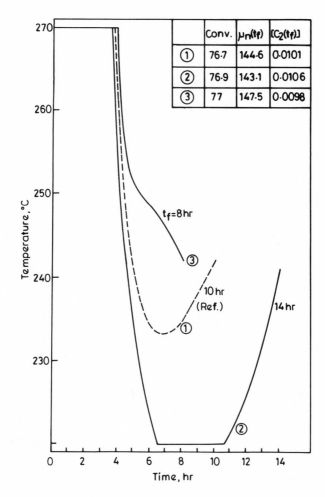

	Conv.	$\mu_n(t_f)$	$[C_2(t_f)]$
①	76.7	144.6	0.0101
②	76.9	143.1	0.0106
③	77	147.5	0.0098

Figure 7.25. Effect of changing t_f on the optimal temperature profile.[51] Parameter values as in text.

The effect of adding P_{1x} and altering the various weightage factors has also been studied.[51] A close study of the values of the objective function for the various cases suggests that the best reactor for obtaining polymer of $\mu_n(t_f)$ close to 140 and with α_1, α_2, and α_3 given by the reference values is when $t_f = 10$ hr, $[P_{1x}]_0 = 0.02$ mole/kg, and $[W]_0 = 0.13$ mole/kg. Under these conditions, the final monomer conversion is 80.6%, $\mu_n(t_f) = 134.2$, and $[C_2]_{t_f} = 0.0114$ mole/kg. If the design engineer can tolerate this deviation of $\mu_n(t_f)$ from the desired value of 140, this represents close to a globally optimized reactor.

Another optimization problem which is more relevant for reactors in *operation* (where there is less flexibility than in the *design* of new reactors) is one where the values of μ_n and $[C_2]$ at $t = t_f$ are specified *exactly*, i.e., there are end-point constraints of the following type:

$$\mu_n(t_f) = \mu_{n,d} \tag{7.8.3a}$$

$$[C_2]_{t_f} = [C_2]_d \tag{7.8.3b}$$

in addition to *local* constraints on temperature as given by Eq. (7.8.2). The objective function to be minimized may be selected as

$$I\{T(t)\} = [C_1]_{t_f}^2 \tag{7.8.4}$$

Figure 7.26 shows some typical results[52] obtained using algorithms suggested by Bryson and Ho.[54] It is observed that the temperature profiles are qualita-

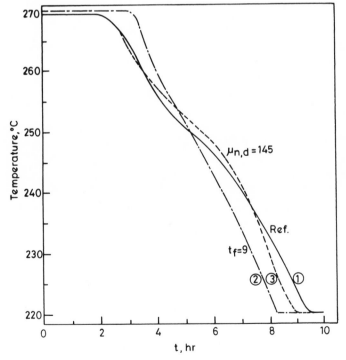

Figure 7.26. Optimal temperature profiles for the end-point constrained problem.[52] Reference conditions: $[C_1]_0 = 8.8$ moles/kg, $[W]_0 = 0.16$ moles/kg, $[P_{1x}]_0 = 0$, $\mu_{n,d} = 140$, $[C_2]_d = 0.01$ moles/kg, $t_f = 10$ hr.

tively different from those in Figs. 7.24 and 7.25. Similar results have been obtained by Mochizuki and Ito,[49] who used a different mechanism for cyclic oligomer formation. Naudin ten Cate[48] also obtained qualitatively similar results as in Fig. 7.26, using Eq. (7.8.3a) as the only end-point constraint, and minimizing both $[C_1]_{t_f}$ and the cyclic oligomers at the end of the reactor. Both these groups of workers, however, have presented semiquantitative results only, for proprietary reasons. More recently,[53] the reaction time, t_f, has been considered as a variable, instead of being specified *a priori*. The optimization algorithm becomes quite complicated in such cases. Optimal temperature profiles using different objective functions, "stopping" conditions, and end-point constraints have been obtained.

Hoftyzer *et al.*[50] have used the kinetic scheme incorporating only the three major reactions, ring-opening, polycondensation, and polyaddition in Table 7.1 and minimized the reaction time t_f for obtaining polymer having $\mu_{n,d} = 140$. They used *two* control variables and obtained the optimal temperature profile (with $T_{max} = 270°C$, $T_{min} = 220°C$) and the optimal water concentration profile (with $[W]_{max} = 1 \text{ mole/kg}$, $[W]_{min} = 0.2 \text{ mol/kg}$) using dynamic programming. Their semiquantitative results indicate that $[W]$ should be maintained at $[W]_{max}$ in the beginning and then should be decreased at $t = t_1^*$ immediately to $[W]_{min}$, while the temperature should be at T_{max} in the beginning ($t \leq t_2^*$) and then be decreased monotonically. Use of the optimization algorithm of Chapter 6 for this case (instead of dynamic programming) where $[W]$ and T are the two control variables, reveals that the results are extremely sensitive to $[W]$ and severe numerical problems are encountered.

In all the optimization studies discussed above, the optimal temperature (and $[W]$) profiles are obtained without consideration of heat (and mass) transfer aspects in the reactor. In other words, the reaction mass is maintained at these computed temperatures somehow by supplying or removing heat (or water) at rates computed by appropriate energy balance (or mole balance) equations. Relatively little work exists on the optimization of nylon 6 tubular reactors with both mass and energy transfer accounted for.

7.9. CYCLIZATION REACTIONS IN NYLON 6 POLYMERIZATION

In the discussion so far, the reactions associated with only the cyclic dimer have been considered. It is known that higher cyclic oligomers C_n are also present in the reaction mass and these need to be removed by an energy-intensive, hot water extraction process. Any simulation or optimization study, therefore, must incorporate reactions associated with these compounds. Relatively little kinetic information exists on these reactions and most optimization studies have used very crude models to account for them.[48,49,55] In this section, the available information on the kinetic and equilibrium constants for cyclization of nylon 6 oligomers is presented.

Semlyen et al.[56,57] and other workers[58-60] have obtained the equilibrium constants for the following cyclization reaction:

$$P_{n+m} \underset{}{\overset{K'_{C_n}}{\rightleftharpoons}} P_m + C_n \qquad (7.9.1)$$

[which can replace Eq. (7.3.1a)], for n up to 6, using gel permeation chromatography measurements. Their values of K'_{C_n} are shown[61] in Fig. 7.27. The data can be curve-fitted by equations of the following types:[1,19]

$$\log_{10}\{K'_{C_n} \text{ (moles/kg)}\} = \frac{a(n)}{T(\text{K})} + b(n), \qquad T > 483 \text{ K} \qquad (7.9.2)$$

$$K'_{C_n} \text{ (moles/kg)} = \frac{a^*(T)}{n^2} \qquad (7.9.3)$$

with $a(n)$, $b(n)$, and $a^*(T)$ obtained empirically.

Recently, some theoretical modeling[62-64] has been attempted to explain the variation of K'_{C_n} with n. Jacobson and Stockmayer[62] have mentioned that in order to obtain C_n by reaction (7.9.1), P_{n+m} must first break into P_n and P_m (reaction 2 of Table 7.1) and then P_n must cyclize. The entropy of the first step, ΔS_1, can be written as

$$\Delta S_1 = \mathcal{k} \ln(V/V_s) \qquad (7.9.4)$$

where $\mathcal{k}$ is the Boltzmann constant. The term in parentheses in Eq. (7.9.4) represents the fact that after reaction, the cleaved chain-ends can be anywhere in the volume V of the reaction mass, whereas just before reaction, they must lie within a small volume V_s. In order to estimate the entropy change for the cyclization of P_n, it is observed that the two chain ends of P_n must turn around and lie within a small volume V_s of each other. Using concepts of chain statistics[35] (also refer to Section 4.7) the fraction of molecular configurations that satisfy this criterion can be found to be

$$\text{Prob} = \left(\frac{3}{2\pi\nu n l^2}\right)^{3/2} V_s \qquad (7.9.5)$$

where ν is the number of backbone *bonds* in a repeat unit, each of length l. The entropy change of the cyclization of P_n is then

$$\Delta S_2 = \mathcal{k} \ln(\text{Prob}/n) \qquad (7.9.6)$$

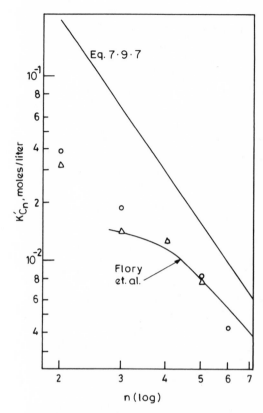

Figure 7.27. Experimental values of K'_{C_n} at 252°C. $\bigcirc$: Data of Semlyen et al.[56,57]; $\triangle$: data of Zahn et al.[58-60] To obtain K'_{C_n} in moles/kg, divide by 0.93. Plots of Eq. (7.9.7) and computed results of Flory et al.[61] also shown. (Reprinted with permission from Ref. 61. Copyright 1976, American Chemical Society, Washington, DC.)

where n must be used because of the n symmetrical $-\mathrm{CONH}-$ groups in C_n. The total entropy for reaction (7.9.1), thus, is the sum of ΔS_1 and ΔS_2 and is $\mathit{k}\ln\{(3/2\pi\nu)^{3/2}V/(l^3n^{5/2})\}$. If the energy change associated with Eq. (7.9.1) is neglected (since the same number of bonds is involved on both sides), the entropy change is related to the change in free energy, and thus to the equilibrium constant, as

$$K'_{C_n} = \exp(\Delta S/\mathit{k}) = \left(\frac{3}{2\pi\nu}\right)^{3/2} \frac{1}{l^3 n^{5/2} N_{\mathrm{Av}}} \qquad (7.9.7)$$

where the Avogadro number N_{Av} has been substituted for $1/V$ to obtain K'_{C_n} in conventional units of moles/liter. Equation (7.9.7) has been plotted in Fig. 7.27 and is found to overestimate K'_{C_n}.

Equation (7.9.7) assumes that the backbone bonds are completely free to rotate about each other. Recently, Flory et al.[61,64,65] have used the rotational isomeric state model to replace Eq. (7.9.5). In their model, they

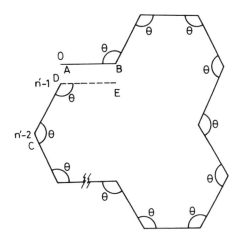

Figure 7.28. The location of bonds during cyclization.

have accounted for the presence of bond angle restrictions and steric hindrances and have *also* incorporated the constraint that for cyclization to occur, the first bond AB (see Fig. 7.28) and the last bond CD must lie at an angle θ, the bond angle (by postulating a hypothetical bond DE which must lie at angle θ to CD and must be "almost" parallel to AB). Their computed results are shown in Fig. 7.27 and are found to explain experimental results much better.

It should be emphasized that K'_{C_n} as used in this section is written in terms of molecular species such that

$$K'_{C_n} = \frac{[P_m][C_n]}{[P_{n+m}]} \tag{7.9.8}$$

If Eq. (7.9.1) is written as

$$P_{n+m} \underset{nk'_{C_n}}{\overset{k_{C_n}}{\rightleftharpoons}} P_m + C_n \tag{7.9.9}$$

such that k_{C_n} and k'_{C_n} represent the rates associated with *functional groups*, the corresponding equilibrium constant is obtained as[18,19]

$$K_{C_n} \equiv \frac{k_{C_n}}{k'_{C_n}} = nK'_{C_n} \tag{7.9.10}$$

Some preliminary simulations using the empirical Eq. (7.9.3) have been carried out.[19] It is observed that the conversion of the monomer is essentially unchanged by the incorporation of the cyclization reactions but that μ_n reduces by almost 10%. Figure 7.29 shows[19] the variation of the cyclic

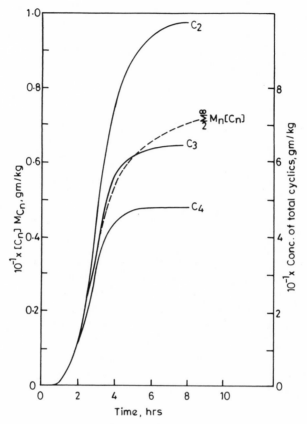

Figure 7.29. Variation of cyclic oligomer concentrations[19] in nylon 6 polymerization in isothermal sealed tubes. $[W]_0 = 0.44$ gmoles/kg, $[P_{1x}]_0 = 0$, $T = 235°C$, $a^*(T) = 0.1692$.

oligomer concentrations for a typical isothermal, nonvaporizing batch reactor. Since k'_{C_n} [Eq. (7.9.9)] has been assumed equal to k_3 of Table 7.1 in this study, the time variation of $[C_2]$, $[C_3]$, ..., etc. may be slightly in error. The total cyclic compounds are observed to be around 6%–7% by weight, which is slightly higher than industrially encountered values. More recently,[9] experimental HPLC data on the equilibrium concentrations of C_3 to C_6 at different values of T and $[W]_0$ have been reported and it is expected that in the near future, more precise rate and equilibrium constants will become available to explain these.

7.10. POLYMERIZATION OF NYLON-6,6

In addition to nylon 6, several other nylons (e.g., nylon 11, 12, 6-6, 6-10, 6-12) are also of commercial importance. Kinetic data on these are generally very scarce[66,67] and very little simulation work has been reported on their polymerization, except on nylon-6,6. This polymer is prepared in a multistage process, and the first step consists of the preparation of nylon-6,6 salt

$$H_2N(CH_2)_6-NH_2 + HOOC-(CH_2)_4-COOH \rightarrow$$

$$H_2N-(CH_2)_6NH_3^{+\ -}O-\overset{\overset{\displaystyle O}{\displaystyle \|}}{C}-(CH_2)_4-COOH \quad (7.10.1)$$

Dry adipic acid (AA) is mixed with an aqueous solution of hexamethylene diamine (HMDA) and the salt (MP 195°C) is precipitated by the addition of methanol. The salt is then centrifuged and an aqueous solution of about 60% salt concentration is sent to the polymerization stage. The use of the salt enables the manufacture of high molecular weight polymer, since equimolar amounts of $-NH_2$ and $-COOH$ groups are thereby ensured. The exact metering of the diamine and the adipic acid, and the subsequent make up of vaporized diamine is extremely cumbersome, even though several recent processes have been commercialized in which molten HMDA and AA are used directly and salt manufacture is avoided.[66,67]

The aqueous salt solution is heated to about 280°C at pressures of about 250 psi in an oxygen-free atmosphere. This promotes rapid polymerization. The number average molecular weight increases to about 44,000 under these conditions. Thereafter, the pressure is reduced to atmospheric and the reactor is held at about 275°C for about 1 hr to obtain nylon-6,6 of molecular weights about 16,000–18,000, which have commercial usage.

The rate of polymerization of the nylon salt has been studied by Ogata,[68,69] and is a typical example of reversible ARB polycondensation. The polymerization is represented by

$$-COOH + -NH_2 \underset{k/K}{\overset{k}{\rightleftharpoons}} -CONH- + H_2O \quad (7.10.2)$$

Ogata has found from sealed tube experiments that both k and K are functions of temperature as well as the *initial* water concentration. He suggests the use of

$$\log_{10} K = \frac{5800}{T(K)} - 8.32 \quad (7.10.3)$$

even though the experimental values of K vary by almost $\pm 20\%$ around the value given by this equation. Recently, his data have been empirically curve fitted[70,71] and it has been found that K can be represented in terms of the *initial* water concentration and temperature while the rate constant,

TABLE 7.8. Rate Constants[70] for
Nylon 6-6 Polymerization[a]

Forward rate constant

$$\log_{10} k = \frac{13.1}{[W]^{0.025}} - \frac{4830}{T}$$

Equilibrium constant

$$K = \exp\{a - b[W]_0\}$$

For $[W]_0 < 3.4$:

T (°C)	200	210	220
a	6.17	6.25	6.28
b	0.22	0.23	0.24

For $[W]_0 \geq 3.4$

T (°C)	200	210	220
a	5.207	5.35	5.41
b	0.01	0.0083	0.01

---

[a] Concentrations of water in mole per mole of repeat unit. k in liters/mol hr.

k, is represented in terms of the water concentration in the reaction mass *at the time of interest* and the temperature. The exact correlations for k and K are given in Table 7.8. Equations (2.9.2), (2.9.4), and (2.9.5) are still applicable to the polymerization of the nylon salt, provided the appropriate values of k and K are used, and these can be integrated numerically to give the MWD and moments. The results are shown in Figs. 7.30–7.32. It is observed that equilibrium is attained at relatively low values of μ_n. More work is required to obtain better correlations for k and K as well as to carry out simulations of batch polymerizations with a temperature cycle and at constant pressures (with water evaporated) to obtain higher molecular weight product.

Considerable attention has been focused on *continuous* polymerizations of nylon-6,6 in the recent past, using molten AA and HMDA. The reactions are carried out at very high pressures (1000 psi) and at 275°C for about 15–30 min, to prevent any boiling. After this, the pressure is decreased (using valves or tubes of increasing diameter[67]) to about 250–350 psi, to prevent freezing during the vaporization of steam (several kinds of equipment have been patented to achieve this[67]). The melt is then polymerized to equilibrium at atmospheric pressure, the usual residence times being about 1 hr for the last finishing stage.

There is very little information available on the design of continuous nylon-6,6 reactors. The kinetic information is untrustworthy (data of Ogata were taken only in the range of 200-220°C) and no information exists on vapor-liquid equilibrium to model the vaporization of water. This information is very crucial since the final degree of polymerization depends on the water concentration in the melt, and this in turn depends on the pressure and temperature, as well as on the concentration of the polymer and the monomers in the liquid phase. Detailed simulation and optimization of

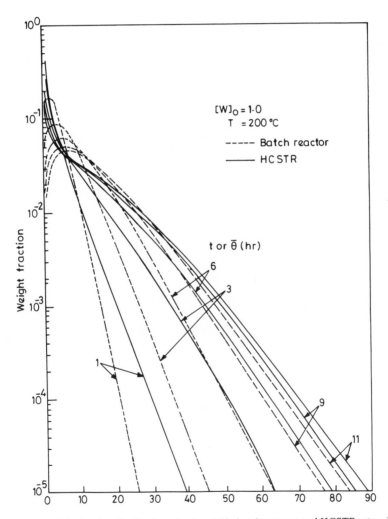

Figure 7.30. Weight fraction distribution of nylon-6,6 in batch reactors and HCSTRs at various reaction times[70,71] $[W]_0$ in moles/mole repeat unit.

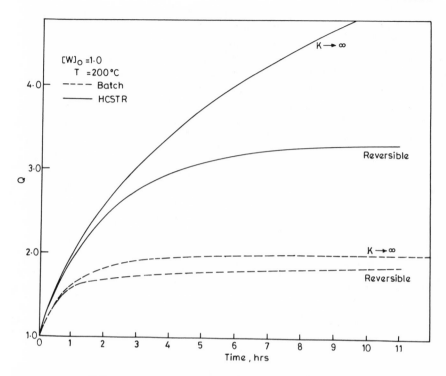

Figure 7.31. PDI[70,71] for nylon-6,6 polymerization for batch reactors and HCSTRs. Results for $K \to \infty$ (irreversible) also shown for comparison. Note that $Q \to 2$ for the irreversible case, even though k varies with time owing to its [W] dependence.

nylon-6,6 reactors must wait till such information becomes available in the published literature.

7.11. CONCLUSIONS

In this chapter, the polymerization of two important nylons has been discussed. It is found that the reaction mechanism is more complex compared to the simpler scheme discussed in previous chapters. Detailed modeling of industrial reactors has been presented in which important side reactions have been included and experimentally determined rate and equilibrium constants have been used. An attempt is made to optimize these reactors and it is observed that different temperature profiles are obtained depending on the objective function used. The modeling of postreactor operations for nylon 6, however, is not discussed since detailed modeling of the solid state polymerization involved is still in its infancy[72] (see Section

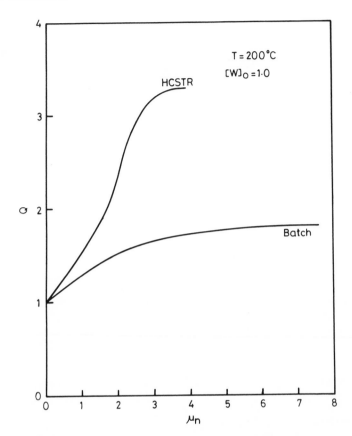

Figure 7.32. Polydispersity index[70,71] for nylon-6,6 polymerization in batch reactors and HCSTRs.

5.4). Similarly, it is seen that detailed information and modeling of nylon-6,6 reactor is not yet available. There exists a need for a considerable amount of research in these areas.

REFERENCES

1. H. K. Reimschuessel, Nylon 6. Chemistry and mechanisms, *J. Polym. Sci. Macromol. Rev.* **12**, 65–139 (1977).
2. H. K. Reimschuessel, in *Ring Opening Polymerization* (K. S. Frisch and S. L. Reegan, Eds.), 1st ed., pp. 303–326, Marcel Dekker, New York (1969).
3. K. Tai and T. Tagawa, Simulation of hydrolytic polymerization of ε-caprolactam in various reactors, *Ind. Eng. Chem., Prod. R&D* **22**, 192–205 (1983).

4. J. Sebenda, Recent progress in the polymerization of lactams, *Prog. Polym. Sci.* **6**, 123-168 (1978).

5. S. K. Gupta and A. Kumar, Simulation and design of nylon 6 reactors, *J. Macromol. Sci. Rev. Macromol. Chem. Phys.* **C26**, 183-246 (1986).

6. W. H. Carothers and G. T. Berchet, Studies on polymerization and ring formation. VIII. Amides from ε-aminocaproic acid, *J. Am. Chem. Soc.* **52**, 5289-5291 (1930).

7. P. H. Hermans, D. Heikens, and P. F. van Velden, On the mechanism of the polymerization of ε-caprolactam. II. The polymerization in the presence of water, *J. Polym. Sci.* **30**, 81-104 (1958).

8. Ch. A. Kruissink, G. M. van der Want, and A. J. Staverman, On the mechanism of the polymerization of ε-caprolactam. I. The polymerization initiated by ε-aminocaproic acid, *J. Polym. Sci.* **30**, 67-80 (1958).

9. K. Tai and T. Tagawa, The kinetics of hydrolytic polymerization of ε-caprolactam. V. Equilibrium data on cyclic oligomers, *J. Appl. Polym. Sci.* **27**, 2791-2796 (1982).

10. H. K. Reimschuessel and G. J. Dege, Polyamides: Decarboxylation and desamination in nylon 6 equilibrium polymer, *J. Polym. Sci. A-1* **8**, 3265-3283 (1970).

11. G. J. Dege and H. K. Reimschuessel, Peroxidation of caprolactam and its effect on equilibrium polymerization of cyclic dimer. *J. Polym. Sci., Polym. Chem. Ed.* **11**, 873-896 (1973).

12. Y. Arai, K. Tai, H. Teranishi, and T. Tagawa, Kinetics of hydrolytic polymerization of ε-caprolactam. 3. Formation, *Polymer* **22**, 273-277 (1981).

13. K. Tai, H. Teranishi, Y. Arai, and T. Tagawa, The kinetics of hydrolytic polymerization of ε-caprolactam, *J. Appl. Polym. Sci.* **24**, 211-224 (1979).

14. K. Tai, H. Teranishi, Y. Arai, and T. Tagawa, The kinetics of hydrolytic polymerization of ε-caprolactam. II. Determination of the kinetic and thermodynamic constants by least-squares curve fitting, *J. Appl. Polym. Sci.* **25**, 77-87 (1980).

15. F. Wiloth, Mechanism and kinetics of the polymerization of ε-caprolactam in the presence of water. X. Comparison of experimental data with the integration results of a completed reaction-kinetic system of differential equations, *Z. Phys. Chem. N.F.* **11**, 78-102 (1957).

16. S. M. Skuratov, A. A. Strepichejev, and E. N. Kanarskaja, Über die wechselseitige umwandlung von zyklischen und linearen polymeren, *Faserforsch. Textiltech* **4**, 390-392 (1953).

17. H. K. Reimschuessel and K. Nagasubramanian, On the optimization of caprolactam polymerization, *Chem. Eng. Sci.* **27**, 1119-1130 (1972).

18. S. K. Gupta, A. Kumar, P. Tandon, and C. D. Naik, Molecular weight distributions for reversible nylon-6 polymerizations in batch reactors, *Polymer* **22**, 481-487 (1981).

19. S. K. Gupta, C. D. Naik, P. Tandon, and A. Kumar, Simulation of molecular weight distribution and cyclic oligomer formation in the polymerization of nylon-6, *J. Appl. Polym. Sci.* **26**, 2153-2163 (1981).

20. S. K. Gupta, A. Kumar, and K. K. Agrawal, Simulation of three-stage nylon-6 reactors with intermediate mass transfer at finite rates, *J. Appl. Polym. Sci.* **27**, 3089-3101 (1982).

21. M. Cave, M.S. dissertation, Imperial College, London, UK (1975).

22. H. M. Hulburt and S. Katz, Some problems in particle technology: A statistical mechanical formulation, *Chem. Eng. Sci.* **19**, 555-574 (1964).

23. K. W. Min, On the application of fractional moments in determining average molecular weight, *J. Appl. Polym. Sci.* **22**, 589-591 (1978).

24. A. Gupta and K. S. Gandhi, Molecular weight distribution in batch hydrolytic polymerization of caprolactam, *J. Appl. Polym. Sci.* **27**, 1099-1104 (1982).

25. A Ramagopal, A. Kumar, and S. K. Gupta, Computational scheme for the calculation of molecular weight distributions for nylon 6 polymerization in homogeneous, continuous flow stirred tank reactors with continuous removal of water, *Polym. Eng. Sci.* **22**, 849-856 (1982).

26. S. K. Gupta and A. Kumar, Simulation of step growth polymerizations, *Chem. Eng. Commun.* **20**, 1-52 (1983).
27. K. Tai, Y. Arai, H. Teranishi, and T. Tagawa, The kinetics of hydrolytic polymerization of ε-caprolactam. IV. Theoretical aspect of the molecular weight distribution, *J. Appl. Polym. Sci.* **25**, 1789 (1980).
28. S. K. Gupta, D. Kunzru, A. Kumar, and K. K. Agrawal, Simulation of nylon 6 polymerization in tubular reactors with recycle, *J. Appl. Polym. Sci.* **28**, 1625-1640 (1983).
29. Allied Chem. Corp., British Patent, 938652 (1963).
30. S. C. Chu and I. C. Twilley, 6th Annual Synthetic Fibers Symp., AIChE, Virginia, 1969.
31. K. Brown, in *Numerical Solutions of Systems of Nonlinear Algebraic Equations* (C. D. Byrne and C. A. Hall, Eds.), 1st ed., pp. 281-348, Academic, New York (1973).
32. K. Nagasubramanian and H. K. Reimschuessel, Caprolactam polymerization: Polymerization in backmix flow systems, *J. Appl. Polym. Sci.* **16**, 929-934 (1972).
33. M. V. Tirrell, G. H. Pearson, R. A. Weiss, and R. L. Laurence, An analysis of caprolactam polymerization, *Polym. Eng. Sci.* **15**, 386-393 (1975).
34. K. Tai, Y. Arai, and T. Tagawa, The simulation of hydrolytic polymerization of ε-caprolactam in various reactors, *J. Appl. Polym. Sci.* **27**, 731-736 (1982).
35. A. Kumar and S. K. Gupta, *Fundamentals of Polymer Science and Engineering*, 1st ed., Tata McGraw-Hill, New Delhi, India (1978).
36. H. Jacobs and C. Schweigman, Mathematical model for the polymerization of caprolactam to nylon-6, Proc. 5th Eur./2nd Intl. Symp. Chem. Rxn. Eng., Amsterdam, 2-4 May, 1972, pp. B7.1-26.
37. Vereinigte Glanzstoff Fabriken, German Patent, 1167021 (1962).
38. O. Fukumoto, Equilibria between polycapramide and water, *J. Polym. Sci.* **22**, 263-270 (1956).
39. A. Gupta and K. S. Gandhi, in *Frontiers of Chem. Rxn. Eng.* (L. K. Doraiswamy and R. A. Mashelkar, Eds.), 1st ed., pp. 667-681, Wiley Eastern, New Delhi, India (1984).
40. A. Gupta, M. Tech. Dissertation, IIT, Kanpur, India (1981).
41. *Encyclopedia of Industrial Chemical Analysis*, Vol. 8, pp. 115, Wiley, New York (1971).
42. *International Critical Tables*, Vol. 3, 233 pp., McGraw-Hill, New York (1928).
43. C. Giori and B. T. Hayes, Hydrolytic polymerization of caprolactam. I. Hydrolysis—polycondensation kinetics; . . . II. Vapor-liquid equilibria, *J. Polym. Sci. A-1* **8**, 335-349, 351-358 (1970).
44. J. P. Roos, Mathematical modeling of the sorption of volatile components in Newtonian-high-viscous liquids with the aid of bubbling, *Adv. Chem. Ser.* **133**, 303-315 (1974).
45. P. Levenspiel, *Chemical Reaction Engineering*, 2nd ed., Chapt. 13, Wiley, New York (1972).
46. J. J. Carberry, *Chemical and Catalytic Reaction Engineering*, 1st ed., McGraw-Hill, New York (1976).
47. K. Nagasubramanian and H. K. Reimschuessel, Diffusion of water and caprolactam in nylon 6 melts, *J. Appl. Polym. Sci.* **17**, 1663-1677 (1973).
48. W. F. H. Naudin ten Cate, Application of the maximum principle of Pontryagin to optimize a nylon 6 continuous polymerization process, Proc. Internl. Cong. Use of Elec. Comp. in Chem. Eng., Paris, April 1973.
49. S. Mochizuki and N. Ito, Optimal polymerization temperature profile for nylon-6 with low cyclic oligomers content, *Chem. Eng. Sci.* **33**, 1401-1403 (1978).
50. P. J. Hoftyzer, J. Hoogschagen, and D. W. van Krevelen, Optimization of caprolactam polymerization, Proc. 3rd Eur. Symp. Chem. Rxn. Eng., Amsterdam, 15-17 Sept. 1964, pp. 247-253.
51. A. Ramagopal, A. Kumar, and S. K. Gupta, Optimal temperature profiles for nylon 6 polymerization in plug-flow reactors, *J. Appl. Polym. Sci.* **28**, 2261-2279 (1983).
52. S. K. Gupta, B. S. Damania, and A. Kumar, Optimization of nylon-6 reactors with end-point constraints, *J. Appl. Polym. Sci.* **29**, 2177-2194 (1984).

53. A. K. Ray and S. K. Gupta, Optimization of nonvaporizing nylon 6 reactors with stopping condition, *J. Appl. Polym. Sci.* **31**, 4529–4550 (1986).

54. A. E. Bryson and Y. C. Ho, *Applied Optimal Control*, 1st ed., Blaisdell, Waltham, Massachusetts (1969).

55. A. Mochizuki and N. Ito, The hydrolytic polymerization kinetics of ε-caprolactam, *Chem. Eng. Sci.* **28**, 1139–1147 (1973).

56. J. A. Semlyen and G. R. Walker, Equilibrium ring concentrations and the statistical conformation of polymer chains. II. Macrocyclics in nylon 6, *Polymer* **10**, 597–601 (1969).

57. J. M. Andrews, F. R. Jones, and J. A. Semlyen, Equilibrium ring concentrations and the statistical conformations of polymer chains. 12. Cyclics in molten and solid nylon-6, *Polymer* **15**, 420–424 (1974).

58. H. Spoor and H. Zahn, Eine methode zur quantitativen papierchromatographischen bestimmung von sekundären aminen und amiden. 17. Mitteilung über oligomere, *Z. Anal. Chem.* **168**, 190–195 (1959).

59. H. Zahn and G. B. Gleitsman, Oligomers and pleionomers of synthetic fiber-forming polymers, *Agnew. Chem.* **75**, 772–783 (1963).

60. M. Rothe, Polymerhomologe ringamide in polycaprolactam, *J. Polym. Sci.* **30**, 227–238 (1958).

61. M. Mutter, U. W. Suter, and P. J. Flory, Macrocyclization equilibria. 3. Poly (6-aminocaproamide), *J. Am. Chem. Soc.* **98**, 5745–5748 (1976).

62. J. Jacobson and W. H. Stockmayer, Intramolecular reaction in polycondensations. I. The theory of linear systems, *J. Chem. Phys.* **18**, 1600–1606 (1950).

63. J. A. Semlyen, Ring-chain equilibria and conformation of polymer chains, *Adv. Polym. Sci.* **22**, 41–75 (1976).

64. P. J. Flory, U. W. Suter, and M. Mutter, Macrocyclization equilibria. I. Theory, *J. Am. Chem. Soc.* **98**, 5733–5739 (1976).

65. P. J. Flory, *Statistical Mechanics of Chain Molecules*, 1st ed., Wiley, New York (1969).

66. D. C. Jones and T. R. White, in *Step Growth Polymerization* (D. H. Solomon, Ed.), 1st ed., pp. 41–94, Marcel Dekker, New York (1972).

67. D. B. Jacobs and J. Zimmerman, in *Polymerization Processes* (C. E. Schildknecht and I. Skeist, Eds.), 1st ed., pp. 424–467, Wiley, New York (1977).

68. N. Ogata, Studies on polycondensation reactions of nylon salt. I. The equilibrium in the system of polyhexamethylene adipamide and water, *Makromol. Chem.* **42**, 52–67 (1960).

69. N. Ogata, Studies on polycondensation reactions of nylon salt. II. The rate of polycondensation reaction of nylon 66 salt in the presence of water, *Makromol. Chem.* **43**, 117–131 (1961).

70. A. Kumar, S. Kuruville, A. R. Raman, and S. K. Gupta, Simulation of reversible nylon-6,6 polymerization, *Polymer* **22**, 387–390 (1981).

71. A. Kumar, R. K. Agarwal, and S. K. Gupta, Simulation of reversible nylon-6,6 polymerization in homogeneous continuous-flow stirred tank reactors, *J. Appl. Polym. Sci.* **27**, 1759–1769 (1982).

72. F. C. Chen, R. G. Griskey, and G. H. Beyer, Thermally induced solid state polycondensation of nylon 6-6, nylon 6-10 and polyethylene terephthalate, *AIChE J.* **15**, 680–685 (1969).

EXERCISES

1. Derive the equations in Table 7.3.

2. Write appropriate balance equations for the two-HCSTR model of the top of the VK column, shown in Fig. 7.20. Use Eq. (7.7.1) for vapor–liquid equilibrium.

3. Model the single HCSTR with vaporization for nylon 6 polymerization on the lines discussed in Chapter 5. Use Raoult's law for the sake of simplicity.

4. Change the kinetic scheme of Exercise 3 in Chapter 5 for nylon 6 polymerization. Also generalize the equations obtained there to a nylon 6 reactor where the pressure first builds up in the vapor space from 1 atm to about 6 atm in about 5 hr. Thereafter, a control valve releases the vapor to maintain the pressure close to 6 atm for a few more hours, till the reaction is almost complete. Use the Flory–Huggins theory to correlate the vapor-liquid equilibrium.

5. In one nylon 6 reactor, the feed flows countercurrently inside a thin coil within the reactor, in order to get preheated (and with some possibility of reaction). Thereafter, it issues out at the top and flows vertically down, as in VK columns. The pressure is maintained high enough to prevent vaporization. Write down appropriate mole and energy balance equations for this reactor and develop an algorithm to solve them.

6. Write down appropriate equations for the reactor model shown in Fig. 7.22.

7. Obtain the Hamiltonian and equations for the adjoint variables for the optimization problem posed in Eqs. (7.8.1) and (7.8.2). Then obtain the conditions for optimality. Confirm your equations with those in Ref. 51.

8. In one post reactor operation, molten product flows down the inner sides of a cylindrical tube, with some amount of stirring, using blades. A vacuum is applied to remove W and some monomer. Using concepts developed for wiped film reactors (with no "bulk") in Chapter 5, model the operation of such a finishing reactor.

POLYESTER REACTORS

8.1. INTRODUCTION

Polyesters are produced either as saturated or unsaturated polyesters. Among the saturated polyesters, polyethylene terephthalate (PET) is produced in the largest quantity, and in view of its commercial importance, a major portion of this chapter is devoted to the design of PET reactors. An unsaturated polyester resin usually consists of a polymerizable ethylenic monomer and a polyester prepared from a glycol and a dibasic acid or an anhydride. In the following, the engineering aspects of PET formation are considered first.

Most of the polyethylene terephthalate (PET) manufactured is used either as film or as fiber in the textile industry. The monomer used in the manufacture of the polyethylene terephthalate is bis-hydroxyethyl terephthalate, having the molecular structure

$$OHCH_2CH_2O-\overset{\overset{\displaystyle O}{\|}}{C}-\underset{}{\bigcirc}-\overset{\overset{\displaystyle O}{\|}}{C}-OCH_2CH_2OH \quad (BHET).$$

The latter can be manufactured through the following two routes:[1] (a) transesterification of dimethyl terephthalate (DMT); and (b) direct esterification of terephthalic acid. Most of the dimethyl terephthalate

$$CH_3O\overset{\overset{\displaystyle O}{\|}}{C}-\underset{}{\bigcirc}-\overset{\overset{\displaystyle O}{\|}}{C}OCH_3$$

for manufacturing PET is produced by the esterification of terephthalic acid with methanol where the former is derived from the air oxidation of p-xylene.

The polymerizable grade DMT can be separated from this mixture through distillation and/or crystallization. The DMT thus produced can be esterified with ethylene glycol ($OHCH_2CH_2OH$) to give bis-hydroxyethyl terephthalate as follows:

$$CH_3O\overset{\displaystyle O}{\overset{\displaystyle \|}{C}}-\!\!\!\bigcirc\!\!\!-\overset{\displaystyle O}{\overset{\displaystyle \|}{C}}OCH_3 + 2\ OHCH_2CH_2OH$$

DMT

$$\rightleftharpoons\ OHCH_2CH_2O\overset{\displaystyle O}{\overset{\displaystyle \|}{C}}-\!\!\!\bigcirc\!\!\!-\overset{\displaystyle O}{\overset{\displaystyle \|}{C}}OCH_2CH_2OH + 2\ CH_3OH$$

BHET   (8.1.1)

The above reaction is known as the transesterification reaction and it is shown later that the product formed from Eq. (8.1.1) is never BHET alone but consists of several oligomers of BHET. The usual procedure of carrying out this reaction is to dissolve DMT in ethylene glycol and react in either batch reactors or HCSTRs at about 180°C and one atmospheric pressure in presence of a suitable catalyst.[2-4] The condensation product methanol evaporates along with ethylene glycol and is continuously distilled to separate the latter, which is recycled to the reactor.[1]

The direct esterification of terephthalic acid

$$TPA,\ COOH-\!\!\!\bigcirc\!\!\!-COOH$$

was generally not preferred earlier because of the difficulties in its purification. This is because it does not melt and has very limited solubility in standard industrial solvents. However, with the improvements in modern technology, this route has been gaining importance. The overall reaction can be represented by

$$OH\overset{\displaystyle O}{\overset{\displaystyle \|}{C}}-\!\!\!\bigcirc\!\!\!-\overset{\displaystyle O}{\overset{\displaystyle \|}{C}}-OH + 2\ OHCH_2CH_2OH$$

G

TPA

$$\rightleftharpoons\ OHCH_2CH_2O\overset{\displaystyle O}{\overset{\displaystyle \|}{C}}-\!\!\!\bigcirc\!\!\!-\overset{\displaystyle O}{\overset{\displaystyle \|}{C}}OCH_2CH_2OH + 2H_2O$$

BHET

(8.1.2)

The rate of this reaction even in the presence of catalysts is small, and to produce BHET in high proportion, it is required that either excess of ethylene glycol is used or the overall reaction is driven in the forward direction through fast removal of the condensation product, water. Industrially, economic conversion levels have been achieved by using high temperatures and pressures where terephthalic acid ionizes and autocatalyzes the formation of BHET.[5,6] Sometimes, to avoid high pressures and yet obtain high reaction temperatures, one of the methods adopted is to dissolve high molecular weight inerts to lower the vapor pressure of the contents of the reaction mass. In fact, some of the patents report the use of polyethylene terephthalate as the inert to obtain a reaction temperature as high as 330°C[7] in the transesterification step.

After the monomer, BHET, is prepared (either through DMT or terephthalic acid route), fiber grade polyethylene terephthalate is manufactured by subjecting BHET to the following two stages of polymerization:

(a) *polycondensation step*, in which the BHET prepolymer is polymerized between 260 and 290°C at high vacuum (0.5–1 torr);
(b) the *final step*, in which special wiped film reactors are used.

Special reactors for these two stages are necessary because the overall reaction is equilibrium controlled in the polycondensation step and is mass transfer controlled in the final stage. In this chapter, each of these stages of polymerization is discussed in detail.

8.2. TRANSESTERIFICATION STAGE OF PET FORMATION FROM DIMETHYL TEREPHTHALATE

8.2a. Industrial Process and Various Side Reactions

The formation of bishydroxyethyl terephthalate from DMT [as in Eq. (8.1.1)] is sometimes called the transesterification reaction, and a typical industrial flow sheet of the process carrying this out is shown in Fig. 8.1. Solid DMT is first melted in tank H and is mixed with ethylene glycol in a mixer M before charging it to the reactor E (which is called an ester interchange reactor). During the reaction, methanol is produced which evaporates along with some ethylene glycol. The latter is separated in the distillation column D and recycled to the reactor. Industrially, batch reactors as well as continuous flow stirred tank reactors are used for carrying out the transesterification reaction.

In this stage, the usual molar ratio of DMT and ethylene glycol in the feed is about 1:2 and the reaction occurs mostly in the temperature range

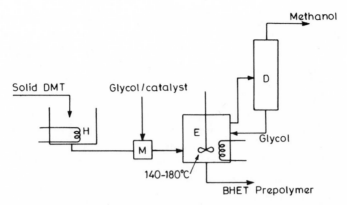

Figure 8.1. Flow sheet of the transesterification stage of PET formation.

of 140–200°C at atmospheric pressure. The use of a catalyst is common so that the reaction can be completed in a reasonable reactor residence time. The transesterification reaction has been proposed to occur through the nucleophilic attack of the hydroxyl groups of ethylene glycol upon the carbon of the ester carbonyl groups of DMT. The reaction intermediate is formed by the coordination of this carbon atom to the metal species of the catalyst. This coordination is expected to lower the electron density on the carbon of the ester carbonyl group and thus facilitate the attack of the OH groups.[4] Based upon this reaction mechanism, various catalysts have been studied in the literature and have been ordered in the literature in terms of either (a) the electronegativity of their metal species or (b) the stability of this coordination complex. A more simple, but empirical, way of grading various catalysts would be in terms of the relative rates of evolution of methanol determined experimentally,[1] as done below:

$$Zn(OAc)_2 \cdot H_2O(4.6) > Zn\ formate(4.5) > Pb(OAc)_2 \cdot 3H_2O(3.5)$$

$$> Cd(OAc)_2 \cdot 2H_2O(2.8) > PbO(2.1) > ZnSO_4(1.1) > Ca(OAc)_2 \cdot H_2O(1.0)$$

where the numbers in parentheses represent the relative rates of methanol evolution. Zinc acetate is found to be the most effective catalyst and is commonly used industrially.

On reaction of DMT with ethylene glycol (G), the $-OCH_3$ group of the former interacts with the OH group of the latter and methanol is formed as the condensation product. The reaction between the functional groups initially leads to the formation of half-esterified DMT; however, as the reaction progresses, the reaction mass consists of BHET which can react not only with itself (polycondensation reaction between two CH_2OH groups) but can esterify the unreacted as well as the half-esterified ethylene glycol.

After the reaction has progressed for some time, the reaction mass consists of oligomers of the following types:

D_n:

$$(8.2.1a)$$

H_n:

$$(8.2.1b)$$

and

P_n:

$$(8.2.1c)$$

where D_1 is DMT and P_1 is BHET. Theoretically, n could take any value, and Reference 1 states that the value of n after the transesterification stage can be as high as 3.

There have been several experimental studies[3-8] which have clearly demonstrated that free ethylene glycol and half-esterified ethylene glycol react at different rates. Keeping this in mind, the main reactions involving methyl ester and hydroxyethyl groups are

Reaction with ethylene glycol:

$$D_n + G \underset{k_0'}{\overset{4k_1}{\rightleftharpoons}} H_n + M, \qquad n = 1, 2, \ldots \qquad (8.2.2)$$

$$H_n + G \underset{k_0'}{\overset{2k_1}{\rightleftharpoons}} P_n + M, \qquad n = 1, 2, \ldots \qquad (8.2.3)$$

Transesterification reaction:

$$D_n + H_m \underset{k_1'}{\overset{2k_2}{\rightleftharpoons}} D_{n+m} + M, \qquad m, n = 1, 2, \ldots \qquad (8.2.4)$$

$$H_n + H_m \underset{\frac{1}{2}k_1'}{\overset{2k_2}{\rightleftharpoons}} H_{m+n} + M, \qquad m, n = 1, 2, \ldots \qquad (8.2.5)$$

$$P_n + D_m \underset{\frac{1}{2}k_1'}{\overset{4k_2}{\rightleftharpoons}} H_{m+n} + M, \qquad m, n = 1, 2, \ldots \qquad (8.2.6)$$

$$P_n + H_m \underset{k_1'}{\overset{2k_2}{\rightleftharpoons}} P_{m+n} + M, \qquad m, n = 1, 2, \ldots \qquad (8.2.7)$$

Polycondensation reaction:

$$P_n + P_m \underset{k_2'}{\overset{4k_3}{\rightleftharpoons}} P_{m+n} + G, \qquad m, n = 1, 2, \ldots \qquad (8.2.8)$$

$$H_n + H_m \underset{k_1'}{\overset{2k_3}{\rightleftharpoons}} D_{m+n} + G, \qquad m, n = 1, 2, \ldots \qquad (8.2.9)$$

In the mechanism given above, k_1 is the reactivity of the free glycol (G) and k_2 is the reactivity of the half esterified glycols ($-OCH_2CH_2OH$ group) in H_m and P_m. M represents a molecule of methanol which can react with a reacted bond in D_m, H_m, and P_m through the reverse reactions as shown.

To find out the reactivity of the reverse reactions involving D_m, H_m, P_m, and methanol (M), a molecule of H_5 is considered as an example, with various carbonyl groups numbered:

$$(8.2.10)$$

If methanol attacks at point 4 of H_5, one would form D_2 and P_3, whereas if it attacks point 5, H_2 and H_3 are formed. If it is assumed that M can attack points 4 and 5 with equal likelihood, the reactivities of the reverse steps of both Eqs. (8.2.5) and (8.2.6) are ($\frac{1}{2}k_1'$) as shown. With this clarification and using the kinetic scheme given in Eqs. (8.2.2)-(8.2.8), it is easy to write the mole balance equations for D_n, H_n, and P_n. This is left as an exercise. In principle, these equations can be solved numerically to get the detailed molecular weight distribution of P_n, D_n, and H_n.

One defines the following groups:

$$E_m = \text{—C}_6\text{H}_4\text{—COCH}_3 \quad (8.2.11a)$$

$$E_g = \text{—C}_6\text{H}_4\text{—COCH}_2\text{CH}_2\text{OH} \quad (8.2.11b)$$

$$Z = \text{—C}_6\text{H}_4\text{—C—OCH}_2\text{CH}_2\text{OC—C}_6\text{H}_4\text{—} \quad (8.2.11c)$$

in order to simplify the reaction scheme. The concentration of these groups can be written in terms of $[D_n]$, $[H_n]$, and $[P_n]$ as follows:

$$[E_m] = \sum_{r=1}^{\infty} (2[D_r] + [H_r]) \quad (8.2.12a)$$

$$[E_g] = \sum_{r=1}^{\infty} (2[P_r] + [H_r]) \quad (8.2.12b)$$

and

$$[Z] = \sum_{r=1}^{\infty} (r-1)([P_r] + [D_r] + [H_r]) \quad (8.2.12c)$$

where Z represents a reacted bond formed through the reaction of functional groups. With the help of mole balance equations for D_n, H_n, and P_n, it is possible to write the balance equations for E_m, E_g, and Z as was done for ARB polymerization in Chapter 2. It can be easily shown that the balance equations for E_m, E_g, and Z involve the concentrations of these groups only. This implies that the polymer formation written in the mechanism of Eqs. (8.2.2)–(8.2.9) can be equivalently written in terms of the reaction of functional groups only[10] as written in Table 8.1. It may be observed that if the polycondensation reaction is ignored, the above mechanism would reduce to the case of induced asymmetry as discussed in Chapter 3.[11-13]

At the usual temperature of the ester interchange reactor, the mechanism given in Eqs. (8.2.2)–(8.2.9) is a considerable idealization and there are several side reactions that are known to occur along with the formation of

TABLE 8.1. Various Reactions of Functional Groups in the Transesterification Stage of PET Formation from the DMT Route

Main reactions:

1. Ester interchange

$$E_m + G \underset{k_1/K_1}{\overset{k_1}{\rightleftharpoons}} E_g + M$$

2. Transesterification

$$E_m + E_g \underset{k_2/K_2}{\overset{k_2}{\rightleftharpoons}} Z + M$$

3. Polycondensation

$$2E_g \underset{k_3/K_3}{\overset{k_3}{\rightleftharpoons}} Z + G$$

Important side reactions:

4. Acetaldehyde formation

$$E_g \xrightarrow{k_4} E_c + A$$

5. Diethylene glycol formation

a.
$$E_g + G \xrightarrow{k_5} E_c + D$$

b.
$$E_g \xrightarrow{k_6} E_c + E_D$$

6. Water formation

a.
$$E_c + G \underset{k_7/K_4}{\overset{k_7}{\rightleftharpoons}} E_g + W$$

b.
$$E_c + E_g \underset{k_8/K_5}{\overset{k_8}{\rightleftharpoons}} Z + W$$

7. Vinyl group formation

a.
$$Z \xrightarrow{k_9} E_c + E_v$$

b.
$$E_g + E_v \xrightarrow{k_3} Z + A$$

Symbols

A: CH_3CHO

D: $OHCH_2CH_2OCH_2CH_2OH$

E_m:

E_g:

E_c:

E_v:

(continued)

TABLE 8.1 (*continued*)

E_D: ⬡—$COCH_2CH_2OCH_2CH_2OH$ (with =O above the C)

M: CH_3OH

W: H_2O

Z: ⬡—$COCH_2CH_2OC$—⬡ (with =O above each C)

BHET. It may be pointed out that these cannot be ignored and must also be considered for any realistic simulation. As a matter of fact, because of these reactions, the condensation product consists of acetaldehyde and water along with methanol. The overall mechanism of BHET formation in terms of the functional groups has been given in Table 8.1. Acetaldehyde (A) is formed when either E_g or a bond (Z) is degraded. In the former case, an acid group (E_c) is formed, whereas in the latter, a vinyl group (E_v) is formed. Water is formed when an acid end group (E_c) reacts either with ethylene glycol G or E_g functional groups. It can be seen in the table that only some of these reactions are reversible.[14,15]

The formation of ether (D and E_D) in the reaction mass is an important side reaction because, as pointed out earlier, even a small amount of these has a great influence on the properties of the final polymer. There exists a considerable controversy over the mechanism of diethylene glycol formation. Earlier, it was believed that it is formed by the dehydration reaction of ethylene glycol as[9,16]

$$2OHCH_2CH_2OH \rightarrow OHCH_2CH_2O-CH_2CH_2OH + H_2O \qquad (8.2.13)$$

It was later shown that only a very small portion of diethylene glycol is formed through this reaction.[17] Reimschuessel[18] has suggested that these are formed through the interaction of E_g and ethylene glycol as

$$E_g + G \nearrow^{E_D + W}_{\searrow D + E_c} \qquad (8.2.14)$$

In addition to this, the degradation of E_g (reaction 4 of Table 8.1) has been postulated to occur via the formation of an intermediate ethylene oxide

$$E_g \longrightarrow E_c + \underset{\underset{O}{\diagdown \diagup}}{CH_2 - CH_2} \qquad (8.2.15a)$$

$$\underset{\underset{O}{\diagdown \diagup}}{CH_2 - CH_2} \longrightarrow CH_3CHO \qquad (8.2.15b)$$

This ethylene oxide can also interact with G and E_g to form D and E_D as

$$\underset{\underset{O}{\diagup \diagdown}}{CH_2 - CH_2} + G \longrightarrow D \qquad (8.2.16a)$$

$$\underset{\underset{O}{\diagup \diagdown}}{CH_2 - CH_2} + E_g \longrightarrow E_D \qquad (8.2.16b)$$

One can, however, ignore the mechanistic details and postulate the formation of D and E_D through reaction 5 of Table 8.1. It can be seen that this is slightly different from those given in Eqs. (8.2.14) and (8.2.16).

8.2b. Rate Constants

The transesterification reaction has been studied by several workers.[3,8,9,19,20] Generally the progress of the reaction is followed by determining the amount of methanol distilled out. There have been a number of models proposed for the transesterification step. Most of these assume the volume of the reaction mass as time invariant, even though methanol formed leaves the reaction mass continuously. Yamanis and Adelman[21] have analyzed experimental data of different investigators and there appears to be quite some uncertainty in the values of k_2 reported. Detailed simulations, however, have shown that the final results are fairly insensitive to the choice of the value of k_2.[13-15] Rate constant k_1, as discussed earlier, depends upon the catalyst used. For the catalyst zinc acetate, k_1 has been reported as a function of the catalyst concentration, which is shown in Fig. 8.2. From the figure, it is evident that the effect of a differential increase in the catalyst concentration on k_1 is considerably larger at small concentrations than at higher values.

To determine the rate constant for the polycondensation step in the laboratory, workers have used sealed tube experiments,[9] semibatch reactors[22-25] and thin stagnant films.[26-28] The literature shows a wide variation

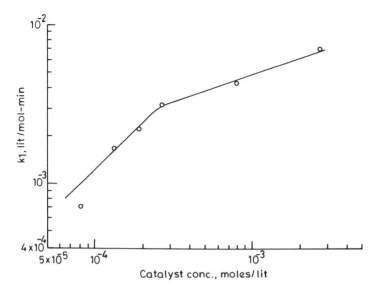

Figure 8.2. Dependence of rate constant k_1 on the concentration of zinc acetate catalyst at 197°C.[14]

in the reported values of k_3. It is generally believed that this arises because the vapor–liquid equilibrium of ethylene glycol required in correctly interpreting the rate data is subject to considerable uncertainty. Recently there has been a simulation study which has used the rate constant estimated from the data of Yokoyama *et al.*[24] This simulation claims to predict the performance of a PET plant in operation,[14,15] and in view of this, the values of the rate constants, as reported in Table 8.2, have been used in the present discussion. The study of Yokoyama *et al.*[24] has been made using Sb_2O_3 catalyst, and for the range studied, it was found to be proportional to the concentration of the catalyst used. Based on the experiments on the model system glycodibenzoate ethylene glycol–glycol monobenzoate, Hovenkamp[29] showed that the hydroxyl group deactivates the Sb_2O_3 catalyst, and the following relation was deduced for the polymerizing system:

$$k_3 = k_{3,0} \frac{1 + 0.9132\mu_n}{1 + 0.913\mu_{no}} \qquad (8.2.17)$$

where $k_{3,0}$ is the polycondensation rate constant for the initial average chain length, μ_{no}, of the polymer and μ_n is the average chain length at the time of interest. In Table 8.3, it is $k_{3,0}$ which has been reported. In their simulation studies, Ravindranath and Mashelkar have used this $k_{3,0}$, even though the catalyst used in the first stage is zinc acetate and not antimony oxide. They

TABLE 8.2. Various Rate and Equilibrium Constants, k_i and K_i Defined in Table 8.1

$$k_i = A_i e^{-E_i/RT}$$

	1	2	3	4	5	6	7	8	9
Activation energy, E_i (kcal/mol)	15.0	15.0	18.5	29.8	29.8	29.8	17.6	17.6	37.8
Frequency factor, A_i (liters/mol min)	6.0×10^4 $4.0 \times 10^{4\,a}$ 2.0×10^4	3.0×10^4 $2.0 \times 10^{4\,a}$ 1.0×10^4	6.8×10^9	$2.17 \times 10^{9\,b}$	2.17×10^9	2.17×10^9	1.0×10^6	1.0×10^6	$3.6 \times 10^{9\,b}$
Equilibrium constant K_i	0.3	0.15	0.5	—	—	—	2.5	1.25	—
k_i at 200°C (liters/mol min)	4.43×10^{-3}	2.22×10^{-3}	1.79×10^{-3}	3.29×10^{-5}	3.29×10^{-5}	3.29×10^{-5}	7.17×10^{-4}	7.17×10^{-3}	1.07×10^{-8}
k_i at 280°C (liters/mol min)	4.49×10^{-2}	2.25×10^{-2}	3.12×10^{-2}	3.29×10^{-3}	3.29×10^{-3}	3.29×10^{-3}	1.09×10^{-1}	1.09×10^{-1}	3.6×10^{-6}

[a] The frequency factors used in calculating the fourth and fifth rows.
[b] Units are min^{-1}.

TABLE 8.3. Mole Balance Equations for Batch Ester Interchange Reactors

$$\frac{1}{V}\frac{de_m}{dt} = -\mathcal{R}_1 - \mathcal{R}_2 \equiv f_1$$

$$\frac{1}{V}\frac{de_g}{dt} = \mathcal{R}_1 - \mathcal{R}_2 - 2\mathcal{R}_3 - \mathcal{R}_4 - \mathcal{R}_5 - \mathcal{R}_6 + \mathcal{R}_7 - \mathcal{R}_8 - \mathcal{R}_{10} \equiv f_2$$

$$\frac{1}{V}\frac{dz}{dt} = \mathcal{R}_2 + \mathcal{R}_3 + \mathcal{R}_8 - \mathcal{R}_9 + \mathcal{R}_{10} \equiv f_3$$

$$\frac{1}{V}\frac{de_c}{dt} = \mathcal{R}_4 + \mathcal{R}_5 + \mathcal{R}_6 - \mathcal{R}_7 - \mathcal{R}_8 + \mathcal{R}_9 \equiv f_4$$

$$\frac{1}{V}\frac{de_v}{dt} = \mathcal{R}_9 - \mathcal{R}_{10} \equiv f_5$$

$$\frac{1}{V}\frac{d}{dt} = \mathcal{R}_5 = f_6'$$

$$\frac{1}{V}\frac{de_D}{dt} = \mathcal{R}_6 = f_6''$$

$$\frac{1}{V}\frac{dg}{dt} = -\mathcal{R}_1 + \mathcal{R}_3 - \mathcal{R}_7 - \mathring{Q}_G'(t)$$

$$\frac{1}{V}\frac{da}{dt} = \mathcal{R}_4 + \mathcal{R}_{10} - \mathring{Q}_A'(t)$$

$$\frac{1}{V}\frac{dw}{dt} = \mathcal{R}_7 + \mathcal{R}_8 - \mathring{Q}_W'(t)$$

$$\frac{1}{V}\frac{dm}{dt} = \mathcal{R}_1 + \mathcal{R}_2 - \mathring{Q}_M'(t)$$

$$f_6 = f_6' + f_6''$$

$$\mathcal{R}_1 = k_1\left(2e_m g - \frac{e_g m}{K_1}\right)\bigg/ V^2$$

$$\mathcal{R}_2 = k_2\left(e_m e_g - \frac{2zm}{K_2}\right)\bigg/ V^2$$

$$\mathcal{R}_3 = k_3\left(e_g^2 - \frac{4zg}{K_3}\right)\bigg/ V^2$$

$$\mathcal{R}_4 = k_4 e_g / V$$

$$\mathcal{R}_5 = 2k_5 e_g g / V^2$$

$$\mathcal{R}_6 = k_6 e_g^2 / V^2$$

$$\mathcal{R}_7 = k_7\left(2e_c g - \frac{e_g w}{K_4}\right)\bigg/ V^2$$

$$\mathcal{R}_8 = k_8\left(e_c e_g - \frac{2zw}{K_5}\right)\bigg/ V^2$$

$$\mathcal{R}_9 = k_9 z / V$$

$$\mathcal{R}_{10} = k_3 e_v e_g / V^2$$

(continued)

TABLE 8.3 (*continued*)

Vapor pressure:
  Ethylene glycol[a]

$$\log P_G^0 (\text{mm Hg}) = 21.61 - \frac{3729}{T} - 4.042 \log(T)$$

  Water[b]

$$\log P_W^0 (\text{mm Hg}) = 8.024 - \frac{1721.35}{(T - 311)}$$

  Methanol[b]

$$\log P_M^0 (\text{mm Hg}) = 8.506 - \frac{1979.38}{T - 256}$$

Molar volumes (in $\text{cm}^3/\text{g mole}$)[c]:

$$v_{\text{DMT}} = 191.5\{1 + 0.0014(T - 413)\}$$

$$v_M = 43.5\{1 + 0.0014(T - 413)\}$$

$$v_G = 60.6\{1 + 0.0014(T - 413)\}$$

$$v_W = 19.422 + 0.00138(T - 413)$$

Dimensionless variables:

$$x_1 = \frac{e_m}{e_{m_0}}, \qquad x_2 = \frac{e_g}{e_{m_0}}, \qquad x_3 = \frac{g}{e_{m_0}}, \qquad x_4 = \frac{z}{e_{m_0}}$$

$$x_5 = \frac{e_c}{e_{m_0}}, \qquad x_6 = \frac{+e_D}{e_{m_0}}, \qquad x_7 = \frac{e_v}{e_{m_0}}$$

$$c_{m_0} = \frac{e_{m_0}}{V_0}, \qquad x_8 = \frac{m}{e_{m_0}}, \qquad x_9 = \frac{w}{e_{m_0}}, \qquad x_{10} = \frac{a}{e_{m_0}}$$

[a] Reference 8.
[b] Reference 31.
[c] T is the temperature in K. v_P is taken to be the same as v_{DMT}.

have justified this by claiming that the final results are fairly insensitive to the choice of the value of $k_{3,0}$.

PET is subjected to temperatures of 270–300°C during the manufacturing and subsequent processing stage which leads to degradation of polymers. Some of the degradation reactions are given in Table 8.1. There is very little information available in the literature on the rate constants of the side reactions, k_4, k_5, k_6, k_7, and k_8 (as defined in Table 8.1). Indeed, good data on side reactions, not only for PET polymerization but also for other systems, are viewed by major producers as proprietary. Hovenkamp and Muting[30] have studied the total DEG formation with a mixture of manganese acetate and antimony trioxide as catalyst and have evaluated k_5 and k_6 at 270°C. Yokoyama *et al.*[24] have studied the degradation of the hydroxyethyl ester end group (E_g), and through a careful simulation they evaluated k_4, k_8, and k_9. Because of the similarity of reactions, k_5 and k_6 were taken to be the same as k_4, k_7, and k_8. The terminal vinyl end group was assumed to behave exactly as E_g, which means the reaction between E_g and E_g (i.e., reaction 3 of Table 8.1) and the reaction between E_g and E_v are identical kinetically. The use of these rate constants has been once again justified on the basis that the simulation studies have predicted the performance of a PET plant in operation quite accurately.

The various equilibrium constants (K_1-K_5) in PET formation are, surprisingly, found to be almost independent of temperature. Challa[9] has found the ester interchange equilibrium constant K_1 to be 0.3. The poly-condensation equilibrium K_3 has been reported by several workers[8,31] to be 0.5. Since

$$K_2 = K_1 K_3 \qquad (8.2.18)$$

K_2 can be calculated to be 0.15, and this has been included in Table 8.2. The esterification equilibrium of benzoic acid–ethylene glycol and benzoic acid–2-hydroxyethyl benzoate have been studied[13] for which the equilibrium constants were found to be 2.15 and 1.5, respectively, independent of temperature. These could be treated as model experiments representing reactions 6a and 6b of Table 8.1 and K_4 and K_5 could be taken equal to these values.

8.2c. Simulation of the Ester Interchange Reactor

In the flow sheet of Fig. 8.1, the ester interchange reactor is operated at constant pressure (usually 1 atmosphere) and at a temperature of 220°C. At these conditions, the vapor pressures of ethylene glycol, methanol, water, and acetaldehyde are more than the total pressure applied, which means that these would continuously flash from the reactor. Keeping this in mind, one can now establish the mole balance of various functional groups and condensation products in batch reactors as given in Table 8.3. In the table e_m, e_g, e_c, e_v, z, δ, etc. represent the total moles (and not the concentrations) of E_m, E_g, E_c, E_v, Z, D, etc. and V is the volume of the reaction mass. $\mathring{Q}'_G(t)$, $\mathring{Q}'_A(t)$, $\mathring{Q}'_W(t)$, and $\mathring{Q}'_M(t)$ represent the net rate (which is time dependent) of flashing of ethylene glycol, acetaldehyde, water, and methanol from the ester interchange reactor, respectively. $\mathscr{R}_1$-$\mathscr{R}_{10}$ are defined as the rates of the individual reactions in Table 8.1 and are also given in Table 8.3.

It may be relevant to point out here that the balance for z is not an independent equation, but can be written as a linear combination of differential equations for e_m, e_g, e_c, and e_v. This can be easily integrated for a known composition of feed.

To determine $\overset{\circ}{Q}_G'(t)$, $\overset{\circ}{Q}_A'(t)$, $\overset{\circ}{Q}_W'(t)$, and $\overset{\circ}{Q}_M'(t)$ it is assumed that all the polymeric species and dimethyl terephthalate are completely involatile and Raoult's law is assumed to hold. If p_G, p_A, p_W, and p_M are the partial pressures of G, A, W, and M, then

$$P_T = P_G^0(T)x_G + P_A^0(T)x_A + P_W^0(T)x_W + P_M^0(T)x_M \qquad (8.2.19)$$

where P_G^0, P_A^0, P_W^0, and P_M^0 are the vapor pressures (given in Table 8.3[8,32]) and x_G, x_A, x_W, and x_M are the mole fractions of ethylene glycol, acetaldehyde, water, and methanol, respectively, in the liquid phase of the ester interchange reactor. For a specified value of the total pressure, P_T, the equilibrium concentrations x_G, x_A, x_W, and x_M can be found through trial and error.

As the transesterification reaction progresses, there is a shrinkage in the volume of the reaction mass as the condensation product flashes out, and therefore V in Table 8.3, assuming the additivity of volumes of the various components, one has

$$V = v_P M_P + V_G g + v_M m + v_A a + v_W w \qquad (8.2.20)$$

Above v_P, v_G, v_M, v_A, v_W are the molar volumes (given in Table 8.3[8,32] and are equal to the reciprocal of molar density) and M_P is the total moles of the polymer in the reaction mass. It is now possible to simulate the reactor using the numerical technique outlined in Section 5.2. As pointed out earlier, (k_2/k_1) has been reported to fall in the range of 0.01[21] to 0.5[8], but the results are relatively insensitive to its value in this range. Results have also not been found to be considerably affected by the variation of k_3. Consequently the use of the value of k_3 determined for antimony oxide catalyst was not regarded as a serious drawback.

It may be argued that by reducing the pressure of the reactor, methanol would flash with a smaller amount remaining in the reaction mass. This would in turn drive the ester interchange and transesterification reactions more in the forward direction, and a larger conversion can thus be achieved for the same residence time of the reactor. To examine this, pressure P_T was systematically changed and the performance of the semibatch reactor computed. The final conversions measured by the release of methanol are found to change by a very small amount. This perhaps indicates that the concentration of methanol in the reaction mass is very small as such because of its high vapor pressure. As a result, reactions 1 and 2 of Table 8.1 may well be assumed to be irreversible.

Among the various design and operating parameters, temperature appears to be an important one. It may be recognized that the reactor temperature cannot be increased indefinitely because higher values favor

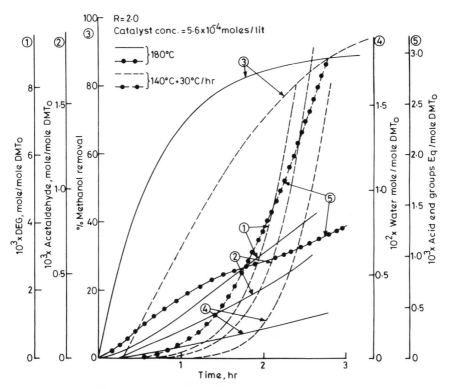

Figure 8.3. Values of m, a, e_{DEG}, w, and e_c versus residence time for isothermal and linearly increasing temperature history. $r \equiv [G]_0/(DMT)_0$.

the degradation of polymer. The usual reactor temperature is 180°C, and results have been computed and plotted in Fig. 8.3.[14] Two temperature policies have been considered: in the first one, the reactor is operated isothermally, whereas in the other, a profile normally encountered in industries, the temperature is increased linearly from the initial value of 140°C at the rate of 30°C/hr. Since every mole of DMT reacted produces two moles of methanol, the extent of reaction can be measured by the amount of methanol released as a percentage of the stoichiometric amount of methanol. The numerical integration of differential equation in Table 8.3 gives Q'_M and it is an easy matter to compute the cumulative methanol formed for a given time of reaction. When the reactor is operated isothermally at 180°C, as seen in Fig. 8.3, methanol begins to flash almost immediately from the reactor and the cumulative methanol evaporated rises quickly to an asymptotic value which is governed by the equilibrium of the first two reactions of Table 8.1. When the temperature of the reactor is increased from 140°C at a rate of 30°C/hr, methanol is found to start flashing from the ester-interchange reactor after about 17 min of reaction. For large times of reaction, the cumulative methanol produced is more than that for the isothermal case. Since the equilibrium constants are independent of temperature, this result at first appears to be surprising, but can be explained as follows. Ethylene glycol in the reaction mass reacts either with E_g or with E_m. In the case of isothermal reaction at 180°C, more ethylene glycol is consumed through reaction 5a of Table 8.1 compared to that for the linearly increasing temperature profile. The equilibrium constant K_1 is defined to be

$$K_1 = \frac{e_g m}{e_m g} \tag{8.2.21}$$

Since g in the reaction mass for the case of linearly increasing temperature history is larger, e_m in it is expected to be smaller, which implies a higher cumulative methanol released as observed in Fig. 8.3.

The effect of temperature history of the ester interchange reactor on the various side products has also been examined in Fig. 8.3. In this figure, e_c, $(\delta + e_D)$, a and w have been plotted for both isothermal and linearly increasing temperature history of the reactor. For small reactor times, e_c, $(\delta + e_D)$, a, and w are more for the isothermal reactor, but these values shoot up very rapidly for large times of reaction for the case of the increasing temperature history. If the ester interchange reactor is desired to be operated such that in addition to obtaining high conversion, the formation of side products is also minimized, information shown in Fig. 8.3 suggests the following strategy. One could use a high reactor temperature initially so that reactions 1 to 3 of Table 8.1 occur rapidly and then the temperature is subsequently lowered to reduce the formation of side products. A detailed optimization of the reactor, as discussed later, indeed shows this.

Another very important variable besides the reactor temperature is the ratio, r, of ethylene glycol to dimethyl terephthalate charged initially. In Fig. 8.4, the effect of r on the performance of the reactor has been examined. For an isothermal reactor, the cumulative moles of methanol flashed increases very rapidly for small times, and this approaches an asymptotic value essentially determined by the equilibrium of reactions 1 and 2 of Table 8.1. As r is increased, unreacted ethylene glycol in the reaction mass at large reaction times is larger and the value of e_m is expected to be smaller. This implies that the cumulative methanol released should increase as r is made larger, and this is confirmed by the results plotted in Fig. 8.4. When a linearly increasing temperature profile is applied in the batch reactor, the cumulative methanol released for a given r is larger than that for an isothermal reactor. In the same figure, $(\delta + e_D)$ and e_c have also been plotted, and unfortunately on increasing r, the total amounts of these are

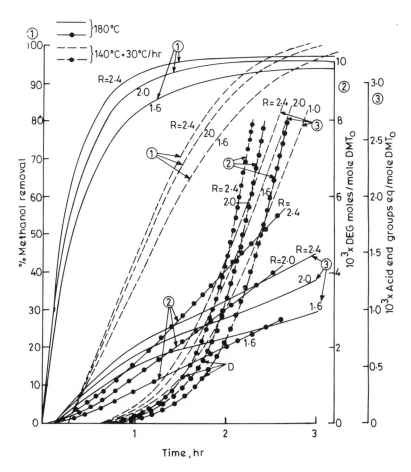

Figure 8.4. Effect of the ratio of DMT and ethylene glycol in feed on m, e_{DEG}, and e_c versus time.[14]

also found to increase. Thus the advantage of having higher rate of reaction is offset by the increased amount of undesirable side products in the reaction mass. One could also add ethylene glycol continuously over the entire reaction time instead of all at the beginning. This has the advantage of reducing $(\delta + e_D)$ slightly without affecting the final conversion of DMT.

Sometimes, transesterification reaction is carried out in a series of homogeneous continuous flow stirred tank reactors (HCSTR) as shown in Fig. 8.5. Vapor flashed from each HCSTR is charged to a single operator. Ethylene glycol is separated and recycled to the first reactor, thus coupling together the equations for all the HCSTRs. This means that the performance of the transesterification process can be obtained only by solving the perfor-

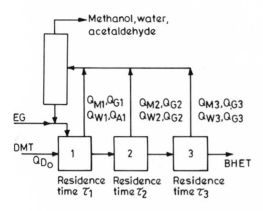

Figure 8.5. Flow sheet for the transesterification process involving a train of HCSTRs.

mance of *all* the HCSTRs simultaneously. The balance equations for a single HCSTR can be easily written, and it can easily be seen that the calculation of the performance of the train of HCSTRs is quite involved. In Fig. 8.6, the performance of one HCSTR with feed consisting of ethylene glycol and DMT has been compared with that of a batch reactor at two temperatures in terms of the conversion of DMT as a function of reactor residence time. At both these temperatures and at any given residence time, the conversion of DMT is higher in batch reactors, which is to be expected. Additionally, in a train of HCSTRs with a finite number of reactors, optimal performance is obtained when the residence times of all the reactors are equal as seen in Table 8.4. In this table, results for (τ_1, τ_2, τ_3) as $(1.5, 1, 0.5)$, $(1, 1, 1)$, and $(0.5, 1, 1.5)$ have been given and the one with $(1, 1, 1)$ is found to give the best results. Calculations have also been made for multiple HCSTRs in series, and in Table 8.4, results have been compared with those for a batch reactor and an HCSTR. It is a well-known result[33] that in the limit as the number of HCSTRs is increased to infinity (keeping the total residence time the same), the performance of a train of HCSTRs approaches that of batch reactors. Thus, the conversion of DMT attained in three HCSTRs is larger than that in a single HCSTR but is lower than that in a batch reactor as expected. In Table 8.4, the amounts of various side products formed have also been given for comparison. For a single HCSTR, the amounts of the side products formed have also been given for comparison. For a single HCSTR, the amounts of the side products formed are the least because of the lowest conversion of DMT attained, whereas the train of HCSTRs gives intermediate performance and the largest amount of the side products are formed in a batch reactor.

The detailed simulation carried out in this section shows that among the various variables, temperature is the single most important parameter. The pressure of the reactor is found to affect the performance only mar-

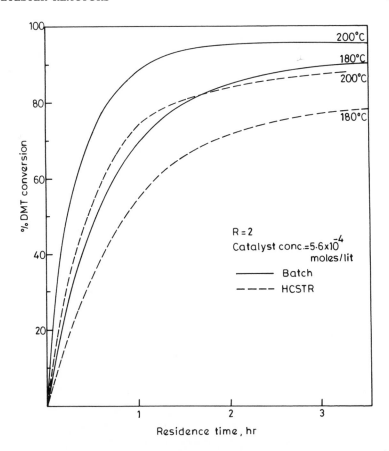

Figure 8.6. Conversion of DMT versus residence time for batch reactors and HCSTRs.[14,15]

ginally. In view of this, in the following optimization study, the optimal temperature profile has been determined.

8.3. OPTIMIZATION OF ESTER INTERCHANGE REACTOR[34]

In this section, temperature has been regarded as the control variable with respect to which the semibatch reactor is optimized. This is because pressure and the ratio of ethylene glycol to DMT in the feed have been shown in the earlier section to play a less important role in the performance of the reactor. The volume of the reaction mass has been assumed to be essentially constant at some average value (which need not be equal to the initial volume of the reaction mass), to simplify the analysis.

TABLE 8.4.[a] Comparison of the Performance of Batch Reactors and a Train of HCSTRs at 180°C

	Batch reactor $\tau = 3$ hr	One HCSTR $\tau = 3$ hr	Three HCSTRs $\tau_1 = 1.5, \tau_2 = 1,$ $\tau_3 = 0.5$ hr	Three HCSTRs $\tau_1 = \tau_2 = \dot{\tau}_3 = 1$ hr	Three HCSTRs $\tau_1 = 0.5, \tau_2 = 1,$ $\tau_3 = 1.5$ hr
DMT conversion (%)	91.70	76.43	85.50	86.20	86.76
Acetaldehyde $10^3 \times$ mole/mole DMT_0	2.12	2.63	2.30	2.21	2.17
$(\delta + e_D)$ $10^2 \times$ mol/mol DMT_0	1.75	2.35	2.18	2.09	2.02
Water $10^2 \times$ mole/mole DMT_0	1.44	2.28	1.99	1.88	1.80
Acid end group $10^3 \times$ mole/mole DMT_0	5.22	3.37	4.18	4.34	4.35

[a] Reference 15.

As the pressure of the semibatch reactor is increased, according to Eq. (8.2.19), the mole fractions of methanol, x_M, water x_W, and acetaldehyde, x_A are expected to increase. Since in Section 8.2 it has been shown that pressure is relatively an unimportant variable, this appears to imply that x_W, x_A, and x_M are small and they are little affected by the change in pressure. As a first approximation, these have been treated as parameters, and on doing this, it is possible to solve the mole balance equations of Table 8.3 without performing flash calculations. This leads to a considerable saving in computer time.

Before attempting to optimize the reactor, it is extremely important to decide an objective function carefully because it is well known that the final results depend considerably upon the choice of this function.[35,36] It may be recognized that the main goal of the ester interchange reactor is to maximize BHET formation, but the kinetic model presented in Table 8.1 does not yield the amount of BHET formed explicitly. It is also not appropriate to state that maximizing BHET is equivalent to the maximization of e_g, for the following reason. In the reaction mechanism of Table 8.1, some of the e_g reacts with itself and forms Z, which amounts to stating that the end product of the ester interchange reactor is not BHET alone, but consists of its oligomers also. But this does not affect the performance of the reactor detrimentally. As a result, it is proposed that the conversion of E_m is maximized, and simultaneously, the amounts of various side products are minimized. If the feed to the semibatch reactor is pure DMT and ethylene glycol, the conversion of E_m is given by $(1 - e_m/e_{m_0})$, where e_{m_0} is the moles of E_m at $t = 0$. Since maximizing conversion is equivalent to minimizing (e_m/e_{m_0}), the objective function, I, can possibly be written as

$$I = \min\left\{\alpha_1\left(\frac{e_m}{e_{m_0}}\right)^2 + \alpha_2\left[\left(\frac{\delta + e_D}{e_{m_0}}\right)^2 + \left(\frac{e_v}{e_{m_0}}\right)^2 + \left(\frac{e_c}{e_{m_0}}\right)^2\right]\right\}_{t=t_f} \quad (8.3.1)$$

where α_1 and α_2 are the weighting factors which are treated as parameters and t_f is the reactor residence time.

The mole balance equations and the objective functions are written in terms of the dimensionless state variables defined in Table 8.3. Differential equations in Table 8.3 can be rewritten in terms of dimensionless variables x_1 to x_7 and these can be arranged in the following vectorial form:

$$\frac{d\mathbf{x}}{dt} = \mathbf{f}(\mathbf{x}, T, x_8, x_9, x_{10}) \quad (8.3.2)$$

In the above equation, $\mathbf{f}(\mathbf{x}, T, x_8, x_9, x_{10})$ are formed by the right-hand sides of the equations in Table 8.3. Since the feed to the reactor consists of a mixture of DMT and ethylene glycol only, the initial condition, x_0, of the vector differential equation is given by

$$x_0 = \{1, 0, x_{3,0}, 0, 0, 0, 0\} \qquad (8.3.3)$$

To be able to find the optimal temperature profile, the Hamiltonian, H (see Chapter 6) is defined as

$$H = \sum_{i=1}^{7} z_i f_i \qquad (8.3.4)$$

where z_1 to z_7 are the adjoint variables whose time dependence can be easily derived. The optimal temperature profile is obtained by

$$\frac{\partial H}{\partial T} = 0 \qquad (8.3.5)$$

Since the equilibrium constants K_1 to K_4 are independent of temperature, $\partial H/\partial T$ can be obtained easily. To obtain the optimum temperature history $T(t)$ Eq. (8.3.5) must be solved simultaneously with the state and adjoint variable equations.[36,27]

The vector iteration method has been used to optimize the ester interchange reactor, and as suggested in Chapter 6, the memory requirements have been reduced by approximating the temperature profile by 100 piecewise continuous linear segments. It takes about 4 min of computer time on a DEC 1090 to find the optimum temperature profile. To check if this approximation was appropriate, the total time domain was divided into 200 piecewise continuous segments and results were found to change only negligibly.

The computer program[34] requires an initial guess (isothermal) of the temperature history for the solution to converge. If the initial guess is more than 530 K, in the second iteration, in some time domain the temperature is changed beyond 580°C and there is an overflow. Similarly when this initial guess is below 480 K, in some part of the time domain, the reactor temperature falls below 420 K and the optimum temperature is never reached. With the initial guess falling in the range of 490–530 K, the optimum solution is found to be independent of this initial guess.

From the vapor pressure data, it can be easily concluded that at 500°C and 1 atm pressure, most of the methanol, water, and acetaldehyde would flash from the reactor leaving only very small amounts of these in the reaction mass. As a first approximation, they have been taken as zero, for which the optimum temperature profile has been calculated and plotted in Fig. 8.7. It is found that high temperature must be used initially but it must

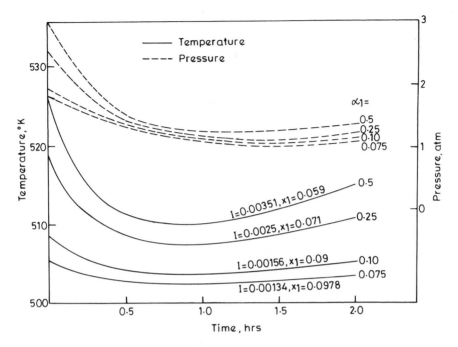

Figure 8.7. Effect of α_1 on the optimum temperature and pressure profile $\alpha_2 = 1$, $[M] = [W] = 0$.

be lowered in the final stages of the reaction. A high initial temperature gives a high rate of consumption of E_m, thus leading to its rapid conversion to E_g as seen from the mechanism of reaction given in Table 8.1. But if the same level of temperature is maintained, the accumulated E_g would react with G and another E_g group leading to the formation of diethylene glycol groups, ultimately lowering the conversion of E_m through reactions 1 and 2 shown in the table. When $\alpha_1 = 1$, at the optimum temperature profile, the term x_1^2 in the objective function [Eq. (8.3.1)] dominates. As α_1 is lowered, the optimum profile in Fig. 8.7 is found to become flatter, and this occurs because of the reduced importance of the conversion in the objective function. As the importance of x_5, x_6, and x_7 increases, the choice of lower temperatures initially is governed by steps 4 and 5 of Table 8.1; however, for large times, the reactor temperature must be increased so as to enhance the forward step of the water formation. If the vapor–liquid equilibrium of ethylene glycol is assumed at the reactor temperature, it is possible to calculate the pressure of the reactor using Eq. (8.2.19), and P_T versus time of reaction has also been plotted in Fig. 8.7. In Fig. 8.8, x_5, x_6, and x_7 have been plotted with α_1 as parameter, and as α_1 is reduced, their values are also found to reduce. Even though the assumptions in generating Fig. 8.8 are different from those in generating Fig. 8.3, they could be compared

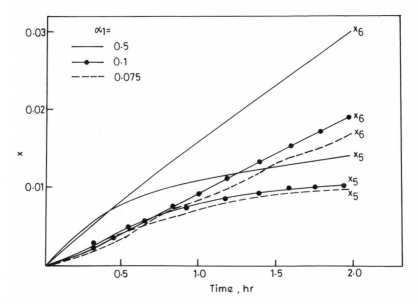

Figure 8.8. Fractions of various side products as a function of time for temperature profiles given in Fig. 8.7.

qualitatively. The use of linearly increasing temperature profiles in Fig. 8.3 gives an almost exponentially increasing magnitude of the various side products and therefore is inferior compared to the temperature profiles shown in Fig. 8.7. The latter not only gives a higher conversion of E_m, but also limits the formation of side products as seen in Fig. 8.8. This result may also be contrasted with the optimum temperature policy derived for step growth polymerization,[38] for which a maximum possible temperature (isothermal) has been suggested in Chapter 6.

If the pressure of the reactor is increased, its immediate effect would be sensed on the mole fractions of acetaldehyde, methanol, and water. Because of its exceptionally high vapor pressure, acetaldehyde can be assumed to be a noncondensible vapor at the reactor temperature, which means that in Eq. (8.2.19)

$$x_A = 0 \qquad\qquad (8.3.6)$$

Since x_G is completely specified by the vapor–liquid equilibrium for a given pressure, x_M and x_W can always be determined uniquely. Consequently, instead of studying the effect of pressure, one could equivalently study the effect of x_M and x_W on the temperature profile as has been done in Fig. 8.9. As x_M and $x_W = 0.1$, a maximum in the profile is obtained. This

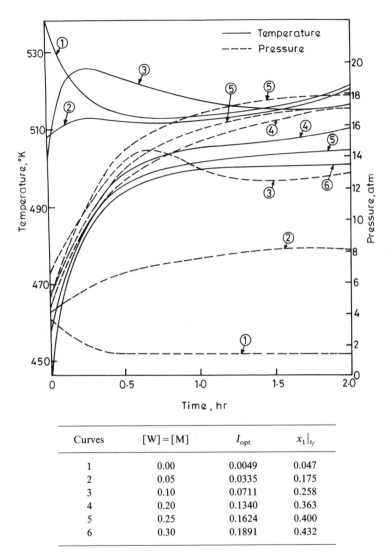

Figure 8.9. Effect of [M] and [W] on the optimum temperature profile for $\alpha_1 = \alpha_2 = 1.0$.

Curves	$[W] = [M]$	I_{opt}	$x_1\vert_{t_f}$
1	0.00	0.0049	0.047
2	0.05	0.0335	0.175
3	0.10	0.0711	0.258
4	0.20	0.1340	0.363
5	0.25	0.1624	0.400
6	0.30	0.1891	0.432

maximum disappears as x_M and x_W are further decreased and the temperature starts from a low value, approaching a higher value asymptotically. This is found because at high values of x_M and x_W, the ester interchange and the transesterification steps of Table 8.1 contribute little to the overall conversion of E_m and the choice of temperature is essentially governed by side reactions 4, 5, and 7 of the table. This means that the temperature should be low initially; however, it should be increased as the reaction proceeds for the following reason: High temperature would give a larger amount of E_g in the reaction mass, which would, in turn, give larger conversion of E_m through the transesterification reaction.

In Fig. 8.10, the effect of ethylene glycol and DMT in the feed [i.e., $x_{3,0}$ in Eq. (8.3.3)] on the optimum temperature profile has been examined.

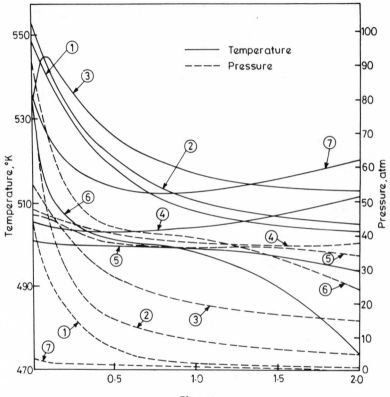

Curves	$[G]_0/[E_m]_0$	I_{opt}	$x_1\vert_{t_f}$
1	0.25	0.37436	0.612
2	0.50	0.10032	0.316
3	0.75	0.01576	0.116
4	1.25	0.00288	0.036
5	1.50	0.00190	0.028
6	1.75	0.00185	0.013
7	1.00	0.0049	0.049

Figure 8.10. Effect of $[G]_0/[E_m]_0$ ratio in the feed on the optimum temperature and pressure profiles for $\alpha_1 = \alpha_2 = 1$ and $[M] = [W] = 0$.

The use of high temperature initially as usual is governed by the consideration that the conversion of E_m by reactions 1 and 2 of Table 8.1 is maximized. When sufficient E_g is formed in the reaction mass, the temperature needs to be lowered to minimize side reactions 4, 5, and 7 of the table. In the presence of excess ethylene glycol in the feed, the temperature needs to be lowered to minimize the formation of side products. The lowering of final temperature in Fig. 8.10 is found to suppress step 5 of Table 8.1.

In this section the optimum temperature history has been calculated for the ester interchange reactor. Various parameters affecting the temperature profile have been identified. In the next section, the direct esterification of terephthalic acid is simulated.

8.4. DIRECT ESTERIFICATION OF TEREPHTHALIC ACID

8.4a. Industrial Process

Industrially, in the direct esterification of terephthalic acid (TPA), the latter is mixed with ethylene glycol and reacted to give BHET in a train of stirred tank reactors as shown in Fig. 8.11. This is done because of the limited solubility of TPA and unfavorable conversion due to high equilibrium constant. The solubility of TPA in ethylene glycol and BHET have been studied by several workers[42,43] and can be correlated by the following empirical equation:

Solubility of TPA in ethylene glycol

$$\ln c = 1.19 - 1240/T \qquad (8.4.1)$$

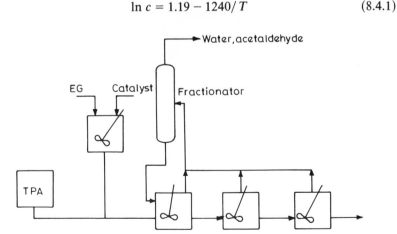

Figure 8.11. Flow sheet for the direct esterification of terephthalic acid for the formation of BHET.

Solubility of TPA in BHET

$$\ln c = 1.90 - 1420/T \qquad (8.4.2)$$

where c is the concentration of TPA in moles/kg of the solution and T is in °K. It can easily be seen from the above equations that the solubility of TPA in BHET is larger than it is in ethylene glycol.

Even though it is desirable to operate the reactor with an excess of ethylene glycol so that the reaction mass is homogeneous, it cannot be done because of the high cost of the latter. Therefore, the reactor producing BHET through this process consists of a continuous phase of ethylene glycol with dissolved TPA having dispersed phase of TPA granules. To maintain the well-mixed condition in this slurry reactor, a stirred tank reactor is normally used industrially. In Section 8.2, it has been shown that a train of HCSTRs gives a lower conversion compared to that obtained in batch reactors but is better than a single HCSTR. For this reason, industrially three HCSTRs in series are used as shown in the process flow sheet of Fig. 8.11. As done in the production of BHET from DMT, in this process the vapors leaving the various reactors are collected and separated in a single fractionator. As a result, the mole balance equations for various species for all these reactors must be solved simultaneously. In addition to this computational difficulty, it is necessary that the heterogeneous nature of the reactor be properly accounted for. In this section, a first-order model has been presented to account for this.

8.4b. Reactions and Their Rate Constants

Before proceeding to the modeling of the reactors, the various reactions and the rate constants in the direct esterification of terephthalic acid are discussed first. If the reactions of end groups alone are considered, then they can be written as

$$(8.4.3)$$

$$(8.4.4)$$

$$\text{~}\underset{\text{benzene ring}}{\bigcirc}-\overset{\overset{\text{O}}{\|}}{\text{C}}\text{OCH}_2\text{CH}_2\text{OH} + \text{OHCH}_2\text{CH}_2\text{O}\overset{\overset{\text{O}}{\|}}{\text{C}}-\underset{\text{benzene ring}}{\bigcirc}\text{~} \rightleftharpoons$$

$$\text{~}\underset{\text{benzene ring}}{\bigcirc}-\overset{\overset{\text{O}}{\|}}{\text{C}}\text{OCH}_2\text{CH}_2\text{O}\overset{\overset{\text{O}}{\|}}{\text{C}}-\underset{\text{benzene ring}}{\bigcirc}\text{~} + \text{H}_2\text{O} \qquad (8.4.5)$$

Reactions (8.4.3), (8.4.4), and (8.4.5) are identical to reactions 6a, 6b, and 3 of Table 8.1, respectively. It is expected that the various side reactions occurring in the formation of BHET through the direct esterification route must be the same as those occurring in the DMT route. This would imply that the overall mechanism of reaction for the direct esterification is identical to that given in Table 8.1 except that the ester interchange and transesterification reactions do not occur. The main reactions are now steps 6 and 3 of Table 8.1 and the side reactions are 4, 5, and 7.

It may be recognized that if reactions (8.4.3) and (8.4.5) between terephthalic acid and ethylene glycol occur, the resultant product is not BHET alone but would once again be a mixture of various oligomers of BHET along with the desired BHET. To simplify the kinetic analysis, in the literature, the direct esterification of TPA has been studied using model compounds like benzoic acid,[31,42,43] in which case the various reactions are

$$\bigcirc-\text{COOH} + \text{OHCH}_2\text{CH}_2\text{OH} \underset{k_7/K_4}{\overset{2k_7}{\rightleftharpoons}}$$

$$\bigcirc-\text{COOCH}_2\text{CH}_2\text{OH} + \text{H}_2\text{O} \qquad (8.4.6)$$

$$\bigcirc-\text{COOH} + \text{OHCH}_2\text{CH}_2\text{OOC}-\bigcirc \underset{2k_8/K_5}{\overset{k_8}{\rightleftharpoons}}$$

$$\bigcirc-\text{COOCH}_2\text{CH}_2\text{OOC}-\bigcirc + \text{H}_2\text{O} \qquad (8.4.7)$$

$$2\bigcirc-\text{COOCH}_2\text{CH}_2\text{OH} \underset{4k_3/K_3}{\overset{k_3}{\rightleftharpoons}}$$

$$\bigcirc-\text{COOCH}_2\text{CH}_2\text{OOC}\bigcirc + \text{H}_2\text{O} \qquad (8.4.8)$$

These reactions have been shown to be catalyzed by the carboxyl end groups as well as the catalyst antimony triacetate $Sb(COOCH_3)_3$. The catalytic effect upon various rate constants can be represented as

$$k_7 = k_7^0 + k_7^c[E_c]$$ (8.4.9)

$$k_8 = k_8^0 + k_8^c[E_c]$$ (8.4.10)

$$k_3' = (k_3/K_3) = k_{3r}^0 + k_{3r}^0[E_c] + k_{3r}^m[E_c] + k_{3r}^m[Sb] + k_{3r}^{cm}[E_c][Sb]$$ (8.4.11)

The temperature dependences of k_7, k_8, and k_{3r} have been correlated by assuming k_7^0, k_7^c, k_8^0, k_8^c, k_{3r}^0, k_{3r}^c, k_{3r}^m, and k_{3r}^{cm} each represented by the Arrhenius equation of the form

$$k_i^0, k_i^c = (A_i^0, A_i^c)\, e^{-(E_i^0, E_i^c)/RT}, \qquad i = 7, 8, 3r$$ (8.4.12)

The temperature dependences of the equilibrium constants, K_3, K_7 and K_8 have been written as

$$K_i = \exp\left(\frac{\Delta S_i}{R} - \frac{\Delta H_i}{RT}\right), \qquad i = 3, 8$$ (8.4.13a)

$$K_7 = K_8/K_3$$ (8.4.13b)

and these are given in Table 8.5.

TABLE 8.5. Rate Constants and Equilibrium Constants for Reactions Given in Eqs. (8.4.6)–(8.4.8)

$$k_{3r}\left(\frac{kg}{mole\ min}\right) = 5.038 \times 10^{17} \exp\left(-\frac{49160}{1.987\,T}\right)$$

$$+ 4.071 \times 10^{14} \exp\left(-\frac{27600}{1.987\,T}\right)[Sb] + 3.325 \times 10^6 \exp\left(-\frac{21860}{1.987\,T}\right)[E_c] + 360[E_c][Sb]$$

$$k_7\left(\frac{kg}{mole\ min}\right) = 2.55 \times 10^{-5} + 8.672 \times 10^5 \exp\left(-\frac{20630}{1.987\,T}\right)[E_c]$$

$$k_8\left(\frac{kg}{mole\ min}\right) = 8.426 \times 10^8 \exp\left(-\frac{29710}{1.987\,T}\right) + 75.64 \exp\left(-\frac{10310}{1.987\,T}\right)[E_c]$$

$$K_7 = \exp\left(-\frac{19.21}{1.987} + \frac{9122}{1.987\,T}\right)$$

$$K_8 = \exp\left(-\frac{10.81}{1.987} + \frac{5666}{1.987\,T}\right)$$

It is assumed that there is negligible vinyl group formation, which means that step 7 of the mechanism given in Table 8.1 does not occur. In the absence of any other information, rate constants k_4, k_5, and k_6 have been taken to be the same as those given in Table 8.2.

8.4c. Simulation of the Reactor

It is necessary to define two kinds of concentrations to be able to account for the slurry nature of the reaction mass. $[E_g]$, $[E_c]$, $[G]$, $[D]$, $[E_D]$, $[Z]$, and $[W]$ are the concentrations (in moles/kg) based on the mass of the continuous liquid phase and $[E_g']$, $[E_c']$, $[G']$, $[D']$, $[E_D']$, $[Z']$, and $[W']$ are the concentrations (in moles/kg) based on the *total* mass of the reaction mass (including the undissolved TPA). If $\mathring{Q}_0$ and $\mathring{Q}$ are total flow rates (in kg/min) in and out of the reactor, then the mole balances of a functional group in a given HCSTR can be easily written and are summarized in Table 8.6. It may be observed that the terms in $\mathcal{R}_3$–$\mathcal{R}_8$ are written in terms of concentrations in the continuous phase only. This is because the

TABLE 8.6. Mole Balance Relations for Various Functional Groups in an HCSTR for Direct Esterification[a]

$$\mathring{Q}_0[E_{c0}'] - \mathring{Q}[E_c'] + \rho_m V_1(-\mathcal{R}_7 - \mathcal{R}_8 + \mathcal{R}_4 + \mathcal{R}_5 + \mathcal{R}_6) = 0$$

$$\mathring{Q}_0[E_{g0}'] - \mathring{Q}[E_g'] + \rho_m V_1(\mathcal{R}_7 - \mathcal{R}_8 - 2\mathcal{R}_3 - \mathcal{R}_4 - \mathcal{R}_5 - \mathcal{R}_6) = 0$$

$$\mathring{Q}_0[Z_0'] - \mathring{Q}[Z'] + \rho_m V_1(\mathcal{R}_8 + \mathcal{R}_3) = 0$$

$$\mathring{Q}_0[G_0'] - \mathring{Q}[G'] + \rho_m V_1(-\mathcal{R}_7 + \mathcal{R}_3) - \mathring{Q}_V[G]^v = 0$$

$$\mathring{Q}_0[W_0'] - \mathring{Q}[W'] + \rho_m V_1(\mathcal{R}_7 + \mathcal{R}_8) - \mathring{Q}_V[W]^v = 0$$

$$\mathring{Q}_0([\delta_0'] + [E_{D0}']) - \mathring{Q}([\delta'] + [E_D']) + \rho_m V_1(\mathcal{R}_5 + \mathcal{R}_6) = 0$$

$$\rho_m V_1 \mathcal{R}_4 - \mathring{Q}_V[A]^v = 0$$

$$\mathcal{R}_3 = k_3\left([E_g]^2 - \frac{4[Z][G]}{K_3}\right)$$

$$\mathcal{R}_4 = k_4[E_g]$$

$$\mathcal{R}_5 = 2k_5[E_g][G]$$

$$\mathcal{R}_6 = k_6[E_g]^2$$

$$\mathcal{R}_7 = k_7\left(2[E_c][G] - \frac{[E_g][W]}{K_1}\right)$$

$$\mathcal{R}_8 = k_8\left([E_c][E_g] - \frac{2[Z][W]}{K_2}\right)$$

[a] $\mathcal{R}_3$–$\mathcal{R}_8$ in moles/kg min; $[E_c']$, $[E_g']$ etc.: concentrations in moles/kg based on reaction mass; $[E_c]$, $[E_g]$ etc.: concentrations in moles/kg based on the continuous phase; ρ_m: density in kg/liter; $\mathring{Q}_V$: vapor flow rate in kg/min; $[G]^v$, $[W]^v$, and $[A]^v$: concentrations of G, W, and A in vapor phase.

reaction is occurring in the continuous medium where ethylene glycol and terephthalic acid are present together.

Kemkecs[44] has observed that the acid end group concentration $[E_c]$, in the continuous medium is constant until the reaction mass becomes homogeneous. If $\mathring{Q}_0$ is the total flow rate and $\mathring{Q}_0^1$ is the flow rate of the continuous medium, then the latter has been correlated by[45]

$$\mathring{Q}_0^1 = \frac{\mathring{Q}_0\{0.127[E_g] + 0.192[Z'] + 0.062[G'] + 0.018[W']\}}{1 - 0.083[E_c]} \quad (8.4.14)$$

Evidently $\mathring{Q}_0^1$ becomes equal to $\mathring{Q}_0$ when the reaction mass becomes homogeneous. With the knowledge of $\mathring{Q}_0$ and $\mathring{Q}_0^1$, the concentrations in the homogeneous phase could be calculated from

$$[E_c] = \frac{\mathring{Q}_0}{\mathring{Q}_0^1}[E_c'] \quad (8.4.15a)$$

$$[E_g] = \frac{\mathring{Q}_0}{\mathring{Q}_0^1}[E_g'] \quad (8.4.15b)$$

$$[Z] = \frac{\mathring{Q}_0}{\mathring{Q}_0^1}[Z], \text{ etc.} \quad (8.4.15c)$$

For a specified reactor pressure, a similar relation as in Eq. (8.2.19) can be used and the flash calculations carried out to determine the amount of ethylene glycol, acetaldehyde, and water evaporated. This specifies terms like $F_V[G]^v$, $F_V[A]^v$, and $F_V[W]^v$ in Table 8.6, and with the knowledge of these, the exit concentrations can be found by solving all balance equations simultaneously.

Computations have been carried out for a three HCSTR sequence with their residence times totaling 3 hr. Even though the final product of the reactor sequence always consists of various oligomers of BHET along with the monomer, as a first approximation for the calculation of the solubility of TPA in the reaction mass, they have taken it as BHET alone. The effect of various choices of the residence times, τ_1, τ_2, and τ_3 of these reactors and temperatures T_1, T_2, and T_3 has been summarized in Table 8.7. On comparing results of rows 2 and 4, it is found that keeping $T_1 < T_2 < T_3$ is a better strategy not only in terms of the final TPA conversion but also in terms of the total DEG formation. In the calculations depicted in rows 5 and 6, residence times τ_1 and τ_3 are varied, and these results suggest that higher residence times should be assigned to the reactor operating at higher temperatures.

TABLE 8.7. Effect of the Choice of T_1, T_2, T_3, τ_1, τ_2, and τ_3 on the Final Conversion of TPA and Side Products ($P_T = 5$ atm, $F \equiv [G]_0/[TPA]_0 = 1.7$)

No.	τ_1 (hr)	τ_2 (hr)	τ_3 (hr)	T_1 (°C)	T_2 (°C)	T_3 (°C)	Final TPA conversion (%)	Acetaldehyde ($10^3 \times$ mole/⬡)	DEG ($10^2 \cdot$ mole/⬡)
1	1	1	1	240	240	240	64.1	2.91	3.90
2	1	1	1	250	250	250	84.6	6.70	7.16
3	1	1	1	260	260	260	90.7	13.40	11.9
4	1	1	1	240	250	260	84.7	7.42	7.05
5	0.5	1	1.5	240	250	260	86.1	8.95	7.68
6	1.5	1	0.5	240	250	260	80.1	6.00	6.38

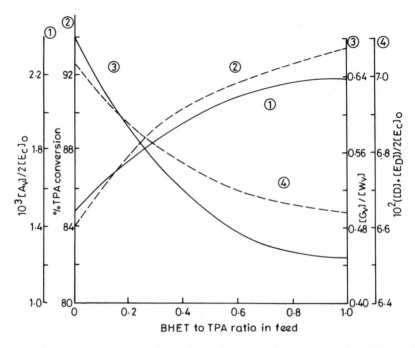

Figure 8.12. Effect of (BHET/TPA) ratio in the feed upon the final conversion of TPA and side products. $T_1 = 240°C$, $T_2 = 250°C$, $T_3 = 260°C$, $P = 5$ atm, $\tau_1 = \tau_2 = \tau_3 = 1$ hr, $F = [G]_0/[TPA]_0 = 1.7$.

In Fig. 8.12, the effect of the recycle of BHET on the final conversion and side product from the three reactor sequence has been examined. As the recycle ratio is increased, the final conversion is found to improve, which is because the solubility of TPA in BHET is larger and more TPA goes into the solution, thus giving higher rates of reaction. In addition to the increase in the solubility of TPA, as the recycle of BHET is increased, the total amounts of ethylene glycol and water evaporated (given on scale 3 of Fig. 8.12) from the reactors reduce as shown. As the BHET recycled is increased, it amounts to increasing E_g groups in the reaction mass, which means that the contribution of reaction 4 of Table 8.1 increases, thus giving higher acetaldehyde formation (scale 1). Similarly with the increase in recyle, the excess E_g can react through reaction 5b as well as 6b, and this way the diethylene glycol group formation (scale 4) is found to be reduced.

In Fig. 8.13, the fraction of ethylene glycol in the feed has been varied and the effect examined for two values of the recycle ratio $F(= BHET/TPA)$. As the fraction of ethylene glycol in the feed is increased, more of TPA goes into the solution, as a result of which the final conversion is found to

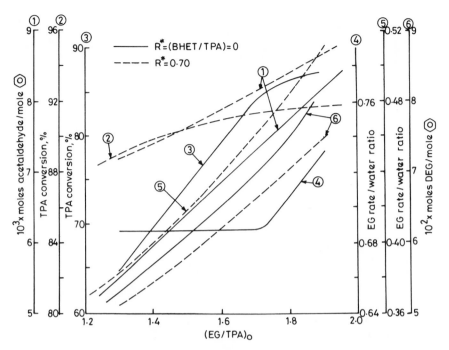

Figure 8.13. Effect of (EG/TPA) in the feed with and without recycle on the concentration of various products in the outlet stream in the direct esterification of TPA. $P = 5$ atm, $T_1 = 240°C$, $T_2 = 250°C$, $T_3 = 260°C$, $\tau_1 = \tau_2 = \tau_3 = 1$ hr.

improve. On scales 2 and 5 of this figure, results for $F = 0.7$ have been plotted, and as can be seen (and expected from the results of Fig. 8.12) higher conversions of TPA and reduced ratio of flow of ethylene glycol and water are obtained. On scales 1 and 6, acetaldehyde and DEG formed have been shown. Naturally as the amount of ethylene glycol in the feed is increased, the total DEG formed would increase. Acetaldehyde is also formed in larger quantities because the moles of E_g groups would increase with the increase in ethylene glycol, and as a result, their degradation to form more acetaldehyde is enhanced.

In Fig. 8.14, the effect of pressure has been examined. On the increase of the pressure of the reactors, the amount of ethylene glycol leaving the reactions mass falls, thus increasing the relative importance of the reverse reactions in Table 8.1. It is thus seen that the conversion of TPA as well as the amount of side products formed fall in magnitude. On adding BHET (shown by dotted lines in the figure), the conversion of TPA improves, the amount of DEG formed is reduced, and the amount of acetaldehyde formed is increased for the reasons explained earlier.

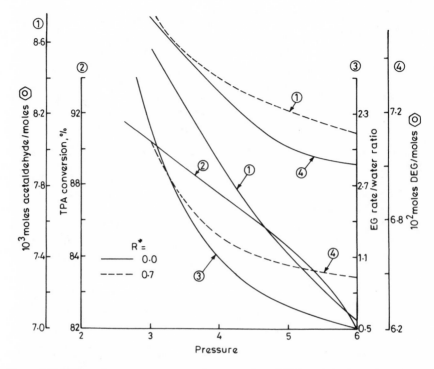

Figure 8.14. Effect of reactor pressure on the conversion of TPA and the formation of side products. $F = 1.7$, $T_1 = 240°C$, $T_2 = 250°C$, $T_3 = 260°C$, $\tau_1 = \tau_2 = \tau_3 = 1$ hr, $F = [G]_0/[TPA]_0 = 1.7$, $R^* = 0.7$.

8.5. POLYCONDENSATION STEP (SECOND STAGE) OF PET FORMATION

8.5a. Side Reactions in Polycondensation Step

The polycondensation step of PET formation consists of the polymerization of BHET. In terms of functional groups, this occurs primarily through the reaction of E_g. Since the various side reactions like the formation of vinyl end groups, acid groups, etc., are the same as those for the transesterification step, the reaction mechanism given in Table 8.1 can easily be adopted. However, as there are very few E_m groups in the reaction mass, steps 1 and 2 of Table 8.1 do not take place. With this simplified kinetic mechanism, it is possible to write down the mole balance equations for semibatch and continuous reactors by substituting $e_m = 0$ in Table 8.3. In the polycondensation step, because of the unfavorable equilibrium, it is necessary to remove ethylene glycol continuously so that high conversion of functional groups is attained. Such an analysis has been done and results for semibatch and continuous reactors calculated.[46,47]

The analysis of the polycondensation step in terms of functional groups alone becomes inadequate because of the following two *additional* reactions:

a. redistribution reaction; and
b. cyclization reaction.

Polyesters which are terminated with hydroxyl and carboxyl groups can in general undergo three kinds of reactions under redistribution. These can be schematically represented by[48]

Intermolecular alcoholysis:

$$\begin{array}{ccccc}
\text{---}O\text{---}CO\text{---} & & & & \\
+ & \rightleftharpoons & \text{---}OH & + & CO\text{---} \\
& & & & | \\
HO\text{---} & & & & O\text{---}
\end{array} \qquad (8.5.1)$$

Intermolecular acidolysis:

$$\begin{array}{ccccc}
\text{---}O\text{---}CO\text{---} & & \text{---}O & & \\
+ & \rightleftharpoons & | & + HO\text{---}CO\text{---} \\
\text{---}COCH & & \text{---}CO & &
\end{array}$$
$$(8.5.2)$$

Transesterification:

$$\begin{array}{ccccc}
\text{---}O\text{---}CO\text{---} & & \text{---}O & & CO\text{---} \\
+ & \rightleftharpoons & | & + & | \\
\text{---}CO\text{---}O\text{---} & & \text{---}CO & & O\text{---}
\end{array}$$
$$(8.5.3)$$

In the studies of the polymerization of BHET in sealed tubes, Challa[9] has followed the concentration of monomer as a function of time. It was assumed that the reaction mass does not have any carboxyl end group $[E_c]$ and there is negligible transesterification reaction occurring. With only intermolecular alcoholysis, the redistribution rate constant was derived.[9] Kotliar[48] points out that the polycondensation and redistribution rates for polyesters are of the same order of magnitude around 254°C but the Arrhenius activation energy for the latter is larger. Since in PET formation the usual reactor temperature is around 282°C, it is implied that the redistribution reaction cannot be ignored. It would be shown in this section that this reaction has no effect on the zeroth and first moments of the MWD, which means that the redistribution reaction cannot be accounted in any analysis involving functional groups alone.

In addition to the redistribution reaction discussed above, commercial polyethylene terephthalate contains approximately 2%–5% cyclic oligomers,[1,49] the amount of which varies with the thermal history of the polymerization. Three mechanisms of cyclic oligomer formations in PET polymerization have been suggested by Goodman and Nesbit[50,51] as follows:

(a) Cyclodepolymerization with the chain end:

$$(8.5.4)$$

(b) Cyclization of short chain oligomers:

$$(8.5.5)$$

(c) Exchange-elimination reactions within or between polymer ester groups:

$$(8.5.6)$$

Peebles *et al.*[52] suggest that among these Eq. (8.5.5) is the main route of cyclization and the studies of Ha and Choun[49] have further confirmed this. In the latter study, commercial polymer was carefully fractionated and the (essentially) monodispersed polymer was exposed to high temperatures (270°C) for 5, 10, 30, 60, and 120 min. If μ_n is the number average chain length of the polymer, then the total concentration, [C], of the cyclic oligomers was found to increase in the polymer sample as

$$C = k_2 t^{0.3} / \mu_n \qquad (8.5.7)$$

where t is the time of exposure to the high temperature. For short times as in these experiments, Ha and Choun point out that the reaction in Eq. (8.5.5) could be assumed to be irreversible. The time-dependent rate constant arises because of the simultaneous degrading of polymer chains at the high temperature of reaction.

8.5b. Rate Constants of Various Reaction Steps[5]

It has already been discussed in Section 8.2, that there is a wide variation in the values of k_3 reported in the literature. In this section, the uncatalyzed rate constants reported by Challa have been used to demonstrate the effect of redistribution and cyclization reactions on the MWD of the polymer formed.

There have been very few studies reported on the cyclization of PET, and in the following discussion, a semiempirical approach has been adopted to find an expression for the cyclization rate constant. The forward step of cyclization as in Eq. (8.5.5) has been the subject of several theoretical studies, and as shown in Section 4.7, the cyclization constant, k_c, is related to the polycondensation rate constant, k_3,[53-59] of (Table 8.1) as

$$k_c = k_3 \left\{ \left(\frac{3}{2\pi \langle r^2 \rangle} \right)^{3/2} \frac{1}{N_{av}} \right\} \qquad (8.5.8)$$

where $\langle r^2 \rangle^{1/2}$ is the radius of gyration of the polymer chain and N_{av} is the Avogadro number.

The cyclization of a linear chain is a reversible reaction and has been represented by Semlyen[56] as

$$P_n \underset{mk_c'(m)}{\overset{k_c(m)}{\rightleftharpoons}} P_{n-m} + C_m, \qquad m < n \qquad (8.5.9)$$

where the reverse reaction can occur between the hydroxyl groups of P_{n-m} and any reacted bond of C_m. Since there are m bonds on C_m, the rate constant for the reverse reaction has been written as $mk'_c(m)$. Semlyen *et al*,[57-59] have measured equilibrium constant $K(m)$ for various values of m at 270°C and attempted to correlate their data using the rotational isomeric state model of the polymer chain. The *empirical* curve-fit of the experimental data of $K(m)$ is found to fall exponentially with n and can be correlated by

$$\frac{1}{K_c(n)} = A_T e^{-B_T/n} \tag{8.5.10}$$

where A_T and B_T are constants which are independent of chain length but are functions of temperature.

The experiments on the cyclization studies discussed above are not directly applicable to the polymerizing mass in the polycondensation reactors for the following reasons. In addition to the reaction in Eq. (8.5.9), the chain ends of P_n can form C_n and in the reverse reaction the latter can interact with ethylene glycol (G) present in the reaction mass as follows:

$$P_n \underset{nk''_c}{\overset{k_c}{\rightleftharpoons}} C_n + G \tag{8.5.11}$$

In addition, as the polymer molecules grow in chain length, its radius of gyration, $\langle r^2 \rangle^{1/2}$, increases, which means that according to Eq. (8.5.9), the probability of its cyclization increases. With the chain-length dependent cyclization rate and equilibrium constant the overall polycondensation step of polymerization cannot be written in terms of functional groups. As a result molecular weight distribution of the polymer must be solved first, from which the number and weight average molecular weights and the polydispersity index of the polymer formed can be calculated.

The reaction mechanism of the polycondensation step of PET formation is summarized in Table 8.8 in terms of the molecular species. In the second stage, the use of monofunctional compounds like cetyl alcohol has also been recommended.[60] As a result there are five major reactions occurring in the reaction mass: polycondensation, reaction with monofunctional compounds, redistribution, cyclization and degradation. Because of these, the reaction mass consists of polymer chains (P_n), monofunctional oligomers (P_{nx}), cyclic oligomers (C_n), and degraded polymers (Z_n). With the mechanism of Table 8.8, it is now possible to write the balance equations for the various species.[61,62]

8.5c. Simulation of the MWD of the Polymer Formed in Polycondensation Reactors

As stated in the beginning of this section, the redistribution reaction is mainly the interaction between the reacted bonds of a polymer molecule

TABLE 8.8. Reaction Mechanism of the Polycondensation Stage
of PET Formation

1. Polycondensation

$$P_m + P_n \underset{k_p'}{\overset{k_p}{\rightleftharpoons}} P_{m+n} + G$$

2. Reaction with monofunctional compounds

$$P_m + P_{nx} \underset{k_m'}{\overset{k_m}{\rightleftharpoons}} P_{(m+n)x} + G$$

3. Redistribution

$$P_m + P_n \overset{k_r}{\longrightarrow} P_{m+n-r} + P_r$$

4. Cyclization

  a.

$$P_m \underset{mk_c'}{\overset{k_c}{\rightleftharpoons}} C_m + G \qquad m \geq 2$$

  b.

$$P_m \underset{nk_c''}{\overset{k_c}{\rightleftharpoons}} C_{m-n} + P_n \qquad m \geq 3$$

5. Degradation

$$P_m \overset{k_d}{\longrightarrow} Z_r + Z_{m-r}$$

$$P_n = \text{OH} - \text{CH}_2\text{CH}_2\text{O} \left[\overset{O}{\overset{\|}{C}} - \bigcirc - \overset{O}{\overset{\|}{C}}\text{OCH}_2\text{CH}_2\text{O} \right]_n \text{H}$$

$$P_{1x} = \text{CH}_3(\text{CH}_2)_{14}\text{OH}$$

$$P_{nx} = \text{CH}_3(\text{CH}_2)_{14} - \text{O} \left[\overset{O}{\overset{\|}{C}} - \bigcirc - \overset{O}{\overset{\|}{C}}\text{OCH}_2\text{CH}_2\text{O} \right]_{n-1} \text{H}$$

$$C_n = \text{O} \boxed{- \overset{O}{\overset{\|}{C}} - \bigcirc - \overset{O}{\overset{\|}{C}}\text{OCH}_2\text{CH}_2\text{O} -}_n$$

$$Z_n = \text{H} \left[\text{OCH}_2\text{CH}_2\text{O}\overset{O}{\overset{\|}{C}} - \bigcirc - \overset{O}{\overset{\|}{C}} \right]_n \text{OCH}=\text{CH}_2$$

and the OH group of another molecule. In view of this, not only P_x and P_y, but P_x, P_{yx}, and C_z can also react through this mechanism. Since the concentrations of monofunctional and cyclic oligomers are small, only P_x and P_y have been assumed to interact in this way. Symbolically, this can be written as

$$P_x + P_y \rightarrow P_{x-r} + P_{y+r} \qquad (8.5.12)$$

where r can take any value except $r = (x - y)$ when the reactants and products become identical.

It is observed that P_1 cannot react with P_1 nor with P_2. If P_3 reacts with another molecule of P_3, the molecular species that can be formed are P_1 and P_5 or P_2 and P_4. Similarly when P_3 reacts with P_4, the molecular species that can be formed are either P_1 and P_6 or P_5. This exercise is repeated and the various reactions between P_x and P_y are recorded. From this list of reactions, all those reactions which form P_n are collected and it is found that P_n can be formed in two distinct ways:

(a) by a *process of elimination* in which a reacted bond of P_x having chain length greater than n undergoes a redistribution reaction such that $(x - r) = n$; and

(b) by a *process of combination* in which a given molecule P_y having chain length less than x combines with a part of a chain of the other molecule such that $(y + r) = n$.

It is further observed that P_n disappears when

(a) the $(n - 1)$ bonds of P_n react with the hydroxyl group of any other molecule; and

(b) the OH groups of P_n react with any of the reacted bonds of any other molecule.

To decide over the reactivity of the reaction in Eq. (8.5.11), as an example, the reaction of the OH groups of P_3 with the various sites of P_5 are considered. The structure of P_5 is

Evidently the redistribution reactions can occur at 1a and b to 4a and b of P_5. If the reaction occurs at 1a, two P_4 molecules are formed, and if it occurs at 1b, P_1 and P_7 are formed. Similarly P_3, P_5 and P_2, P_6 are formed due to the reactions at 2a and 2b, P_2, P_6 and P_3, P_5 by reactions at 3a and 3b, and P_1, P_7 and $2P_4$ by reactions at 4a and 4b, respectively. It can thus be seen that there are two sites which give P_2, two sites which give P_3, and so on. If k_r is the rate constant involving an OH group and a reacted bond, the overall reactivity of P_3 and P_5 would therefore be $2 \times 2k_r$ or $4k_r$ because P_3 has two OH groups which can participate in the redistribution reaction.

With the help of the above discussion it is now possible to derive an expression for the rate of formation of P_n through this step. As an example, the rate of formation of P_5 is considered. The total concentration of hydroxyl groups in the reaction mass is equal to $2 \sum_{i=1}^{\infty} [P_i]$. P_5 is formed by the process of elimination whenever any OH group $(=2 \sum_{i=1}^{\infty} [P_i])$ in the reaction mass reacts with the appropriate bond of P_6, P_7, P_8 etc., which occurs at a rate $4k_r$ $(\sum_{i=1}^{\infty} [P_i])(\sum_{j=6}^{\infty} [P_j])$. P_5 is formed by a process of combination when P_1 reacts with P_5, P_6, $P_7, \ldots$; P_2 reacts with P_4, P_5, $P_6, \ldots$; P_3 reacts with P_3, P_4, $P_5, \ldots$; or P_4 reacts with P_2, P_3, $P_4, \ldots$. Therefore the rate of formation of P_5 through combination is equal to $4k_r \sum_{i=1}^{4} \sum_{j=5-i+1}^{\infty} [P_i][P_j]$. It is removed when four of its reacted bonds $(=4[P_5])$ react with the hydroxyl groups $(=2 \sum_{i=1}^{\infty} [P_i])$, therefore the rate of removal is equal to $4k_r(4[P_5]) \sum_{i=1}^{\infty} [P_i]$. P_5 can also be removed when its OH group reacts with the reacted bonds of any other molecule $\{= \sum_{j=2}^{\infty} (j-1)[P_j]\}$, i.e., the rate of reaction is $4k_r[P_5] \sum_{j=2}^{\infty} (j-1)[P_j]$. Following similar arguments, one can write the formation of P_n, $\dot{P}_{n,\mathrm{III}}$ through the redistribution reaction as

$$\dot{P}_{1,\mathrm{III}} = -4k_r[P_1] \sum_{i=2}^{\infty} (i-1)[P_i] + 4k_r \left(\sum_{i=1}^{\infty} [P_i] \right) \left(\sum_{j=2}^{\infty} [P_j] \right) \qquad (8.5.13a)$$

$$\dot{P}_{n,\mathrm{III}} = -4k_r(n-1)[P_n] \sum_{i=1}^{\infty} [P_i] - 4k_r(P_n) \sum_{i=2}^{\infty} (i-1)[P_i]$$

$$+ 4k_r \left(\sum_{i=1}^{\infty} [P_i] \right) \left(\sum_{j=n+1}^{\infty} [P_j] \right) + 4k_r \sum_{i=1}^{(n-1)} \sum_{j=n-i+1}^{\infty} [P_i][P_j], \qquad n \geq 2$$

$$(8.5.13b)$$

The rate of formation for P_1 is different because P_1 does not have any reacted bond and it cannot be formed through the combination step.

Looking at Eq. (8.5.11), it is found that in the redistribution reaction, the total number of polymer molecules and the total number of phenyl rings must remain unchanged. In the fourth reaction of Table 8.8, P_n can either undergo a cyclization reaction or it can react with a cyclic compound.

Similarly P_n can be formed if a C_n reacts with G or C_r $(r < n)$ reacts with P_{n-r}. Therefore the rate of formation, $\dot{P}_{n,\text{IV}}$, of P_n is given by

$$\dot{P}_{2,\text{IV}} = -k_c[P_2] - 2[P_2] \sum_{m=2}^{\infty} mk''_{cm}[C_m] + 4k'_c[G][C_2] \qquad (8.5.14a)$$

$$\dot{P}_{n,\text{IV}} = -k_c[P_n] - k_c(n-2)[P_n] - 2[P_n] \sum_{m=2}^{\infty} mk''_{cm}[C_m]$$

$$+ 2nk'_c[G][C_n] + 2 \sum_{m=2}^{n-1} mk''_{cm}[C_m][P_{n-m}], \qquad n \geq 3 \qquad (8.5.14b)$$

The expression for $\dot{P}_{2,\text{IV}}$ is different because for $n = 2$, the second and fourth terms in Eq. (8.5.14b) vanish because the possibility of formation of C_1 has been excluded. One can now easily write the mole balance relations for various reaction species in batch reactors, and results have been summarized in Table 8.9.[61-65] Various rate constants have been written in dimensionless parameters R_1-R_9.

The mole balance equations above involve a number of rate constants. Since the reaction with monofunctional compounds involve the same functional groups as in the polycondensation step, it can be reasonably assumed that it has the same reactivity as the latter, i.e.,

$$R_2 = 1 \qquad (8.5.15a)$$

and

$$R_3 = R_1 \qquad (8.5.15b)$$

According to Eq. (8.5.8), R_4 is some fraction less than unity depending upon the radius of gyration of the cyclizing polymer molecule. As the relation between the radius of gyration and chain length of short chains is a complex function, as a first approximation, it has been taken as an adjustable parameter such that after the end of polymerization, the total amount of cyclic oligomer formed is no more than about 5% by weight. Various parameter values used in the simulation have been listed in Table 8.10. Before the mass transfer resistance for the removal of the condensation product becomes important (which is beyond 90% conversion), the entire reaction mass can be assumed to be at some uniform concentration given by the vapor-liquid equilibrium existing at the pressure applied. Therefore the effect of the application of vacuum has been simulated using Tables 8.9 and 8.10 by systematically varying the value of y_g in the reaction mass.

TABLE 8.9. The Mole Balance Equations for Various Species in Batch Reactors[61-65]

$$\frac{d[P_n]}{dt} = -4[P_n]k_p \sum_{m=1}^{\infty} [P_m] + 2k_p \sum_{m=1}^{n-1} [P_m][P_{n-m}] + 2k'_p[G](n-1)[P_n]$$

$$+ 2k'_p[G] \sum_{m=n+1}^{\infty} [P_m] - 2k_m[P_n] \sum_{m=1}^{\infty} [P_{mx}] + 2k'_m[G] \sum_{m=n+1}^{\infty} [P_{mx}] - k_c[P_n]$$

$$+ 2k'_c n[G][C_n] + k_r \left\{ -4[P_n] \sum_{m=2}^{\infty} (m-1)[P_m] - 4[P_n](n-1) \sum_{m=1}^{\infty} [P_m] \right.$$

$$\left. + 4 \sum_{m=1}^{\infty} [P_m] \sum_{m'=n+1}^{\infty} [P_{m'}] + 4 \sum_{j=1}^{n-1} \sum_{m=n-i+1}^{\infty} [P_j][P_m] \right\} - k_c(n-2)[P_n]$$

$$+ 2 \sum_{m=2}^{n-1} m[C_m][P_{n-m}]k''_{cm} - k_d(n-1)[P_n] - 2[P_n] \sum_{m=2}^{\infty} m[C_m]k''_{cm}$$

$$\frac{d[P_{nx}]}{dt} = 2k_m[P_{nx}] \sum_{n=1}^{\infty} [P_n] + 2k_m \sum_{m=1}^{n-1} [P_m][P_{n-m}] - 2k'_m[G](n-1)[P_{nx}]$$

$$+ 2k'_m[G] \sum_{m=n+1}^{\infty} [P_{mx}]$$

$$\frac{d[C_n]}{dt} = k_c[P_n] - 2nk'_c[G][C_n] + 2k_c \sum_{m=n+1}^{\infty} [P_m] - 2nk''_{cm}[C_n] \sum_{m=1}^{\infty} [P_m]$$

$$\frac{d[Z_n]}{dt} = 2k_d \sum_{m=n+1}^{\infty} [P_m]$$

$$\frac{d[P_1]}{dt} = -4[P_1]k_p \sum_{m=1}^{\infty} [P_m] + 4k'_p[G] \sum_{m=n+1}^{\infty} [P_m] - 2k_m[P_1] \sum_{m=1}^{\infty} [P_{mx}]$$

$$+ 2k'_m[G] \sum_{m=n+1}^{\infty} [P_{mx}] + k_r \left\{ -4[P_1] \sum_{m=2}^{\infty} (m-1)[P_m] \right.$$

$$\left. + 4 \sum_{j=1}^{\infty} [P_j] \sum_{m=n+1}^{\infty} [P_m] \right\} + 2k_c \sum_{m=2}^{\infty} m[C_m]k''_{cm}$$

$$R_1 = \frac{k'_p}{k_p}, \qquad R_2 = \frac{k_m}{k_p}, \qquad R_3 = \frac{k'_m}{k_p}, \qquad R_4 = \frac{k_c}{k_p[P_1]_0}$$

$$R_5 = \frac{nk'_c}{k_p}, \qquad R_6 = \frac{k_r}{k_p}, \qquad R_8 = \frac{nk''_c}{k_p}, \qquad R_9 = \frac{k_d}{k_p}$$

$$\tau = k_p[P_1]_0 t$$

$$y_{P_{ix}} = \frac{[P_{ix}]}{[P_1]_0}, \qquad i = 1, 2, \ldots$$

$$y_{P_i} = \frac{[P_i]}{[P_1]_0}, \qquad i = 1, 2, \ldots$$

$$y_{C_i} = \frac{[C_i]}{[P_1]_0}, \qquad i = 1, 2, 3, \ldots$$

$$y_g = \frac{[G]}{[P_1]_0}$$

TABLE 8.10. Various Rate Parameters Used in Simulating Noncatalytic
Polycondensation Reactors[a]

Temperature (°C)	R_1	R_2	R_3	R_4	R_6	$10^4 \times R_7$	A_T	B_T
282	2.631	1.0	2.63	0.01	4.78	0.41	2440.6	14
300	2.55	1.0	2.55	0.01	5.76	6.54	2321.6	13.75
320	2.63	1.0	2.63	0.01	7.51	1.03	2208.4	12.5
340	2.71	1.0	2.71	0.01	9.64	1.59	2100.6	11.1
360	2.77	1.0	2.77	0.01	12.20	2.14	1998.2	10.6

$$\ln k_p = -\frac{22.00 \times 10^3}{1.987\,T} + 17.850 \frac{\text{liter}}{\text{mole hr}}$$

$$\ln k_p' = -\frac{23.00 \times 10^3}{1.987\,T} + 19.724 \frac{\text{liter}}{\text{mole hr}}$$

$$\ln k_r = -\frac{30.88 \times 10^3}{1.987\,T} + 27.397 \frac{\text{liter}}{\text{mole hr}}$$

[a] Rate constants have been taken from Reference 9.

In Fig. 8.15, R_4 has been varied with various uniform values of y_g and the weight fraction of the total cyclic polymers plotted as a function of the dimensionless time τ.[65] It may be observed that in the presence of large amounts of ethylene glycol, reaction 4a of Table 8.8 may be more favored than reaction 4b. In Fig. 8.15, results for both reactions 4a and b (shown by solid lines) and reaction 4a only (shown by dashed lines) have been shown. For a given value of R_4 and specified vacuum level (i.e., a given value of y_g), the amount of cyclic polymer by reactions 4a and b is considerably higher compared to when only reaction 4a occurs. As the vacuum is increased, y_g in the reaction mass is lowered and the weight fraction of the cyclic polymer increases and it is the highest for irreversible polymerization (i.e., $y_g = 0.0$). As R_4 is lowered, as expected, the total fraction of total cyclics also falls.

Since the cyclic polymer formed is usually removed from the product, and the monofunctional polymer behaves the same way kinetically as P_n, in the information reported in Figs. 8.16 and 8.17, the polymer is assumed to be a mixture of bifunctional and monofunctional oligomers. The molecular weight distribution has been calculated using

$$W_i = \frac{i[P_i] + i[P_{ix}]}{\sum_{i=1}^{\infty} \{i[P_i] + i[P_{ix}]\}} \tag{8.5.16}$$

In Fig. 8.16, the effect of vacuum has similarly been studied by systematically varying y_g. The curves marked "closed" in Fig. 8.16 indicate the case where ethylene glycol is not removed from the reaction mass, and for this case the MWD does not even undergo a maximum. This is because the conversion of the functional groups even for $\tau = 1.6$ is small owing to the unfavorable equilibrium. As vacuum is applied, y_g takes on smaller and smaller values, thus reducing the contribution of the reverse reactions. The final conversion increases with the reduction in y_g, and at $y_g = 0.1$, the MWD curves just begin to have a maximum. On the same figure the results for $R_4 = 0.1$ with only reaction 4a contribution to the cyclization have been plotted for comparison. It is found that reaction 4b is an important reaction and must also be considered. In Fig. 8.17, the number average molecular weight and the polydispersity index Q have been plotted as a function of the dimensionless time, τ. For a given y_g, the curves rise quickly, almost approaching an asymptotic value for large times of polymerization. As y_g is reduced, these curves settle at higher average molecular weights and the approach towards the asymptote becomes slower and slower.

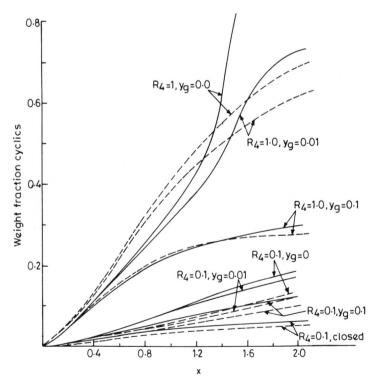

Figure 8.15. Effect of vacuum and R_4 on the total amount of cyclic compound formed in the reaction mass. Dashed line is the result for the case when reaction 4(a) in Table 8.8 alone is considered.

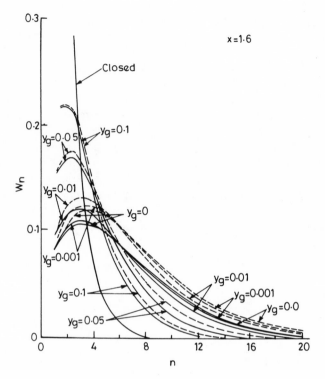

Figure 8.16. Effect of vacuum on weight fraction distribution at $\tau = 1.6$, $R_1 = R_3 = 2.636$, $R_4 = 0.01$, $R_2 = 1.0$, $R_6 = 4.781$, $R_9 = 0.4104 \times 10^{-4}$, $[P_{1x}]_0 = 0.01$.

In Fig. 8.18, the effect of temperature on the weight fractions of degraded polymer and cyclics has been examined. As the temperature of polymerization is increased from 300°C to 360°C, there is about a fivefold increase in the fraction of the degraded polymer at $\tau = 2$. As opposed to this, the fraction of cyclics formed is found to decrease with the increase in temperature; however, this decrease is only small as seen in the figure. The effect of the amount of cetyl alcohol in the feed on the number average molecular weight and polydispersity index has been found to be marginal in the range studied.

8.6. WIPED FILM REACTORS FOR THE FINAL STAGES OF PET FORMATION

In the previous section it has been shown that the formation of PET is not favored because of the high value of the equilibrium constant. As a result, the removal of ethylene glycol is necessary so that the polymerization

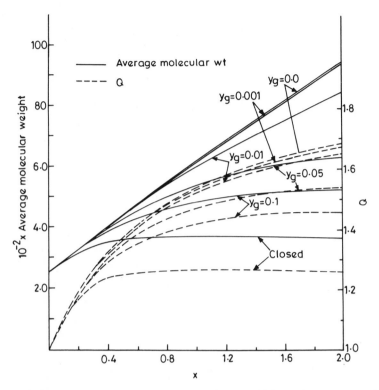

Figure 8.17. Effect of vacuum on number average molecular weight and polydispersity index of the PET formed in batch reactors. $R_1 = R_3 = 2.631$, $R_2 = 1.0$, $R_4 = 0.01$, $R_6 = 4.781$, $R_9 = 0.4104 \times 10^{-4}$, $[P_{1x}]_0 = 0.01$.

is driven more in the forward direction to form PET of high molecular weight. Mass transfer resistance for the removal of ethylene glycol becomes important in industrial semibatch reactors after the average chain length increases beyond 30, and the polymerization is then usually carried out in special wiped film reactors.[66]

As discussed in Chapter 5, there are several geometries of the wiped film reactor. A schematic diagram of a reactor commonly employed industrially for PET is shown in Fig. 8.19. It can be seen that the bulk of the polymer flows in the axial direction, from which some of the polymer is spread on the cylindrical wall as a thin film. This film is scraped after a certain exposure time and mixed with the bulk of the polymer fluid. Fluid transport in spreading and scraping the film has been analyzed by McKelvey,[67] who found that there is a bow wave formation on the edge of the blade. Fluid motion within the bow wave on the blade tends to mix the fluid within it and therefore each of the blades serves as a mixer. As pointed

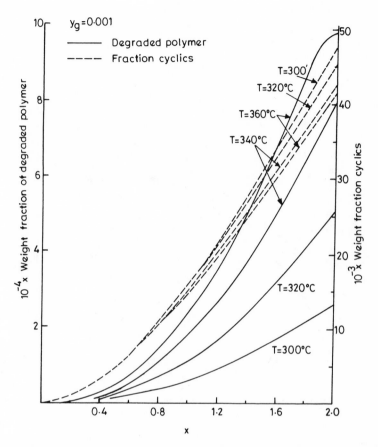

Figure 8.18. Effect of temperature on amount of polymer degraded and cyclized.

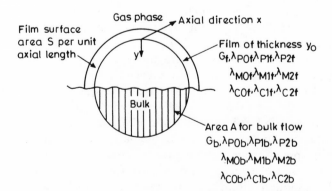

Figure 8.19. Cross section of the wiped film reactor at position x.

out, the axial transport of the fluid has to be assisted by the use of blades, ribbons, screws, etc., and in designing these, it is desired to minimize axial mixing but simultaneously having infinite radial mixing.[68]

In the analysis of wiped film reactors presented in this section, it is assumed that there is no axial mixing but there is infinite radial mixing. It is also assumed that a small amount of polymeric material of the bulk is spread as a film, so that its mixing (after it is scraped) with the bulk contributes negligibly to the change in molecular weight of the polymer. The wiped film reactor has been analyzed in this section for the polymerization mechanism of Table 8.8 after ignoring reactions 4b and 5. The reaction mass consists of polymer molecules (P_n), monofunctional polymer (P_{nx}), and cyclic polymer (C_n), and the analysis of Denson[69] presented in Chapter 5 can easily be extended to describe this complex reaction mass. One starts writing transport equations for each of the molecular species for the film as well as the bulk.[70]

Usually about 300 species are required to calculate the molecular weight distribution of the polymer. This would mean that about 300 equations in the bulk and 300 equations in the film must be solved simultaneously, which is extremely time consuming on even the fastest computer. An equivalent representation would be to combine the sets of the MWD equations in film and bulk into the various moments (λ_{Pi} for polymer, λ_{xi} for monofunctional polymer, and λ_{ci} for cyclic polymer), and usually the first three moments are adequate to describe the MWD.

Using the rate expressions given in Table 8.9, the rates of generation of these moments can be easily derived by summing terms appropriately, and results have been summarized in Table 8.11. As done in Chapter 5, the various transport equations can be similarly combined to determine relations involving moments, and these are given in Table 8.12. If $[P_0]$ is the concentration of the polymer entering the wiped film reactor, equations of Table 8.12 can be easily nondimensionalized using variables given in the table.

TABLE 8.11. Zeroth, First, and Second Moment Generation Equations for Batch Reactors

$$\frac{\mathscr{R}_{P_1}}{k_p C_{p0}^2} \equiv \dot{P}_1 = 4P_1 m_{P0} + 4R_1(m_{p0} - P_1) - 2R_2 P_1 m_{x0} + 2R_3 G(m_{x0} - m_{x1})$$

$$- 4R_6 P_1(m_{P1} - m_{P0}) + 4R_6 m_{P0}(m_{P0} - P_1)$$

$$\frac{\mathscr{R}_{P_{1x}}}{k_p C_{p0}^2} \equiv \dot{P}_{1x} = -2R_2 P_{1x} m_{P0} + 2R_3 G(m_{x0} - P_{1x})$$

$$\frac{\mathscr{R}_{m_{P0}}}{k_p C_{p0}^2} = \dot{m}_{P0} = -m_{P0}^2 + 2R_1 G(m_{P1} - m_{P0}) - 2R_2 m_{x0} m_{P0} + 2R_3 G(m_{x1} - m_{x0})$$

$$+ 2R_5 G m_{c1} - R_4(m_{P0} - P_1)$$

(*continued*)

TABLE 8.11 (*continued*)

$$\frac{\mathcal{R}_{m_{x0}}}{k_p C_{p0}^2} = \dot{m}_{x0} = 0$$

$$\frac{\mathcal{R}_{m_{c0}}}{k_p C_{p0}^2} = \dot{m}_{c0} = R_4(m_{P0} - P_1) - R_5 G m_{c1}$$

$$\frac{\mathcal{R}_{m_{P1}}}{k_p C_{p0}^2} \equiv \dot{m}_{P1} = -2R_2 m_{P1} m_{x0} + R_3 G(m_{x2} - m_{x1}) + 2R_5 G m_{c2} - R_4(m_{c1} - P_1)$$

$$\frac{\mathcal{R}_{m_{x1}}}{k_p C_{p0}^2} \equiv \dot{m}_{x1} = 2R_2 m_{P1} m_{x0} - R_3 G(m_{x2} - m_{x1})$$

$$\frac{\mathcal{R}_{m_{c1}}}{k_p C_{p0}^2} \equiv \dot{m}_{c1} = R_4(m_{P1} - P_1) - R_5 G m_{c2}$$

$$\frac{\mathcal{R}_{m_{P2}}}{k_p C_{p0}^2} \equiv \dot{m}_{P2} = 4m_{P1}^2 - 2R_1 G(m_{P3} - m_{P2}) + 2R_1 G(2m_{P3} - 3m_{P2} + m_{P1})$$

$$- 2R_2 m_{P2} m_{x0} + \tfrac{1}{3}R_3 G(2m_{x3} - 3m_{x2} + m_{x1})$$

$$+ 2R_5 G m_{c3} - R_4(m_{P2} - P_1) + 4R_6\{-m_{P0}(m_{P3} - m_{P2}) - m_{P2}(m_{P1} - m_{P0})$$

$$+ \tfrac{1}{6}m_{P0}(2m_{P3} - 3m_{P2} + m_{P1}) + (m_{P1} - m_{P0})(m_{P2} - 2m_{P1} + m_{P0})$$

$$+ (m_{P1} + m_{P0})(m_{P2} - 3m_{P1} + 2m_{P0}) + \tfrac{1}{6}m_{P0}(2m_{P3} - 9m_{P2} + 13m_{P1} - 6m_{P0})\}$$

$$\frac{\mathcal{R}_{m_{x2}}}{k_p C_{p0}^2} \equiv \dot{m}_{x2} = 2R_2(m_{P1} m_{x0} + 2m_{P1} m_{x1}) - 2R_3 G(m_{x3} - m_{x2}) + \tfrac{1}{3}R_3 G(2m_{x3} - 3m_{x2} + m_{x1})$$

$$\frac{\mathcal{R}_{m_{c2}}}{k_p C_{p0}^2} \equiv \dot{m}_{c2} = R_4(m_{P2} - P_1) - 2R_5 m_{c3}$$

Definitions of variables
Moments

$$\lambda_{Pk} = \sum_{n=1}^{\infty} n^k[P_n], \qquad k = 0, 1, 2, 3$$

$$\lambda_{xk} = \sum_{n=1}^{\infty} n^k[P_{nx}], \qquad k = 0, 1, 2, 3$$

$$\lambda_{ck} = \sum_{n=1}^{\infty} n^k[C_n], \qquad k = 0, 1, 2, 3$$

$$m_{Pk} = \frac{\lambda_{Pk}}{C_{p0}}, \qquad k = 0, 1, 2, 3$$

$$m_{xk} = \frac{\lambda_{xk}}{C_{p0}}, \qquad k = 0, 1, 2, 3$$

$$m_{ck} = \frac{\lambda_{ck}}{C_{p0}}, \qquad k = 0, 1, 2, 3$$

TABLE 8.12. Transport Equations in Film and Bulk of the Wiped-Film Reactors for PET

1. *Film*

$$\frac{\partial G_f}{\partial \tau} = \left(\frac{\mathscr{D}}{y_0^2} t_f\right)\frac{\partial^2 G_f}{\partial \xi^2} + (k_p C_{p0} t_f)\dot{G}_f \tag{1}$$

$$\frac{\partial m_{P0f}}{\partial \tau} + (k_p t_f C_{p0})\dot{m}_{P0f} = 0 \tag{2}$$

$$\frac{\partial m_{P1f}}{\partial \tau} + (k_p t_f C_{p0})\dot{m}_{P1f} = 0 \tag{3}$$

$$\frac{\partial m_{P2f}}{\partial \tau} + (k_p t_f C_{p0})\dot{m}_{P2f} = 0 \tag{4}$$

$$\frac{\partial m_{x1f}}{\partial \tau} + (k_p t_f C_{p0})\dot{m}_{x1f} = 0 \tag{5}$$

$$\frac{\partial m_{x2f}}{\partial \tau} + (k_p t_f C_{p0})\dot{m}_{x2f} = 0 \tag{6}$$

$$\frac{\partial m_{c0f}}{\partial \tau} + (k_p t_f C_{p0})\dot{m}_{c0f} = 0 \tag{7}$$

$$\frac{\partial m_{c1f}}{\partial \tau} + (k_p t_f C_{p0})\dot{m}_{c1f} = 0 \tag{8}$$

$$\frac{\partial m_{c2f}}{\partial \tau} + (k_p t_f C_{p0}) + \dot{m}_{c1f} = 0 \tag{9}$$

$$\frac{\partial P_{1xf}}{\partial \tau} + (k_p t_f C_{p0})\dot{P}_{1xf} = 0 \tag{10}$$

$$\frac{\partial P_{1f}}{\partial \tau} + (k_p t_f C_{p0})\dot{P}_{1f} = 0 \tag{11}$$

$$\bar{N}^*(x) \equiv C_{p0}y_0\bar{N}_G(x) = (C_{p0}y_0)\frac{1}{t_f}\int_0^1 \{(m_{P0f} + m_{c0f} + G_f)_{\tau=1} - (m_{P0f} + m_{c0f} + G_f)_{\tau=0}\}\,d\xi \tag{12}$$

2. *Bulk*

$$-\frac{dG_b}{d\zeta} + (k_p C_{p0}\bar{\theta})\dot{G}_b = \frac{y_0}{Qt_f}S\bar{N}^*(x) \tag{13}$$

$$\frac{dm_{P0b}}{d\zeta} + (k_p C_{p0}\bar{\theta})\dot{m}_{P0b} = 0 \tag{14}$$

$$\frac{dm_{P1b}}{d\zeta} + (k_p C_{p0}\bar{\theta})\dot{m}_{P1b} = 0 \tag{15}$$

$$\frac{dm_{P2b}}{d\zeta} + (k_p C_{p0}\bar{\theta})\dot{m}_{P2b} = 0 \tag{16}$$

(*continued*)

TABLE 8.12 (*continued*)

$$\frac{dm_{x1b}}{d\zeta} + (k_p C_{p0} \bar{\theta}) \dot{m}_{x1b} = 0 \tag{17}$$

$$\frac{dm_{x2b}}{d\zeta} + (k_p C_{p0} \bar{\theta}) \dot{m}_{x2b} = 0 \tag{18}$$

$$\frac{dm_{c0b}}{d\zeta} + (k_p C_{p0} \bar{\theta}) \dot{m}_{c0b} = 0 \tag{19}$$

$$\frac{dm_{c1b}}{d\zeta} + (k_p C_{p0} \bar{\theta}) \dot{m}_{c1b} = 0 \tag{20}$$

$$\frac{dm_{c2b}}{d\zeta} + (k_p C_{p0} \bar{\theta}) \dot{m}_{c2b} = 0 \tag{21}$$

$$\frac{dP_{1b}}{d\zeta} + (k_p C_{p0} \bar{\theta}) \dot{P}_{1b} = 0 \tag{22}$$

$$\frac{dP_{1xb}}{d\zeta} + (k_p C_{p0} \bar{\theta}) \dot{P}_{1xb} = 0 \tag{23}$$

Initial and boundary conditions

a.
$$G_f(y, 0) = G_b(x), \qquad m_{P0f}(y, 0) = m_{P0b}(x)$$
$$m_{P1f}(y, 0) = m_{P1b}(x), \qquad m_{P2f}(y, 0) = m_{P2b}(x)$$
$$m_{x1f}(y, 0) = m_{x1b}(x), \qquad m_{x2f}(y, 0) = m_{x2b}(x)$$
$$m_{c0f}(y, 0) = m_{c0b}(x), \qquad m_{c1f}(y, 0) = m_{c1b}(x)$$

and
$$m_{c2f}(y, 0) = m_{c2b}(x), \qquad P_{1f}(y, 0) = P_{1b}(x), \qquad P_{1xf}(y, 0) = P_{1xf}(x)$$

b.
$$G_f(y_0, t) = C^* \text{ (a constant)}$$

c.
$$\partial G_f / \partial y \big|_{y=0} = 0$$

d.
$$G_b(0) = g_0$$
$$m_{P0b}(0) = p_0, \qquad m_{c0b}(0) = C_0$$
$$m_{P1b}(0) = p_1, \qquad m_{x1b}(0) = m_1, \qquad m_{c1b}(0) = C_1$$
$$m_{P2b}(0) = p_2, \qquad m_{x2b}(0) = m_2, \qquad m_{c2b}(0) = C_2$$

Input conditions taken for the computation

$$g_0 = 0, \qquad p_0 = 0.1, \qquad p_1 = 1.0, \qquad m_2 = 19.0$$
$$m_0 = 0.001, \qquad m_1 = 0.01, \qquad m_2 = 0.019$$
$$C_0 = 0.0, \qquad C_1 = 0.0, \qquad C_2 = 0.0$$
$$P_{10} = 0.01, \qquad P_{1x0} = 1 \times 10^{-6}, \qquad C_{p0} = 0.005$$

(*continued*)

TABLE 8.12 (*continued*)

Dimensionless variables

$$G = \frac{[G]}{[P]_0}, \qquad P_{1x} = \frac{[P_1]}{[P]_0}, \qquad [P_{1x}] = \frac{[P_{1x}]}{[P]_0}$$

$$\xi = y/y_0, \qquad \zeta = x/L, \qquad \tau^* = t/t_f, \qquad \bar{\theta} = LA_b/\dot{Q}$$

$$m_{Pk} = (\lambda_{Pk}/[P]_0), \qquad k = 0, 1, 2, 3$$

$$m_{xk} = \frac{\lambda_{xk}}{[P]_0}, \qquad k = 0, 1, 2, 3$$

$$m_{ck} = \frac{\lambda_{Ck}}{[P]_0}, \qquad k = 0, 1, 2, 3$$

To solve the performance of the wipe film reactor, it is necessary that the concentrations of P_1 and P_{1x} also be known. In addition to this, these equations involve the third moments of P_i, P_{ix}, and C_i distributions, for which some moment closure approximation needs to be made. The method suggested by Tai[71] assumes the following relation between the third moments and the zeroth, first, and second moments:

$$m_{P3} = \frac{m_{P2}(2m_{P2}m_{p0} - m_{P1}^2)}{m_{P0}m_{P1}} \qquad (8.6.1a)$$

$$m_{x3} = \frac{m_{x2}(2m_{x0}m_{x2} - m_{x1}^2)}{m_{x0}m_{x1}} \qquad (8.6.1b)$$

$$m_{03} = \frac{m_{c2}(2m_{c0}m_{c2} - m_{c1}^2)}{m_{c0}m_{c1}} \qquad (8.6.1c)$$

Orthogonal collocation technique (described in Appendix 8.1) has been used to solve eleven partial differential equations in the film and eleven ordinary differential equations in the bulk. G_f is represented in terms of a series[72,73]

$$G_f(\xi) = \sum_{i=1}^{N+1} a_i\{\xi^2\}^{i-1} \qquad (8.6.2)$$

where a_i are constants, independent of ξ and the degree of the polynomial, N, is chosen so as to obtain stable numerical solutions. If ξ_j, $j = 1, 2, \ldots, N + 1$ are the collocation points, then

$$G_{fj} \equiv G_f(\xi_j) = \sum_{i=1}^{N+1} a_i\xi_j^{2i-2}, \qquad j = 1, 2, \ldots, (N + 1) \qquad (8.6.3)$$

These are $(N + 1)$ equations in terms of $(N + 1)$ constants $a_1, a_2, \ldots, a_{N+1}$ and can be written in the following matrix notation:

$$\mathbf{G}_f = \mathbf{QA} \qquad (8.6.4)$$

where

$$Q_{ji} = \xi_j^{2i-2} \qquad (8.6.5a)$$

$$\mathbf{A} = [a_1, a_2, \ldots, aN + 1]^T \qquad (8.6.5b)$$

Similarly

$$\left. \frac{d\mathbf{G}_f}{d\xi} \right|_{\xi_j} = \mathbf{CA} \qquad (8.6.6a)$$

and

$$\left. \frac{d^2\mathbf{G}_f}{d\xi^2} \right|_{\xi_j} = \mathbf{HA} \qquad (8.6.6b)$$

where

$$C_{ji} = \left\{ \frac{d}{d\xi^2} \xi^{2i-2} \right\}_{\xi_j} \qquad (8.6.7a)$$

$$H_{ji} = \left\{ \frac{d^2}{d\xi^2} \xi^{2i-2} \right\}_{\xi_j} \qquad (8.6.7b)$$

From Eq. (8.6.3),

$$\mathbf{A} = \mathbf{q}^{-1}\mathbf{G}_f$$

and $\mathbf{A}$ can be eliminated from Eq. (8.6.6b) to obtain

$$\frac{d^2\mathbf{G}_f}{d\xi^2} = \mathbf{HQ}^{-1}\mathbf{G}_f \equiv \mathbf{VG}_f \qquad (8.6.8)$$

On substituting this in Eq. (1), of Table 8.12, one has

$$\frac{dG_{fj}}{d\xi^2} = \left(\frac{\mathscr{D} t_f}{y_0^2} \right) \sum_{i=1}^{N+1} V_{ji}G_i + (k_p C_{p0} t_f)\dot{G}_{fj}, \qquad j = 1, 2, \ldots, N + 1 \qquad (8.6.9)$$

It can be seen from Eq. (8.6.7a) that the boundary condition (c) of Table 8.12 is automatically satisfied. The boundary condition (b) of the table gives

$$\sum_{i=2}^{N+1} a_i = C^* \qquad (8.6.10)$$

This way the partial differential equation governing the concentration of ethylene glycol in the film is converted into a set of ordinary differential equations [Eq. (8.6.9)] by the orthogonal collocation technique.

To solve the various partial differential equations in the film, the dependence of m_{Pkf}, m_{xkf}, m_{ckf}, etc. on ξ is assumed to be given by similar equations as in Eq. (8.6.1) as

$$m_{Pkf} = \sum_{i=1}^{N+1} \pi_{ki} \xi^{2i-2}, \qquad k = 0, 1, 2; \qquad j = 1 \text{ to } N+1 \qquad (8.6.11a)$$

$$m_{xkf} = \sum_{i=1}^{N+1} \nu_{ki} \xi^{(2i-2)}, \qquad k = 1, 2; \qquad j = 1 \text{ to } N+1 \qquad (8.6.11b)$$

$$m_{ckf} = \sum_{i=1}^{N+1} C_{ki} \xi^{2i-2}, \qquad k = 0, 1, 2; \qquad i = 1, 2, 3, \dots, (N+1) \qquad (8.6.11c)$$

$$P_{1f} = \sum_{i=1}^{N+1} t_i \xi^{2i-2} \qquad (8.6.11d)$$

$$P_{1xf} = \sum_{i=1}^{N+1} w_i \xi^{2i-2} \qquad (8.6.11e)$$

With the help of these, the film equations, other than that for ethylene glycol, can similarly be reduced to ordinary differential equations at the collocation points as follows:

$$\left\{ \frac{dm_{Pkf}}{d\tau} + k_p t_f C_{p0} \dot{m}_{Pkf} \right\}_{\xi_j} = 0, \qquad k = 0, 1, 2; \qquad j = 1 \text{ to } N+1 \qquad (8.6.12a)$$

$$\left\{ \frac{dm_{xkf}}{d\tau} + k_p t_f C_{p0} \dot{m}_{xkf} \right\}_{\xi_j} = 0, \qquad j = 1, 2; \qquad j = 1 \text{ to } N+1 \qquad (8.6.12b)$$

$$\left\{ \frac{dm_{ckf}}{d\tau} + k_p t_f C_{p0} \dot{m}_{ckf} \right\}_{\xi_j} = 0, \qquad k = 0, 1, 2; \qquad j = 1 \text{ to } N+1 \qquad (8.6.12c)$$

$$\left\{\frac{dP_{1f}}{d\tau} + k_p t_f C_{p0} \dot{P}_{1k}\right\}_{\xi_j} = 0, \qquad j = 1 \text{ to } N+1 \tag{8.6.12d}$$

$$\left\{\frac{dP_{1xf}}{d\tau} + k_p t_f C_{p0} \dot{P}_{1xf}\right\}_{\xi_j} = 0, \qquad j = 1 \text{ to } N+1 \tag{8.6.12e}$$

The solution of the set of ordinary differential equations (8.6.16) and (8.6.12) can be solved using the method of Runge and Kutta of fourth order. This would given the various moments m_{Pk}, m_{xk}, and m_{ck}, $k = 0, 1, 2$ at the collocation points. These can now be substituted into Eq. (8.6.9) and constants π_{ki}, ν_{ki}, and c_{ki} can be determined at $\tau = 1$ as

$$\boldsymbol{\pi}_k = \mathbf{Q}^{-1}\mathbf{m}_{Pkf}\big|_{\tau=1}, \qquad k = 0, 1, 2 \tag{8.6.13a}$$

$$\boldsymbol{\nu}_k = \mathbf{Q}^{-1}\mathbf{m}_{xkf}\big|_{\tau=1}, \qquad k = 1, 2 \tag{8.6.13b}$$

$$\mathbf{c}_k = \mathbf{Q}^{-1}\mathbf{m}_{ckf}\big|_{\tau=1}, \qquad k = 0, 1, 2 \tag{8.6.13c}$$

To calculate the flux $\bar{N}^*(x)$ of ethylene glycol from the film it is necessary that the ξ profiles of m_{P0f} and m_{C0f} be known. These can be determined once π_0 and c_0 are known. It is observed that the integral, I, of the ξ profile of m_{P0} gives

$$I = \int_0^1 \sum_{i=1}^{N+1} \pi_{0i} \xi^{2i-2} \, d\xi$$

$$= \sum_{i=1}^{N+1} \frac{\pi_{0i}}{(2i-1)} \tag{8.6.14}$$

Thus, $\bar{N}_G(x)$ at axial position x is given by

$$\bar{N}_G(x) = \frac{C_{p0}y_0}{t_f}\left\{\left[\sum_{i=1}^{N+1} \frac{\pi_{0i} + c_{0i} + a_i}{(2i-1)}\right]_{\tau=1} - [m_{P0b} + m_{c0b} + G_b]_{\tau=0}\right\} \tag{8.6.15}$$

The calculation procedure can now be summarized as follows:

(a) At a given position x, the film equations (8.6.9) and (8.6.12) are solved from $\tau = 0$ to $\tau = 1$ using the Runge–Kutta method of fourth order.

(b) The values of m_{P0f}, m_{C0f}, and G_f at $\tau = 1$ at the various collocation points are used to calculate the coefficients π_{0i}, c_{0i}, and a_i ($i = 1$ to $N+1$).

(c) The flux, $\bar{N}_G(x)$, of ethylene glycol is calculated using Eq. (8.6.15).

(d) This $\bar{N}_G(x)$ is substituted in the bulk equations and the next incremental values of moments at position $(x + \Delta x)$ the bulk are calculated using the Runge-Kutta method of fourth order.

Successive repetition of steps (a)-(d) gives the axial dependence of the various moments in the wiped film reactor. In this calculation, the feed to the reactor has been assumed to be the output of the second stage of polymerization where 90% conversion of functional groups has already been achieved. Assuming the MWD to be approximately given by Flory's distribution, the inlet values of the various moments have been calculated and are given in Table 8.12.

Denson has pointed out that the choice of k_p appears to be an important parameter which determines the final molecular weight of the polymer. The value of k_p for the calculations in wiped film reactors was taken for the catalyzed polymerization and is the same as k_3 of Table 8.3. In the calculation of Fig. 8.15, R_4 was treated as a parameter and was adjusted in such a way that at the end of the reactor no more than 5% cyclic polymer is formed. This artificial method of finding R_4 makes this choice dependent not only upon the reactor residence time but also upon other design parameters of the reactor. As a result of this, two values of R_4, viz., 0.01 and 0.001, have been examined in this simulation. The values of the kinetic parameters R_1-R_6 are given in Table 8.13, and the difference between this table and Table 8.10 arises mainly because of the value of k_p used.

To be able to obtain stable numerical solutions, at least ten collocation points (i.e., $N = 10$) need to be used. In Fig. 8.20, the effect of the film surface areas upon the average chain length, μ_n, of the polymer at the end of the reactor has been examined. When S is increased, μ_n increases very rapidly at some critical value and quickly attains an asymptotic value. Similar behavior is observed for the polydispersity index as shown in Fig.

TABLE 8.13. Rate Parameters R_1-R_6 at Various Temperatures[a]

Temperature (°C)	R_1	R_2	R_3	R_4	R_5	R_6
280	2.0	1.0	2.0	0.01/0.001	0.02/0.002	0.248
290	2.0	1.0	2.0	0.01/0.001	0.02/0.002	0.303
300	2.0	1.0	2.0	0.01/0.001	0.02/0.002	0.368
320	2.0	1.0	2.0	0.01/0.001	0.02/0.002	0.531
340	2.0	1.0	2.0	0.01/0.001	0.02/0.002	0.747
360	2.0	1.0	2.0	0.01/0.001	0.02/0.002	1.03

[a] $\ln k_p = -(18.5/1.987)/T \times 10^3 + 13.4298$ liter/mole min;[15]
$\ln k_r = -(30.88/1.987)/T \times 10^3 + 27.397$ liter/mole hr.[9]

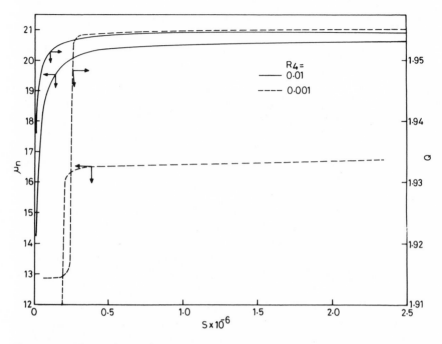

Figure 8.20. Effect of film surface area S on μ_n and Q of the polymer at $\zeta = 1$. $y_0 = 1$ cm, $t_f = 1.0$ sec, $\bar{\theta} = 2000$ sec, $C^* = 0.002$, $T = 280°C$.

8.20 for two values of R_4. For lower values of S, the overall polymerization is mass transfer controlled, as a result of which μ_n is found to improve with increasing S. However, for large S, the overall polymerization is reaction controlled and μ_n becomes independent of S. From Fig. 8.20, it may be observed that to get a $\mu_n = 100$, a very high value of surface area should be used. To have such a large value of S, film must have a large concentration of small bubbles, which in reality is the case. A better model for the film is the diffusion of G from the film to these bubbles as discussed in Chapter 5. In Figs. 8.21 and 8.22, the effects of reactor residence times and temperature have been examined. If the goal is to achieve high molecular weights of the polymer, they appear to be as important parameters as the film surface area, S. In Fig. 8.23, the weight fraction of cyclic compounds formed at $\zeta = 1$ is plotted for the variation of T and $\bar{\theta}$. The weight fraction of cyclic polymer is found to increase tenfold, whereas the average chain length increases by a factor of 3 as the temperature of polymerization is increased from 260 to 360°C. As opposed to the increase in T, when $\bar{\theta}$ is increased from 100 to 6000 sec, the weight fraction of cyclics increases only threefold and μ_n is found to increase by the same amount. Thus it appears

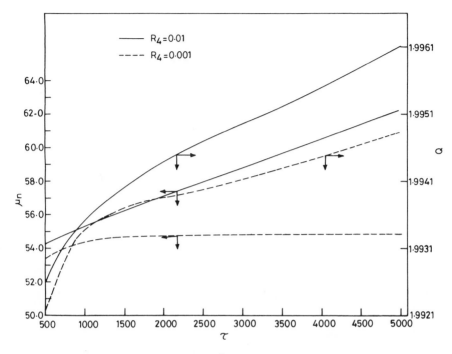

Figure 8.21. Effect of reactor residence time $\bar{\theta}$ on μ_n and Q of PET formed at $\zeta = 1$. $y_0 = 1$ cm, $t_f = 1.0$ sec, $C^* = 0.0002$, $S = 3 \times 10^5$ cm^2, $T = 280°$C.

that the increase in T has less effect on the formation of cyclic polymer. To obtain high μ_n, it appears that both T and $\bar{\theta}$ need to be monitored.

The variation of C^* has been examined in Fig. 8.24 and it is found that μ_n and Q fall drastically as C^* is increased from 10^{-5} to 10^{-2}. In this range of C^*, the total amount of cyclic polymer formed was negligibly affected. The increase in μ_n with the fall in C^* can be explained as follows. With the reduction in C^*, the flux $\bar{N}_G(x)$ of ethylene glycol increases, and with larger removal of ethylene glycol from the film, μ_n of the polymer goes up as observed in Fig. 8.24. As the rotational speed of the drum in the wiped film reactor is increased, the exposure time t_f is reduced. The fraction cyclic polymer formed in the range of t_f lying between 0.5 and 3 sec was found to be negligibly affected. There is a slight fall in μ_n as t_f is increased, but when t_f is reduced to very low values, μ_n shoots up because of the following reason: As t_f is reduced, flux $N_G(x)$ goes up because the former appears in the denominator and the polymerization in the bulk proceeds as if it is irreversible.

There have been several attempts in the literature to model the third stage of PET formation[74-77] but none of these studies account for the side

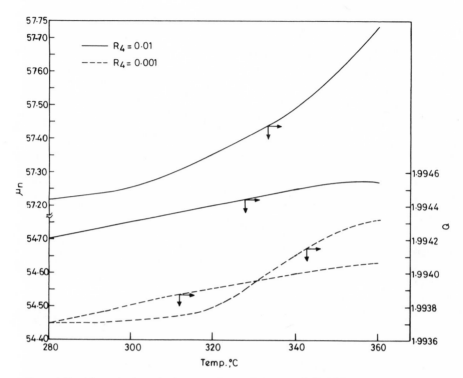

Figure 8.22. Effect of polymerization temperature T on μ_n and Q of PET at $\zeta = 1$. $y_0 = 1$ cm, $t_f = 1$ sec, $\bar{\theta} = 2000$ sec, $C^* = 0.0002$, $S = 3 \times 10^5$ cm^2.

reactions. In these kinetic schemes, only the polycondensation step has been considered. Ault and Mellichamp[74] have ignored the polymerization in the bulk, and the chain growth has been assumed to occur mainly in the film. Since the removal of ethylene glycol is diffusion controlled, transport equations for this static film have been written and solved. These are nonlinear partial differential equations and the nonlinearity arises because of the rate terms. In an alternate approach,[75-77] transport equations have been linearized and the penetration theory solution, so developed, has been tested against available (but limited) experimental results.[25-27,78-80]

8.7. UNSATURATED POLYESTER

The commercially available unsaturated polyester resin usually contains the following ingredients:[81]

a. A polyester;
b. Polymerizable monomer, such as, for example, styrene or a mixture of styrene and methyl methacrylate which serves as a solvent;

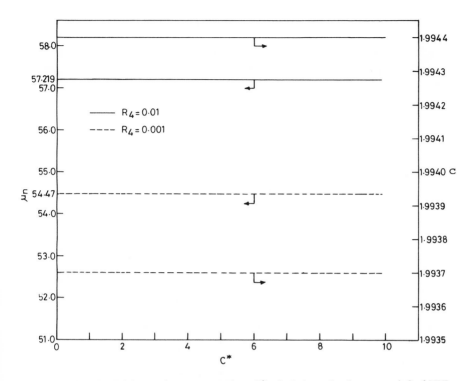

Figure 8.23. Effect of film surface concentration, C^*, of ethylene glycol on μ_n and Q of PET at $\zeta = 1$. $y_0 = 1$ cm, $t_f = 1.0$ sec, $\bar{\theta} = 2000$ sec, $S = 3 \times 10^5$ cm^2, $T = 280°C$.

c. An inhibitor to give a long storage life to the resin under ambient conditions and to moderate the curing of the resin so as to prevent the formation of cracks; hydroquinones, catechols in small amounts serve the purpose of an inhibitor.

The polyesters are prepared by the polymerization of glycols either with anhydrides or acids. Unsaturated polyesters are prepared by using unsaturated anhydride (for example, maleic anhydride) or a diacid (fumaric acid). The polyesterification in presence of ethylene glycol occurs as

$$OH-CH_2CH_2-OH + \underset{\substack{\displaystyle HC-C\diagup \\ \|\\ O \\ \text{Ethylene}\\ \text{maleate}}}{\overset{\displaystyle HC-C\diagdown}{}} \overset{O}{\underset{O}{\diagup}} \xrightarrow{225-240°C} OH(CH_2CH_2OC\underset{\|}{\overset{H}{\underset{O}{C}}}=\underset{\|}{\overset{H}{\underset{O}{C}}}CO)_nH + nH_2O$$

(8.7.1)

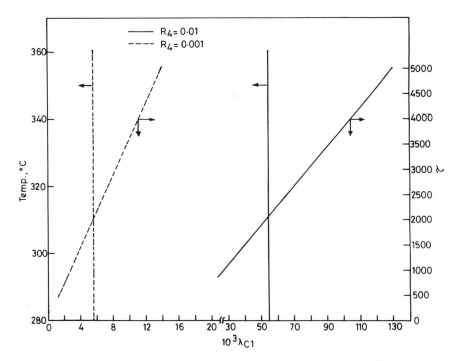

Figure 8.24. Effect of polymerization temperature T and reactor residence time $\bar{\theta}$ on the weight fraction of cyclic polymer $\zeta = 1$. $y_0 = 1$ cm, $t_f = 1$ sec, $C^* = 0.0002$, $S = 3 \times 10^5$ cm^2.

The polymerization is usually carried out batchwise in the temperature range of 227–250°C in which the double bond of the polymer is also found to react to some extent with ethylene glycol. There are very few simulation studies existing for the formation of polyester, which accounts for the reaction of double bonds also. If this reaction is neglected, the polyester formation from ethylene maleate can be approximately represented by the polymerization of cyclic monomers, which is discussed in Chapter 10 in detail. The MWD of an unsaturated polyester formed by reaction of ethylene glycol with the mixture of maleic anhydride and phthalic anhydride and terminated by cyclohexanol has been studied by fractionation as well as GPC.[82] It was observed that there is a greater amount of oligomers of smaller molecular weight than that predicted by the Flory distribution.

Polyester thus formed is dissolved in styrene and a suitable inhibitor is mixed to increase the shelf life of the unsaturated polyester resin. The normal method of curing the polyester resin is to use a suitable initiator like, for example, peroxides. These can be roughly divided as low, medium, and high temperature initiators and are listed in Table 8.14. Sometimes an

TABLE 8.14. Peroxide Catalysts and Temperature of Use in
Unsaturated Polyester Resins

Temperature range of 20–60°C	
t-Butyl hydroperoxide[a]	Pinane hydroperoxide[a]
cumene hydroperoxide[a]	Methyl ethyl ketone peroxide
Temperature range of 60–120°C	
Benzoyl peroxide	Methyl ethyl ketone peroxide
1-Naphthyl peroxide	Lauroyl peroxide
Temperature range of 120–150°C	
t-Butyl perbenzoate	Di-t-butyl peroxide
Dicumyl peroxide	Di-t-butyl diperphthalate

[a] Normally used with an accelerator.

accelerator is also used; this is a compound in the presence of which the initiator decomposes to give free radical at considerably lower temperature.

There are several uses to which the polyester resin can be put, and most of it is consumed in coating and molding industries. Glass-fiber-reinforced sheets are prepared by impregnating a mat of glass fibers with the polyester resin mixed with an oxide of alkaline earth metals (for example, calcium, magnesium, or zinc). The polyester resin used in the sheet molding should be acid group terminated, and the reaction that occurs between the polyester resin and the metal oxide is called a thickening reaction. During this reaction, the viscosity of the resin increases greatly till the impregnated glass sheet becomes nontacky. In the following, a kinetic model has been presented which accounts for the increase in the viscosity of the resin.[83]

It has been demonstrated that the metal oxide does not react either with styrene or with the unsaturation of the polyester molecules[84] and the growth of polyester chains has been suggested to occur in two steps. In the first one, a mole of metal oxide (denoted by MO) adds on to a mole of acid terminated polyester resin. In the final step (sometimes called the neutral salt formation step), the compound so formed reacts with a polyester molecule to give the chain growth as shown below:

Basic salt formation:

$$HOOC\text{\small ∿}COOH + MO \rightleftarrows HOMOOC\text{\small ∿}COOH \qquad (8.7.2)$$

Neutral salt formation:

$$HOOC\text{\small ∿}COOH + HOMOOC\text{\small ∿}COOH \rightleftarrows$$

$$HOOC\text{\small ∿}COOMOOC\text{\small ∿}COOH + H_2O \qquad (8.7.3)$$

It may be recognized that the reaction mass is heterogeneous in nature. Solid metal oxide first gets into solution and the dissolved MO reacts with the acid functional groups to give the basic salt, which reacts further with another acid group. These are represented by:

$$MO(solid) \rightarrow MO(solution) \qquad (8.7.4a)$$

$$MO(solution) + HOOC\backsim \overset{K_1}{\rightleftharpoons} HOMOOC\backsim \qquad (8.7.4b)$$

$$\backsim COOH + HOMOOC\backsim \overset{K_2}{\rightleftharpoons} \backsim COOMOOC\backsim + H_2O \quad (8.7.4c)$$

Above K_1 and K_2 are the equilibrium constants of reactions (8.7.4b) and (8.7.4c), respectively. Since the basic salt formation is a reaction between a strong acid and a base, it is therefore assumed to be fast in comparison to other reactions and hence in equilibrium all the time, i.e.,

$$K_2 = \frac{b}{m_s[COOH]} \qquad (8.7.5)$$

where b is the concentration of the basic salt, m_s is the saturation concentration of the metal oxide, and [COOH] is the concentration of ester molecules (which is also equal to the concentration of functional groups).

One defines S_e and S_w as the moles of metal oxide soluble in dry polyester and water, in terms of which m_s can be written as

$$m_s = S_e[COOH] + S_w[W] \qquad (8.7.6)$$

where [W] is the concentration of water in the liquid phase. If $[COOH]_0$ is the initial concentration, by stoichiometry, the number of moles of $\backsim COOMOOC\backsim$ formed is given by

$$[\backsim COOMOOC\backsim] = [COOH]_0 - [COOH] - b \qquad (8.7.7)$$

Since for every mole of $\backsim COOMOOC\backsim$ formed, one mole of water is produced, the concentration of water, [W], in the liquid phase, at any time is given by

$$[W] = [W]_0 + [COOH]_0 - [COOH] - b \qquad (8.7.8)$$

where $[W]_0$ is the initial concentration of water. In terms of these variables, Eq. (8.7.5) can be rewritten as

$$b = \frac{K_1\{S_e[COOH]^2 + S_w[COOH]([W]_0 + [COOH]_0 - [COOH])\}}{1 + K_1 S_w[COOH]} \qquad (8.7.9)$$

The rate of polymerization is written with the help of Eq. (8.7.4c) as follows. The forward reaction is taken as proportional to the product of reactants, whereas the reverse reaction is assumed to be proportional to $[W][\text{~~COOMOOC~~}]/\mu_n$, where μ_n is the average chain length of the polymer. In this, $[\text{~~COOMOOC~~}]/\mu_n$ is the fraction of molecules containing metal carboxylate bond and μ_n at any given time is equal to the initial number of molecules (i.e., $[COOH]_0$) divided by the number of molecules at the time of interest (i.e., $[COOH] + b$). Therefore

$$\frac{d[COOH]}{dt} = k_2 \left\{ [COOH]b - \frac{1}{K_2} [([W]_0 + [COOH]_0 - [COOH] - b)] \right.$$

$$\left. \times \left[\frac{([COOH]_0 - [COOH] - b)([COOH] + b)}{[COOH]_0} \right] \right\} \qquad (8.7.10)$$

and

$$\mu_n = \frac{[COOH]_0}{[COOH] + b} \qquad (8.7.11)$$

Equation (8.7.10) can be easily integrated since b can be found using Eq. (8.7.9) and with the knowledge of $[COOH]$ versus time, the average chain length μ_n can be calculated with the help of Eq. (8.7.11).

To find the weight average molecular weight, the MWD of the polymer is assumed to be given by the Flory's distribution. One rewrites Eq. (8.7.11) as

$$\mu_n = \frac{1}{1 - \left\{ 1 - \frac{[COOH] + b}{[COOH]_0} \right\}} \qquad (8.7.12)$$

and by analogy p in the Flory's distribution $\{ W_n = (1 - p)^2 p^{n-1} \}$ is taken as

$$p = 1 - \frac{[COOH] + b}{[COOH]_0} \qquad (8.7.13)$$

The weight average length of the polymer, μ_w, is taken to be

$$\mu_w = \frac{2 - \dfrac{[COOH] + b}{[COOH]_0}}{\dfrac{[COOH] + b}{[COOH]_0}} \qquad (8.7.14)$$

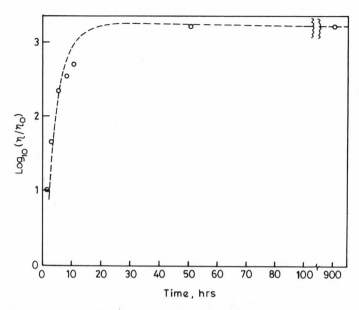

Figure 8.25. Comparison of calculations with experimental data $S_w = 2.8 \times 10^{-6}$ moles of oxide/mole of water, $S_e = 2 \times 10^{-7}$, moles of oxide/moles of polyester, $C_0 = 10^{-1}$ moles/liter, $K_1 = 10^6$ liter/mol, $K_2 = 17.5$ $k_1 = 10^{-2}$ moles/liter sec. (○) Data of Alvey.

The viscosity η (in poise) of the polyester can now be found from the following standard correlations:

$$\log(\eta/\eta_0) = 3.4 \log \mu_w \tag{8.7.15}$$

where η_0 is the viscosity (in poise) of the resin at $t = 0$. Or,

$$\log(\eta/\eta_0) = 3.4 \log \left(\frac{2 - \dfrac{[COOH] + b}{[COOH]_0}}{\dfrac{[COOH] + b}{[COOH]_0}} \right) \tag{8.7.16}$$

This expression for the change in viscosity has been used to test against the experimental data of Alvey, and it can be seen from Fig. 8.25 that Eq. (8.7.16) predicts the thickening reaction very well.

8.8. CONCLUSIONS

Polyesters can be roughly divided into saturated and unsaturated polymers, and in the former category, polyethylene terephthalate is the most important. Commercially, PET is manufactured in three stages, of which the first stage consists of the synthesis of monomer bis-hydroxyethyl terephthalate (BHET). BHET could be prepared either through the transesterification of dimethyl terephthalate (DMT) or the direct esterification of terephthalic acid (TPA) by ethylene glycol. It has been shown that the design of the reactor is not only governed by the high conversion of DMT or TPA, but also by an effort to minimize the formation of side products. A detailed kinetic scheme has been presented for both these routes of the monomer synthesis and these reactors have been carefully simulated and optimized. These studies suggest that a high temperature should be used initially to obtain high conversions of DMT, but the reactor temperature should be lowered later on to minimize the formation of side products.

The kinetic model presented for the first stage of PET formation gives only the concentrations of functional groups. There are several reactions like redistribution and cyclization which are extremely important and cannot be ignored in the second and third stages of PET formation. However, to account for them, it is necessary to consider molecular weight distribution as a whole. These reactions have been modeled mathematically and the MWD derived for batch reactors and HCSTRs in the second stage. To solve for the third stage of the PET formation, moment equations have been derived from the rate expressions and transport equations have been written for the bulk and the film. The partial differential equations for the film have been converted into ordinary differential equations at the collocation points using the orthogonal collocation technique. Using a suitable moment closure technique to break the hierarchy of equations, these have been solved on a computer and a considerable time saving has been found to occur compared to the finite difference method of solution. Solution of the final stages of PET reactors suggests that there is relatively little advantage in increasing the film surface area after some critical value. Among the various design variables, reactor residence time, $\bar{\theta}$, and temperature, T, are found to affect the average chain length of the polymer the maximum. However, increasing $\bar{\theta}$ to obtain higher μ_n is a better strategy compared to increasing T because the latter is found to give a considerably larger amount of side products.

Unsaturated polyesters are usually prepared by polymerization of an anhydride (for example, maleic anhydride) with a diol (for example, propylene glycol) and the polyester resin is obtained by mixing this with styrene and a suitable inhibitor to improve its shelf-life. The resin is mainly used in coating and molding. In sheet molding, a mat of glass fibers is prepared, which is impregnated with the resin mixed with alkaline earth metal oxides. Viscosity of the resin increases due to polymerization, which occurs in two steps. In the first step, the metal oxide reacts with the acid group ended polyester molecules giving rise to the formation of basic salts. In the second step, these react with another polymer molecule giving rise to the increase in molecular weight through the formation of metal carboxylate bonds. A kinetic model has been presented which yields molecular weight of the polymer, and using standard viscosity correlations, viscosity of the resin as a function of time has been predicted.

APPENDIX 8.1. ORTHOGONAL COLLOCATION TECHNIQUE FOR THE SOLUTION OF PARTIAL DIFFERENTIAL EQUATIONS

Several numerical procedures have been suggested for the solution of nonlinear partial differential equations in the literature. Among these the method of weighted residuals has been gaining importance recently. In this method, one takes a set of basis function $\phi_n(z)$ such that these are preferably orthogonal with some weighting function $\rho(z)$:

$$\int_0^1 \rho(z)\phi_n(z)\phi_m(z)\, dz = 0 \qquad (A8.1.1)$$

In addition to the satisfaction of the orthogonality condition, it should also satisfy the boundary conditions. The approximate solution of the partial differential equations is then expressed in terms of N basis functions. In the method of weighted residuals, if the partial differential equations are represented by

$$\frac{\partial x}{\partial t} - A_2\frac{\partial^2 x}{\partial z^2} - A_1\frac{\partial x}{\partial z} - A_0 x - f(x, z, t) = 0 \qquad (A8.1.2)$$

and $\hat{x}(z, t)$ represents the approximate solution in terms of basis functions, the *residual* $R(z, t)$ is computed as

$$R(z, t) = \frac{\partial \hat{x}}{\partial t} - A_2\frac{\partial^2 \hat{x}}{\partial z^2} - A_1\frac{\partial \hat{x}}{\partial z} - A_0\hat{x} - \hat{f} = 0 \qquad (A8.1.3)$$

The goodness of approximation of the solution in terms of basis functions is measured by the extent to which $R(z, t)$ is made small

$$\int_0^1 w_i(z)R(z, t)\, dz = 0, \qquad i = 1, 2, \ldots, N+1 \qquad (A8.1.4)$$

Above $w_i(z)$ are the set of weighting functions to be chosen. In the orthogonal collocation method, w_n are chosen to be delta functions

$$w_n = \delta(z - z_n), \qquad n = 1, 2, \ldots, (N+1) \qquad (A8.1.5)$$

which on substitution in Eq. (A8.1.4) leads to

$$R(z_n, t) = 0, \qquad n = 1, 2, \ldots, (N + 1) \tag{A8.1.6}$$

and the differential equation is required to be solved exactly at N points, called collocation points.

In a recent version of the collocation method, orthogonal polynomials are used as basis functions and $\hat{x}$ is written as

$$\hat{x}(z) = \sum_{i=1}^{N+1} a_i (z^2)^{i-1} \tag{A8.1.7}$$

where a_i are the coefficients yet to be determined. The N collocation points are determined by computing the roots of $P_N(z^2) = 0$, where $P_N(z^2)$ is a polynomial of degree N in z^2, which are extensively tabulated in the literature. If the values of x at the collocation points are given by $\hat{x}_1, \hat{x}_2, \ldots,$ then one can write

$$\hat{\mathbf{X}} \equiv \begin{bmatrix} \hat{x}_1 \\ \hat{x}_2 \\ \vdots \\ \hat{x}_{N+1} \end{bmatrix} = \begin{bmatrix} 1 & z_1^2 & z_1^4 & \cdots & z_1^{2N} \\ \vdots & \vdots & \vdots & & \vdots \\ 1 & z_{N+1}^2 & z_{N+1}^4 & \cdots & z_{N+1}^{2N} \end{bmatrix} \begin{bmatrix} a_1 \\ a_2 \\ \vdots \\ a_{N+1} \end{bmatrix} \equiv \mathbf{QA} \tag{A8.1.8}$$

The first and second derivatives of $\hat{x}$ with respect to z at the collocation points can similarly be written as

$$\frac{d\hat{\mathbf{X}}}{dz} = \begin{bmatrix} \dfrac{d\hat{x}}{dz}\Big|_{z_1} \\ \dfrac{d\hat{x}}{dz}\Big|_{z_2} \\ \vdots \\ \dfrac{d\hat{x}}{dz}\Big|_{z_{N+1}} \end{bmatrix} = \begin{bmatrix} 0 & \dfrac{d(z^2)}{dz}\Big|_{z_1} & \cdots & \dfrac{d}{dz}(z^{2N})\big|_{z_1} \\ \vdots & \vdots & & \vdots \\ 0 & \dfrac{dz^2}{dz}\Big|_{z_{N+1}} & \cdots & \dfrac{dz^{2N}}{dz}\Big|_{z_{N+1}} \end{bmatrix} \begin{bmatrix} a_1 \\ a_2 \\ \vdots \\ a_{N+1} \end{bmatrix} \equiv \mathbf{CA}$$

$$\tag{A8.1.9}$$

$$\left(\frac{d^2\hat{\mathbf{X}}}{dz^2}\right) = \begin{bmatrix} \dfrac{d^2\hat{x}}{dz^2}\Big|_{z_1} \\ \vdots \\ \dfrac{d^2\hat{x}}{dz^2}\Big|_{z_{N+1}} \end{bmatrix} = \begin{bmatrix} 0 & \dfrac{d^2z^2}{dz^2}\Big|_{z_1} & \cdots & \dfrac{d^2}{dz^2}(z^{2N})\Big|_{z_1} \\ \vdots & \vdots & & \vdots \\ 0 & \dfrac{d^2z^2}{dz^2}\Big|_{z_{N+1}} & & \dfrac{d^2z^{2N}}{dz^2}\Big|_{z_{N+1}} \end{bmatrix} \begin{bmatrix} a_1 \\ a_2 \\ \vdots \\ a_{N+1} \end{bmatrix} \equiv \mathbf{HA}$$

$$\tag{A8.1.10}$$

It is thus seen that for a given N and choice of basis functions [as for example in Eq. (A8.1.7)], it is an easy matter to generate $\mathbf{Q}$, $\mathbf{C}$, and $\mathbf{H}$ matrices.

REFERENCES

1. M. Katz in *Polymerization Processes* (C. E. Schildknecht and I. Skeist, Eds.), 1st ed., Wiley Interscience, New York (1977).
2. K. Yoda, K. Timoto, and T. Toda, Continuous process for the production of polyester, *J. Chem. Soc. Jpn. Ind. Chem. Sec.* **67**, 907-920 (1964).
3. K. Tomita and H. Ida, Studies on the formation of P.E.T.: 2. Rate of transesterification of DMT with ethylene glycol, *Polymer* **14**, 55-60 (1973).
4. K. Tomita and H. Ida, Studies on the formation of PET: 3. Catalytic activity of metal compounds in transesterification of D.M.T. with ethylene glycol, *Polymer* **16**, 185-190 (1975).
5. N. Munro and D. MacClean, (I.C.I.), U.S. Patent, 3050533.
6. I. Vansco-Szmercsanyi and E. Makay-Body, Studies on the kinetics of polyesterification, *J. Polym. Sci.* **16C**, 3709-3717 (1968).
7. D. A. Mellichamp, (DuPont), U.S. Patent, 3496146.
8. C. M. Fontana, Polymerization equilibria and kinetics of the catalysed transesterification in the formation of P.E.T., *J. Polym. Sci. A-1* **6**, 2343-2358 (1968).
9. G. Challa, Formation of P.E.T. by ester interchange, *Makromol. Chem.* **38**, 105, 123, 138-146 (1960).
10. J. W. Ault and D. A. Mellichamp, Complex linear polycondensation. I. Semi-batch reactor, *Chem. Eng. Sci.* **27**, 2219-2232 (1972).
11. L. C. Case, Molecular distributions in polycondensation involving unlike reactants. II. Linear distribution, *J. Polym. Sci.* **29**, 455-495 (1958).
12. K. S. Gandhi and S. V. Babu, Kinetics of step polymerization with unequal reactivities, *AIChE J.* **25**, 266-272 (1979).
13. N. Somu, Simulation of the MWD of PET from CSTRs, M.Tech. Dissertation, Department of Chemical Engineering, IIT-Kanpur (1982).
14. K. Ravindranath and R. A. Mashelkar, Modeling of P.E.T. reactors. I. A semibatch ester interchange reactor, *J. Appl. Polym. Sci.* **26**, 3179-3204 (1981).
15. K. Ravindranath and R. A. Mashelkar, Modeling of P.E.T. reactors II. A continuous transesterification process, *J. Appl. Polym. Sci.* **27**, 471-487 (1982).
16. J. R. Kirby, A. J. Baldwin, and R. H. Heidner, Determination of diethylene glycol in P.E.T., *Anal. Chem.* **37**, 1306-1309 (1965).
17. L. H. Buxbaum, The degradation of P.E.T., *Angew. Chem. Int. Ed.* **7**(3), 182-190 (1968).
18. H. K. Reimschuessel, P.E.T. formation–Mechanistic and kinetic aspects of the direct esterification process, *Ind. Eng. Chem. Prod. R&D* **19**, 117-125 (1980).
19. L. H. Peebles and W. S. Wagner, The kinetic analysis of a distilling system and its application to preliminary data on the transesterification of D.M.T. by E.G., *J. Phys. Chem.* **63**, 1206-1212 (1959).
20. W. Griehl and G. Schnock, Zur kinetic der polyesterbildung durch umestermy, *J. Polym. Sci.* **30**, 413-422 (1958).
21. J. Yamanis and M. Adelman, Significance of oligomerization reactions in the transesterification of D.M.T. with ethylene glycol, *J. Polym. Sci. Polym. Chem. Ed.* **14**, 1945-1961 (1976).
22. K. Tomita, Studies on the formation of P.E.T.: 1. Propagation and degradation reactions in the polycondensation of bis(2-hydroxyethyl) terephthalate, *Polymer* **14**, 50-54 (1973).

23. L. Q. Yu, Z. C. Qun, and H. R. Wen, PET polymer formation, *J. Chem. Ind. Eng. (China)* 1, 95-102 (1980).

24. H. Yokoyama, T. Sano, T. Chijiiwa, and R. Kajiya, Studies on the reactions of polycondensation, *J. Jpn Petrol Inst.* 21, 58, 77, 194 (1978).

25. M. Droscher and F. G. Schmidt, Kinetics of the ester-interchange reaction of PET, *Polym. Bull.* 4, 261-266 (1981).

26. V. V. Korshak and S. V. Vinogradova, *Polyesters*, 1st ed., Pergamon, New York (1965).

27. T. M. Pell and T. G. Davis, Diffusion and reaction in polyester melts, *J. Polym. Sci. Polym. Phys. Ed.* 11, 1671-1682 (1973).

28. E. Sunderland and C. W. Andrews, The film as a reaction site and its application to continuous production of paint media, *J. Oil Colour Chemists Assoc.* 32, 511-529 (1949).

29. S. G. Hovenkamp, Kinetic aspects of catalyzed reactions in the formation of P.E.T., *J. Polym. Sci. A-1* 9, 3617-3625 (1971).

30. S. G. Hovenkamp and J. P. Munting, Formation of diethylene glycol as a side reaction during production of PET, *J. Polym. Sci. A1* 8, 679-682 (1970).

31. A. S. Chegolya, V. V. Shevchenko, and G. D. Mikhailov, The formation of P.E.T. in the presence of dicarboxylic acids, *J. Polym. Sci. Polym. Chem. Ed.* 17, 889-904 (1979).

32. R. H. Perry and C. H. Chilton, *Chemical Engineers Handbook*, 5th Ed., McGraw-Hill, New York (1973).

33. R. Aris, *Introduction to the Analysis of Chemical Reactions*, 1st ed., Prentice-Hall, Englewood Cliffs, New Jersey (1965).

34. A. Kumar, V. K. Sakthankar, C. P. Vaz, and S. K. Gupta, Optimization of the transesterification stage of PET reactors, *Polym. Eng. Sci.* 24, 185-204 (1984).

35. L. Lapidus and R. Luus, *Optimal Control of Engineering Processes*, 1st ed., Blaisdell, Waltham, Massachusetts (1967).

36. W. H. Ray and J. Szekeley, *Process Optimization*, 1st ed., Wiley, New York (1973).

37. M. M. Denn, *Optimization by Variational Methods*, 1st ed., McGraw-Hill, New York (1969).

38. J. Hicks, A. Mohan, and W. H. Ray, Optimal control of polymerization reactors, *Can. J. Chem. Eng.* 47, 590-597 (1969).

39. P. J. Flory, *Principles of Polymer Chemistry*, 1st ed., Cornell University Press, Ithaca, New York (1953).

40. A. Kumar, S. N. Sharma, and S. K. Gupta, Optimization of the polycondensation step of PET reactors, *J. Appl. Polym. Sci.* 29, 1045-1061 (1984).

41. A. Kumar, S. N. Sharma, and S. K. Gupta, Optimization of polycondensation step of PET reactors with continuous removal of ethylene glycol, *Polym. Eng. Sci.* 24, 1205-1214 (1984).

42. H. K. Reimschuessel, B. T. Debona, and A. K. S. Murthy, Kinetics and mechanism of the formation of glycol esters. Benzoic acid–ethylene glycol system, *J. Polym. Sci. Polym. Chem. Ed.* 17, 3217-3239 (1979).

43. H. K. Reimschuessel and B. T. Debona, Terephthalic acid esterification kinetics: 2-(2-Methoxyethoxy) ethyl terephthalates, *J. Polym. Sci. Polym. Chem. Ed.* 17, 3241-3254 (1979).

44. J. F. Kemkecs, Direct esterification of TPA, *J. Polym. Sci. C* 22, 713-720 (1966).

45. K. Ravindranath and R. A. Mashelkar, Modeling of P.E.T. reactors: 4. A continuous esterification process, *Polym. Eng. Sci.* 22, 610-618 (1982).

46. K. Ravindranath and R. A. Mashelkar, Modeling of P.E.T. reactors. III. A semibatch prepolymerization process, *J. Appl. Polym. Sci.* 27, 2625-2652 (1982).

47. K. Ravindranath and R. A. Mashelkar, Modeling of P.E.T. reactors: 5. A continuous prepolymerization process, *Polym. Eng. Sci.* 22, 619-627 (1982).

48. A. M. Kotliar, Interchange reactions involving condensation polymers, *J. Polym. Sci. Macromol. Rev.* 16, 367-395 (1981).

49. W. S. Ha and Y. K. Choun, Kinetic study on the formation of cyclic oligomers in P.E.T., *J. Polym. Sci. Polym. Chem. Ed.* 17, 2103-2118 (1979).

50. I. Goodman and B. F. Nesbitt, Equilibrium of cyclic oligomers from PET, *Polymer* 1, 384 (1960).
51. I. Goodman and B. F. Nesbitt, The structures and reversible polymerization of cyclic oligomers from P.E.T., *J. Polym. Sci.* 48, 423–433 (1960).
52. L. H. Peebles, M. W. Huffman, and C. T. Ablett, Isolation and identification of the linear and cyclic oligomers of P.E.T. and the mechanism of cyclic oligomer formation, *J. Polym. Sci. A-1* 7, 479–496 (1969).
53. H. Jacobson and W. H. Stockmayer, Intramolecular reaction in polycondensation. I. The theory of linear systems, *J. Chem. Phys.* 18, 1600–1606 (1950).
54. R. F. T. Stepto, Discussions of the Faraday Society, Gels and Gelling Processes, Norwich (1974).
55. N. A. Plate and O. V. Noah, A theoretical consideration of the kinetics and statistics of reactions of functional groups of macromolecules, *Adv. Polym. Sci.* 32, 133–173 (1979).
56. J. A. Semlyen, Ring chain equilibria and the conformation of polymer chains, *Adv. Polym. Sci.* 22, 41–75 (1976).
57. G. R. Walker and J. A. Semlyen, Equilibrium ring concentrations and the statistical conformations of polymer chains: Part 4. Calculation of cyclic trimer content of P.E.T., *Polymer* 11, 472–485 (1970).
58. D. R. Cooper and J. A. Semlyen, Equilibrium ring concentration and the statistical conformations of polymer chains: Part II. Cyclics in P.E.T., *Polymer* 14, 185–192 (1973).
59. U. W. Suter and M. Mutter, Calculation of microcyclization equilibria for P.E.T., *Makromol. Chem.* 180, 1761–1773 (1979).
60. J. Bjorksten, *Polyesters and Their Applications*, 1st ed., Reinhold, New York (1960).
61. B. Gupta, Modelling and simulation of polycondensation step in polyester formation, M.Tech. Dissertation, IIT-Kanpur (1981).
62. M. V. S. Rao, Investigation of PET batch reactors, M.Tech. Dissertation, IIT-Kanpur (1982).
63. A. Kumar, S. K. Gupta, B. Gupta, D. Kunzru, Modelling of reversible P.E.T. reactors, *J. Appl. Polym. Sci.* 27, 4421–4438 (1982).
64. A. Kumar, S. K. Gupta, and N. Somu, MWD of PET in HCSTRs, *Polym. Eng. Sci.* 22, 314–323 (1982).
65. A. Kumar, S. K. Gupta, M. V. S. Rao, and N. Somu, Simulation of cyclics and degradation product formation in P.E.T. reactors, *Polymer* 24, 449–456 (1983).
66. F. Widmer, A new reactor for continuous manufacture of polyester melt, *Dachema Monograph.* 66, 143–150 (1971).
67. J. M. McKelvey and G. V. Sharps, Fluid transport in thin film polymer processor, *Polym. Eng. Sci.* 19, 651–659 (1979).
68. D. B. Todd and H. F. Irving, Axial mixing in a self-wiping reactor, *Chem. Eng. Prog.* 65(9), 84–89 (1969).
69. M. Amon and C. D. Denson, Simplified analysis of the performance of wiped film polycondensation reactors, *Ind. Eng. Chem. Fundam.* 19, 415–420 (1980).
70. A. Kumar, S. Madan, N. G. Shah, and S. K. Gupta, Solution of final stages of PET reactors, 128 IUPAC Macromolecular Symposium, Amherst, 1982.
71. K. Tai, Y. Arai, H. Teranishi, and T. Tagawa, *J. Appl. Polym. Sci.* 25, 1789–1792 (1980).
72. B. A. Finlayson, *The Method of Weighted Residuals and Variational Principles*, 1st ed., Academic, New York (1972).
73. J. Villadsen, *Selected Approximation Methods for Chemical Engineering Problems*, Inst. for Kemiteknek Numer. Inst., Danmarks Tekniske Hjskole (1970).
74. J. W. Ault and D. A. Mellichamp, A diffusion and reaction model for simple polycondensation, *Chem. Eng. Sci.* 27, 1441–1448 (1972).

75. P. J. Hoftyzer and D. W. Van Krevelen, The influence of evaporation of a volatile product on the rate of conversion in polycondensation processes, Proc. 4th Europ. Symp. Chem. Reaction, *Chem. Eng. Sci.* 123-129 (1971).

76. P. J. Hoftyzer, Kinetics of polycondensation of ethylene glycol terephthalate, *Appl. Polym. Symp.* **26**, 349-363 (1975).

77. K. Ravindranath and R. A. Mashelkar, Modeling of PET reactors 6. A continuous process for final stages of polycondensation, *Polym. Eng. Sci.* **22**, 628-636 (1982).

78. F. Widmer, in *Polymerization Kinetics and Technology* (N. A. J. Platzer, ed.), Advances in Chemistry Series, No. 128, p. 51 (1973).

79. T. Shima, T. Urasaki, and I. Oka in *Polymerization Kinetics and Technology* (N. A. J. Platzer, Ed.), Advances in Chemistry Series, No. 128, p. 183 (1973).

80. G. A. Campbell, E. F. Elton, and E. G. Bobaleck, The kinetics of thin film polyesterification, *J. Appl. Polym. Sci.* **14**, 1025-1035 (1970).

81. E. E. Parker and J. R. Peffer, in *Polymerization Processes* (C. E. Schildknecht and I. Skeist, Ed.), 1st ed., Wiley Interscience, New York (1977).

82. A. Kastanek, J. Zelanka, and K. Hajek, M.W.D. of an unsaturated polyester, *J. Appl. Polym. Sci.* **26**, 4117-4124 (1981).

83. K. S. Gandhi and R. Burns, Studies on the thickening reaction of polyesters resins employed in sheet molding compounds, *J. Polym. Sci. Polym. Chem. Ed.* **14**, 793-811 (1976).

84. F. Alvey, Study of the reaction of polyester resins with magnesium oxide, *J. Polym. Sci. A-1* **9**, 2233-2245 (1971).

EXERCISES

1. From the reactions given in Eqs. (8.2.2)-(8.2.9), write down the mole balance equations for various species in batch reactors. From this show that one can equivalently represent the overall polymerization by the reaction of functional groups.

2. How would you modify Eq. (8.2.19) if it is assumed that acetaldehyde does not remain in the liquid phase?

3. In the analysis presented in Section 8.2, it is assumed that G, A, W, and M flash only at the end of time incremental interval Δt. Modify this analysis in light of Chapter 5.

4. Derive the set of mole balance relations for a single ester interchange HCSTR.

5. Derive adjoint variable equations for the optimization of batch ester interchange reactors, assuming concentrations of methylene glycol, etc. as constant. We can relax this approximation if we use Eq. (8.2.19). How do we do this, considering temperature as the control variable?

6. Repeat the analysis of Exercise 8.5 assuming pressure and temperature as control variables.

7. Assume that the objective function, I, for the polycondensation reactor is

$$I = \alpha_1(e_D - e_D^*) + \int_0^{t_f} \left\{ \alpha_2\left(\frac{1}{e_g} - \mu_n^*\right) + \alpha_3(e_c^2 + e_v^2) \right\} dt$$

where e_D^* and μ_n^* are the desired values of e_D and the average chain lengths of the polymer formed. Derive adjoint equations for the two assumptions given in Exercise 8.5.

8. Using Eq. (8.5.13), show that

$$\sum_{n=1}^{\infty} \dot{P}_{n,\text{III}} = \sum_{n=1}^{\infty} n \dot{P}_{N,\text{III}} = 0.$$

9. Derive the set of equations of Table 8.11.

10. To simplify the analysis of wiped film reactor for ARB polymerization, we have decided to do the following:

(a) We observe that the active film thickness δ is given by

$$\delta = \mathscr{D}_w^{1/2} t^{1/2}$$

where t is the exposure time.

(b) The profile of condensation product, w, in the film is given by

$$W = W_0 + \bar{w}_f(t)\left(1 - \frac{y^2}{\delta^2}\right)$$

where $\bar{w}_f(t)$ is the average film concentration of the condensation product and changes with the film exposure time.

(c) To determine $\bar{W}_f(t)$, treat the entire film as the control volume and make the mole balance for this. Take the rate of formation of W, $\mathscr{R}_w$, as

$$\mathscr{R}_w = \frac{k_p}{K}(\lambda_{10} - P_{b0})\left(\bar{w}_f - \frac{KP_{b0}^2}{\lambda_{10} - P_{b0}}\right)$$

Integrate this differential equation by observing that the solution of $dx/dt + a(t)x = b(t)$ is given by

$$x = \exp\left[-\int a(t)\,dt\right]\left\{\int b(t)\left[\exp\left(\int a(t)\,dt\right)\right]dt + C\right\}$$

Find $\bar{W}_f(t)$ in the film.

(d) Determine $\bar{N}_w$ and use this to establish the concentration of the condensation product in the bulk along the reactor length.

URETHANE POLYMERS AND REACTION INJECTION MOLDING

9.1. INTRODUCTION

Urethane is a polymer that has the characteristic linkage

$$-NH\overset{\overset{\displaystyle O}{\displaystyle \|}}{C}O-$$

and is formed by reaction of a dissocyanate and a polyol.[1-3] The polymerization can be schematically represented as

$$\text{\textasciitilde\textasciitilde NCO} + \text{OH\textasciitilde\textasciitilde} \rightarrow \text{\textasciitilde\textasciitilde NH}\overset{\overset{\displaystyle O}{\displaystyle \|}}{C}\text{O\textasciitilde\textasciitilde} \qquad (9.1.1)$$

Some of the diisocyanates commonly used industrially are toluene diisocyanates

$$\text{TDI, } CH_3\text{—}\hexagon\text{—NCO} \quad \text{or} \quad CH_3\hexagon$$

or diphenyl methane diisocyanate

$$\text{MDI, OCN}\hexagon\text{—}CH_2\text{—}\hexagon\text{—NCO}$$

and both of these are bifunctional monomers.

Polyols used in the manufacture of urethanes have alcoholic functional groups and are rarely small molecules like ethylene glycol. They are usually either polyethers or polyesters with the $-OH$ groups either at the chain ends or on the chains. Polyethers are usually formed by polymerizing propylene oxide with water in alkaline conditions. Polyester polyols are usually prepared by reacting adipic or phthalic acid with excess of glycols. The use of excess glycol is to ensure that the polymer formed has alcohol endings and is bifunctional. Multifunctional polyester polyols can be obtained using multifunctional compounds like glycerine and pentaerythritol. The formation of urethane polymers is usually fast enough without the use of catalysts, but recent applications like reaction injection molding (RIM) require very fast reactions, for which catalysts must be used. Examples include tertiary amines (triethylene diamine, triethyl amine, triethanol amine, etc.) or organo silane compounds (stannous octate, dibutyl tin dilaurate, etc.). The usual ingredients of forming urethane polymer are silicone surfactant (sometimes called a releasing agent) and a flame retardant in addition to a polyol, a diisocyanate, and a suitable catalyst. In producing flexible foams, a blowing agent, like water, is used which, on reaction with an isocyanate group, produces carbon dioxide as follows:

$$-NCO + H_2O \rightarrow -NH_2 + CO_2 \qquad (9.1.2)$$

The CO_2 thus liberated first escapes the reaction mass, but with the progress of reaction, the viscosity increases and the gas is trapped, giving a cellular structure to the final polymer.

The process of manufacturing polyurethane elastomers can roughly be classified into two-shot and single-shot systems. In the former, an excess of a diisocyanate is reacted with a diol and the formation of isocyanate prepolymer can be represented as

$3n$OCNRNCO + $2n$OH∿OH

$\rightarrow$ OCNRNHCOO∿OOCNHRNHCOO∿OOCNHRNCO

Isocyanate prepolymer

$$(9.1.3)$$

This extended diisocyanate can be converted into long chain polymer by using chain extenders like low molecular weight glycols or dimamines as

3OCN∿NCO + OHR'OH

$\rightarrow$ OHR'OOCNH∿NHCOOR'OOCNH∿NHCOOR'OOCNH∿NCO

$$(9.1.4)$$

where OCN⌁NCO represents the molecule of isocyanate prepolymer shown in Eq. (9.1.3). Some of the isocyanate groups in this stage are used in the chain extension step as in Eq. (9.1.4) and some of them are used up to form allophanate and biuret branch points as shown below:

$$\text{⌁NHCOOR'OOCNH⌁} + \text{OCN⌁} \rightarrow \underset{\begin{array}{c} | \\ \text{C=O} \\ | \\ \text{⌁NH} \end{array}}{\text{⌁NCCONH⌁}} \qquad (9.1.5)$$

Allophanates

$$\text{⌁NHCOOR'OOCNH⌁} + \text{OCN⌁} \rightarrow \underset{\begin{array}{c} \S \\ \S \end{array}}{\overset{\begin{array}{cc} O & O \\ || & || \end{array}}{\text{⌁NHCNCNH}}} \qquad (9.1.6)$$

Biurets

Even though the reactions are quite complex, mechanistically a simplified picture can be considered which explains the basic features of polymerization. In this and the next section, only the kinetics of polymerization is modeled, ignoring the physical aspects like mixing, fluid flow, etc., which will be analyzed later. The use of water and diamines in place of glycols in reaction (9.1.4) leads to similar chain extension, except that there is a preponderance of biuret linkages in the network compared to allophanates. In the single-shot process, the diisocyanate, diol, polyol, and the chain extender (glycol, diamine, or water) are mixed together and reacted. In this method, prepolymer formation and chain extension [reactions (9.1.3) and (9.1.4)] occur simultanueously and these must be carefully monitored.

9.2. MODELING OF CROSS-LINKED POLYURETHANE FORMATION[3]

Urethane polymers are produced using a diisocyanate (denoted by A_1A_2), a diol (denoted by BB), a polyol (denoted by B_f) of functionality f, and water as the blowing agent in addition to the usual flame retardants and surfactants. In diisocyanates the two isocyanates groups A_1 and A_2 have different reactivities in general, and this is called structural asymmetry. In aromatic diisocyanates, as in toluene diisocyanate (TDI), one isocyanate group can modify the activity of the other and the activity of both can be modified by the substituents of the aromatic ring. For a mixture of 2,4- and 2,6-isomers of TDI, in Table 9.1[4,5] twelve reactions with the primary and

TABLE 9.1.[a] Rate Constants for Various Reactions of Toluene Diisocyanate
with Polyols

Reactants			Lit equivalent^{-1} sec^{-1} $10^4 k_i$	
Nature of OH	Location of isocyanate	Product	25°C	60°C
Primary hydroxyl	Monomeric para k_1	Urethane	0.613	4.17
	Monomeric ortho k_2		0.230	1.67
	Polymeric para k_3		0.161	1.10
	Polymeric ortho k_4		0.0605	0.439
Secondary hydroxyl	Monomeric para k_5	Urethane	0.204	1.67
	Monomeric ortho k_6		0.0273	0.333
	Polymeric para k_7		0.0538	0.439
	Polymeric ortho k_8		0.00717	0.0877
Urethane	Monomeric para k_9	Allophanate	0.00307	0.0208
	Monomeric ortho k_{10}		0.00409	0.00417
	Polymeric para k_{11}		0.000805	0.00548
	Polymeric ortho k_{12}		0.000108	0.00110

[a] References 4, 5.

secondary hydroxyl groups of polyols have been identified and experimental
values of the rate constants given. Isocyanates during the polymerization
are thus seen to exhibit a first-shell substitution effect, which means that
the A_1 and A_2 groups on polymer chains have different reactivities as
compared to those on diisocyanate *monomers*. The entire mechanism of
polymerization is given in Table 9.2, and for this reason, A_1 and A_2 on
polymer chains have been distinguished by an asterisk.

The reaction of an isocyanate group with water produces an amine
group as shown in Eq. (9.1.2). This can also react with an isocyanate group

TABLE 9.2. Mechanism of Urethane Formation

a. Structural asymmetry of monomer

$$A_2A_1 + B\text{ww} \xrightarrow{k_1} {}^*A_2A_1B$$

$$A_1A_2 + B\text{ww} \xrightarrow{k_2} {}^*A_1A_2B$$

b. First-shell substitution effect

$$\text{ww}A_1^* + B\text{ww} \xrightarrow{k_1^*} \text{ww}A_1B\text{ww}$$

$$\text{ww}A_2^* + B\text{ww} \xrightarrow{k_2^*} \text{ww}A_2B\text{ww}$$

(*continued*)

TABLE 9.2 (*continued*)

---

c. Reactions with blowing agent CD

$$A_2A_1 + CD \xrightarrow{R_1 k_1} {}^*A_2A_1CD$$

$$A_1A_2 + CD \xrightarrow{R_1 k_2} {}^*A_1A_2CD$$

$$\text{\textasciitilde}A_1^* + CD \xrightarrow{R_1 k_1^*} \text{\textasciitilde}A_1CD$$

$$\text{\textasciitilde}A_2^* + CD \xrightarrow{R_1 k_2^*} \text{\textasciitilde}A_2CD$$

$$A_2A_1 + D\text{\textasciitilde} \xrightarrow{R_2 R_1 k_1} {}^*A_2D\text{\textasciitilde}$$

$$A_1A_2 + D\text{\textasciitilde} \xrightarrow{R_2 R_1 k_2} {}^*A_1D\text{\textasciitilde}$$

$$\text{\textasciitilde}A_1^* + D\text{\textasciitilde} \xrightarrow{R_2 R_1 k_1^*} \text{\textasciitilde}A_1D\text{\textasciitilde}$$

$$\text{\textasciitilde}A_2^* + D\text{\textasciitilde} \xrightarrow{R_2 R_1 k_2^*} \text{\textasciitilde}A_2D\text{\textasciitilde}$$

d. Formation of allophanate linkages (M):

$$A_2A_1 + E \xrightarrow{R_3 k_1} M + {}^*A_2\text{\textasciitilde}$$

$$A_1A_2 + E \xrightarrow{R_3 k_2} M + {}^*A_1\text{\textasciitilde}$$

$$\text{\textasciitilde}A_1^* + E \xrightarrow{R_3 k_1^*} M$$

$$\text{\textasciitilde}A_2^* + E \xrightarrow{R_3 k_2^*} M$$

e. Formation of biuret groups (G):

$$A_2A_1 + F \xrightarrow{R_4 k_1} {}^*A_2\text{\textasciitilde} + G$$

$$A_1A_2 + F \xrightarrow{R_4 k_2} {}^*A_1\text{\textasciitilde} + G$$

$$\text{\textasciitilde}A_1^* + F \xrightarrow{R_4 k_1^*} G$$

$$\text{\textasciitilde}A_2^* + F \xrightarrow{R_4 k_1^*} G$$

---

producing a urea linkage as

$$\text{\textasciitilde}NCO + NH_2\text{\textasciitilde} \rightarrow \text{\textasciitilde}NHCONH\text{\textasciitilde} \qquad (9.2.1)$$

The reactions in Eqs. (9.1.2) and (9.2.1) are the same as those in the polymerization of cyclic monomers (for example, between phthalic anhydride and ethylene glycol). If a water molecule is denoted as CD, then the C group reacts with isocyanate first to produce D (an amine group). There is no D on water molecules as such; it is produced only after the reaction. This also implies that the C group cannot exist at the end of a polymer chain.

Allophanate linkages shown in Eq. (9.1.5) are formed by reaction of urethane with isocyanates. The amine groups, D, react with an isocyanate group as in Eq. (9.2.1) and form urea groups. These are included in Table 9.2. With this kinetic model, it is possible to write down the governing mole balance equations for various species in the reaction mass as done in Table 9.3. To do this, it is first necessary to distinguish the A_1 and A_2 functional groups on monomers. One defines $[A_1^r]$ and $[A_2^r]$ as the moles per unit volume of monomeric A_1 and A_2 groups that have been consumed by direct reaction. For convenience, various terms used in this section and the next have also been included in Table 9.3. In this table, $[A]$ is the concentration of the monomer whose time variation is given by $(d[A_1^r]/dt + d[A_2^r]/dt)$.

The stoichiometric relation existing between the various species can be written by observing that all A groups reacted can show up either as B,

TABLE 9.3. Mole Balance for Various Species in Urethane Formation

$$\frac{d[A_1^r]}{dt} = k_1[A]Y$$

$$\frac{d[A_2^r]}{dt} = k_2[A]Y$$

$$\frac{d[A_1^*]}{dt} = \frac{d[A_2^r]}{dt} - k_1^*[A_1^*]Y$$

$$\frac{d[A_2^*]}{dt} = \frac{d[A_2^r]}{dt} - k_2^*[A_2]Y$$

$$\frac{d[C]}{dt} = R_1[C]Z$$

$$\frac{d[D]}{dt} = -R_1R_2[D]Z$$

$$\frac{d[B]}{dt} = -[B]Z$$

$$\frac{d[E]}{dt} = -[B]Z$$

$$\frac{d[F]}{dt} = \{R_1R_2[D] - R_4[F]\}Z$$

$$Y = [B] + R_1[C] + R_1R_2[D] + R_3[E] + R_4[F]$$

$$Z = (k_1 + k_2)[A] + k_1^*[A_1^*] + k_2^*[A_2]$$

(*continued*)

TABLE 9.3 (*continued*)

$$p_A = 1 - \frac{[A]}{[A]_0}$$

$$p_B = 1 - \frac{[B]}{f[B_f]_0 + 2[B_2]_0}$$

$$p_C = \frac{[C]_0 - [C]}{[C]_0}, \qquad p_{C1} = \frac{[C]_0 - [C] - [D]}{[C]_0}$$

$$p_E = \frac{[E]_0 - [E]}{[B]_0}, \qquad p_F = \frac{[F]_0 - [F]}{[C]_0}$$

$$\alpha_1 = \frac{[A_1^r]}{[A] + [A_1^r]}, \qquad \alpha_2 = \frac{[A_2^r]}{[A] + [A_2^r]}$$

$$\alpha_1^* = \frac{[A_2^r] + [A_1^*]}{[A_2^r]}, \qquad \alpha_2^* = \frac{[A_1^r] - [A_2^*]}{[A_1^r]}$$

$$r_{B_f} = \frac{f[B_f]_0}{f[B_f] + 2[B_2]_0 + 2[C]_0}$$

$$r_{B_2} = \frac{2[B_2]_0}{f[B_f]_0 + 2[B_2]_0 + 2[C]_0}, \qquad r = \frac{2[A]_0}{f[B_f] + 2[B_2]_0 + 2[C]_0}$$

$$r_C = (1 - r_{B_f} - r_{B_2}), \qquad r_{B_f} = (1 - r_{B_2})(1 - r_C)$$

$$R_5 = (k_2/k_1)$$

$$R_6 = (k_1^*/k_1)$$

$$R_7 = (k_2^*/k_1)$$

C, D, E, or F groups. This can be written in terms of the variables defined in Table 9.3 as

$$r \frac{\alpha_1}{1 + R_5\alpha_1}\{1 + \alpha_2^* + R_5(1 + \alpha_1^*)\} = r_{B_f}(p_B + p_E) + (1 - r_{B_f})(p_C + p_{C1} + p_F) \tag{9.2.2}$$

It is possible to relate conversions p_C and p_{C1} of Table 9.3 in terms of p_B as follows. By dividing $d[C]/dt$ by $d[B]/dt$ and then integrating the resulting equation, one obtains

$$(1 - p_C) = (1 - p_B)^{R_1} \tag{9.2.3}$$

Similarly dividing $d[D]/dt$ by $d[C]/dt$ and integrating gives

$$p_{C1} = p_C - \frac{1 - p_C}{R_2 - 1}\{1 - (1 - p_C)\}^{R_2 - 1} \tag{9.2.4}$$

Proceeding in the similar way, one has

$$p_B - p_E = \frac{1}{1 - R_3}\{(1 - p_B) - (1 - p_B)^{R_3}\} \qquad (9.2.5a)$$

$$p_F = p_{C1} + \frac{R_2}{(R_2 - 1)}\left\{\frac{(1 - p_C)^{R_4} - (1 - p_C)}{R_4 - 1} - \frac{(1 - p_C)^{R_4} - (1 - p_C)^{R_2}}{(R_4 - R_3)}\right\}$$
$$(9.2.5b)$$

$$[A_2^r] = R_5[A_1^r]$$

$$\frac{[A_1^*]}{[A]_0} = \frac{R_5}{R_6 - 1 - R_5}\left\{\frac{[A]}{[A]_0} - \left(\frac{[A]}{[A]_0}\right)^{R_6/(R_5+1)}\right\} \qquad (9.2.5c)$$

$$\frac{[A_2^*]}{[A]_0} = \frac{1}{(R_5 R_7 - R_5 - 1)}\left\{\frac{[A]}{[A]_0} - \left(\frac{[A]}{[A]_0}\right)^{R_5 R_7/(R_5+1)}\right\} \qquad (9.2.5d)$$

Equations (9.2.5c) and (9.2.5d) can be written in terms of dimensionless variables defined in Table 9.3 as

$$(1 - \alpha_1^*) = \frac{R_5 \alpha_1 + 1}{\alpha_1(R_6 - R_5 - 1)}\left\{\frac{1 - \alpha_1}{R_5 \alpha_1 + 1} - \left(\frac{1 - \alpha_1}{R_5 \alpha_1 + 1}\right)^{R_6/(R_5+1)}\right\} \qquad (9.2.6a)$$

$$(1 - \alpha_2^*) = \frac{R_5 \alpha_1 + 1}{\alpha_1(R_5 R_7 - R_5 - 1)}\left\{\frac{1 - \alpha_1}{(R_5 \alpha_1 + 1)} - \left(\frac{1 - \alpha}{R_5 \alpha_1 + 1}\right)^{R_5 R_7/(R_5+1)}\right\}$$
$$(9.2.6b)$$

If p_B is known, p_C, p_{C1}, p_E, and p_F can be calculated from Eq. (9.2.5). Similarly, if α_1 is known, α_2, α_1^*, and α_2^* can be found from Eq. (9.2.6). But α_1 is related to p_B through the stoichiometric relation in Eq. (9.2.2). It is thus seen that the various concentrations in the reaction mass can be easily found in terms of p_B, which, itself, is determined as a function of time by integrating balance equation for B given in Table 9.3.

After solving for the concentrations of various species, it is of considerable interest to predict when the gelation occurs. In the following section, the gelation criterion has been developed.

9.3. GELATION

In this section Macosko's expectation theory[6,7] (see Chapter 4) is extended to determine the weight average molecular weight of the

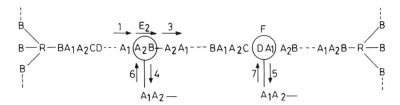

Figure 9.1. A typical urethane molecule.

polyurethane formed. In Fig. 9.1, a polymer chain is shown and it can be seen that branching can occur at polyol, B_f, E, and F (see Table 9.2).

Whenever a polymer chain $\sim\!\!\!\sim$B reacts with monomer A_1A_2, there is a formation of either $\sim\!\!\!\sim$$BA_1A_2^*$ (structure 1) or $\sim\!\!\!\sim$$BA_2A_1^*$ (structure 2). The expectation values $E(M_{A_1}^{out})$ and $E(M_{A_2}^{out})$ due to these structures are evidently zero because they are the chain ends. The polymer chain $\sim\!\!\!\sim$B can also react with $^*A_1A_2\!\!\sim\!\!\!\sim$ and $^*A_2A_1\!\!\sim\!\!\!\sim$ to give $\sim\!\!\!\sim$$BA^*A_1A_2\!\!\sim\!\!\!\sim$ (structure 3) and $\sim\!\!\!\sim$$BA_2^*A_1\!\!\sim\!\!\!\sim$ (structure 4). Thus, for expectation values $E(M_{A_2}^{out})$ and $E(M_{A_1}^{out})$ to have nonzero values, it is necessary that $\sim\!\!\!\sim$$A_1^*$ and $\sim\!\!\!\sim$$A_2^*$ be reacted as in structures 3 and 4. Therefore, to determine the expected values of A_1 and A_2 reacted, $E(M_{A_1}^{out})$ and $E(M_{A_2}^{out})$, looking out in the direction 1 in Fig. 9.1, it is observed it could be reacted to B group with or without branching at A_1 (or A_2), to a C group, to a D group with or without branching at D, to an E group or to an F group. Thus

$$E(M_{A_1}^{out}) = \text{Prob}(A_2^* \text{ reacted}) U \qquad (9.3.1a)$$

$$E(M_{A_2}^{out}) = \text{Prob}(A_1^* \text{ reacted}) U \qquad (9.3.1b)$$

where U is defined in Table 9.4. In this table, Prob(B reacted) is the probability of finding those B groups that would lead to branching and Prob(C) the probability of finding reacted C that is connected to a branch, etc. $E(M_B^{in})$, $E(M_D^{in})$, etc. are the expectation values having the same meaning as in Chapter 4. To evaluate these probabilities, it is observed that the total B groups reacted, B_E, is given by

$$B_E = [B]_0(p_B + p_E) + [C]_0\{p_C + p_{C1} + p_F\}$$

$$= \{[B]_0 + 2[C]_0\}\{r_{B_f}(p_B + p_E) + \tfrac{1}{2}(1 - p_{B_f})(p_C + p_{C1} + p_F)\} \qquad (9.3.2)$$

In terms of this, the various probabilities can be evaluated and are also given in Table 9.4. To find $E(M_{A_1}^{out})$ and $E(M_{A_2}^{out})$, it is once again observed that structures 1 and 2 above have zero expectation values but chains $\sim\!\!\!\sim$$BA_1A_2^*B\!\!\sim\!\!\!\sim$ (structure 5) and $\sim\!\!\!\sim$$BA_2A_1^*B\!\!\sim\!\!\!\sim$ (structure 6) only have nonzero

TABLE 9.4. Various Relations from Application of Expectation Theory to Urethane Network Formation

$$E(M_{A_1}^{out}) = \text{Prob}(A_2^* \text{ reacted})U$$

$$E(M_{A_2}^{out}) = \text{Prob}(A_1^* \text{ reacted})U$$

$$E(M_{A_1^*}^{out}) = U$$

$$E(M_{A_2^*}^{out}) = U$$

$$E(M_B^{out}) = \text{Prob}(B \text{ reacted})Z_1$$

$$E(M_C^{out}) = \text{Prob}(C \text{ reacted})Z_1$$

$$E(M_D^{out}) = \text{Prob}(D \text{ reacted})Z_1$$

$$E(M_E^{out}) = \text{Prob}(E \text{ reacted})Z_1$$

$$E(M_F^{out}) = \text{Prob}(F \text{ reacted})Z_1$$

$$\text{Prob}(A_1) = \frac{1}{1 + \alpha_2^* + R_5(1 + \alpha_1^*)}$$

$$\text{Prob}(A^*) = \frac{R_5 \alpha_1^*}{1 + \alpha_2^* + R_5(1 + \alpha_1^*)}$$

$$\text{Prob}(A_2) = \frac{R_5}{1 + \alpha_2^* + R_5(1 + \alpha_1^*)}$$

$$\text{Prob}(A_2^*) = \frac{\alpha_2^*}{1 + \alpha_2^* + R_5(1 + \alpha_1^*)}$$

$$\text{Prob (B reacted)} = \frac{r_{B_f} p_B}{B_E}$$

$$\text{Prob (C reacted)} = \frac{(1 - r_{B_f}) p_C}{2 B_E}$$

$$\text{Prob (E reacted)} = \frac{p_E}{p_B}$$

$$\text{Prob (F reacted)} = \frac{p_F}{p_{C_1}}$$

$$U = \text{Prob}(B \text{ reacted}) \text{ Prob}(E \text{ not reacted}) E(M_B^{in})$$
$$+ \text{Prob}(B \text{ reacted}) \text{ Prob}(E \text{ reacted})\{E(M_B^{in}) + E(M_E^{in})\}$$
$$+ \text{Prob}(C \text{ reacted}) E(M_C^{in})$$
$$+ \text{Prob}(D \text{ reacted}) \text{ Prob}(F \text{ not reacted}) E(M_D^{in})$$

$$Z_1 = \text{Prob}(A_1) E(M_{A_*}^{in}) + \text{Prob}(A_1^*) E(M_{A_1}^{in})$$
$$+ \text{Prob}(A_2) E(M_{A_2}^{in} + \text{Prob}(A_2^*) E(M_{A_2^*}^{in})$$

$$B_E = ([B]_0 + 2[C]_0)\{r_{B_f}(p_B + p_E) + \tfrac{1}{2}(1 - p_{B_f})(p_C + p_{C_1} + p_F)\}$$

values. Since in these structures, group B occurs with probability one, each of these is equal to U.

The expectation parameters $E(M_B^{in})$, $E(M_C^{in})$, and $E(M_D^{in})$ can be related to the molecular weights of the repeat units formed by the functional groups B, C, and D, i.e.,

$$E(M_B^{in}) = M_B + (f - 1)E(M_B^{out}) \qquad (9.3.3a)$$

$$E(M_C^{in}) = M_C - M_{CO_2} + E(M_D^{out}) \qquad (9.3.3b)$$

$$E(M_D^{in}) = M_C - M_{CO_2} + E(M_C^{out}) \qquad (9.3.3c)$$

where M_B and M_C are the molecular weights between branch points involving B and C groups. Substituting these in Eq. (9.3.1), one has

$$E(M_{A_1}^{out}) = \alpha_2^*(S + QZ_1) \qquad (9.3.4)$$

where

$$S = \frac{r_{B_f} M_B(p_B + p_E) + [(1 - r_{B_f})/2](M_C - M_{CO_2})(p_C + p_{C1} + p_F)}{r_{B_f}(p_B + p_E) + [(1 - r_{B_f})/2](p_C + p_{C1} + p_F)} \qquad (9.3.5a)$$

$$Q = \frac{(f - 1)r_{B_f} p_B(p_B + p_E) + [(1 - r_{B_f})/2]\{2p_F + 2p_{C1} + p_F p_C\} + 2r_{B_f p_E}}{r_{B_f}(p_B + p_E) + [(1 - r_{B_p})/2](p_C + p_{C1} + p_F)} \qquad (9.3.5b)$$

Similarly,

$$E(M_{A_1^*}^{out}) = S + QZ_1 \qquad (9.3.6a)$$

$$E(M_{A_2}^{in}) = \alpha_1^*(S + QZ_1) \qquad (9.3.6b)$$

$$E(M_{A_2^*}^{in}) = S + QZ_1 \qquad (9.3.6c)$$

It is further observed that

$$E(M_{A_1}^{in}) = M_A + E(M_{A_1}^{out}) = M_A + \alpha_2^*(S + QZ_1) \qquad (9.3.7a)$$

$$E(M_{A_1^*}^{in}) = M_A + E(M_{A_1^*}^{out}) = M_A + \alpha_2^*(S + QZ_1) \qquad (9.3.7b)$$

$$E(M_{A_2}^{in}) = M_A + E(M_{A_2}^{out}) = M_A + S + QZ_1 \qquad (9.3.7c)$$

$$E(M_{A_2^*}^{out}) = M_A + E(M_{A_2^*}^{out}) = M_A + \alpha_1^*(S + QZ_1) \qquad (9.3.7d)$$

Substituting these in the expression for Z_1, the following equation is obtained:

$$Z_1 = M_A + \frac{2S(\alpha_2^* + R_5\alpha_1^*)}{1 + \alpha_2^* + R_5(1 + \alpha_1^*)} + 2QZ_1\frac{\alpha_2^* + R_5\alpha_1^*}{1 + \alpha_2^* + R_5(1 + \alpha_1^*)} \qquad (9.3.8)$$

Solving for Z_1,

$$Z_1 = \frac{M_A + \dfrac{2S(\alpha_2^* + R_5\alpha_1^*)}{1 + \alpha_2^* + R_5(1 + \alpha_1^*)}}{1 - \dfrac{2Q(\alpha_2^* + R_5\alpha_1^*)}{1 + \alpha_2^* + R_5(1 + \alpha_1^*)}} \qquad (9.3.9)$$

The molecular weight of the polymer would go to infinity if Z_1 goes to infinity. The gel criterion is therefore given by

$$\frac{2Q(\alpha_2^* + R_5\alpha_1^*)}{1 + \alpha_2^* + R_5(1 + \alpha_1^*)} \geq 1 \qquad (9.3.10)$$

subject to the satisfaction of the stochiometric relation written in Eq. (9.2.2). If there are no side reactions, $p_E = p_F = 0$, in which case

$$Q = \frac{(f - 1)r_{B_f}p_B^2 + [(1 - r_{B_f})/2]2p_{C1}}{r_{B_f}p_B + [(1 - r_{B_f})/2](p_C + p_{C1})} \qquad (9.3.11)$$

and the gel criterion Eq. (9.3.10) reduces to

$$\frac{2\{(f - 1)r_{B_f}p_B^2 + (1 - r_{B_f})p_{C1}\}(\alpha_2 + R_5\alpha_1^*)}{\{r_{B_f}p_B[(1 - r_{B_f})/2](p_C + p_{C1})\}\{1 + \alpha_2^* + R_5(1 + \alpha_1^*)\}} \geq 1 \quad (9.3.12)$$

subject to the satisfaction of the stoichiometric relation given by

$$\frac{1}{1 + R_5\alpha_1}\{1 + \alpha_2^* + R_5(1 + \alpha_1^*)\}$$

$$= \frac{1}{r(1 - r_{B_2})}\{2r_{B_f}p_B + (1 - r_{B_2} - r_{B_f})(p_C + p_{C1})\} \quad (9.3.13)$$

For a given reaction system, the rate constants are fixed and Eq. (9.3.12) can predict the fractional conversions at which gelation will occur. It can be shown that gelation does not occur under all conditions and the region separating those conditions in which gelation occurs is called the gelation envelope. It can be derived as follows.

To demonstrate how to calculate the gelation envelope, the equal reactivity of functional groups is considered as an example for which $R_5 = 1$, $R_6 = 1$, and $R_7 = 1$. It is further assumed that the CD monomer behaves like B_2 monomer; i.e., $R_1 = 2$ and $R_2 = 0.5$. Then

$$p_C = 2p_B - p_B^2 \qquad (9.3.14a)$$

$$pC_1 = p_B^2 \qquad (9.3.14b)$$

$$\alpha_1 = \alpha_2 \qquad (9.3.14c)$$

$$\alpha_1^* = \alpha_2^* \qquad (9.3.14d)$$

$$(1 - \alpha_1^*) = -\frac{1 + \alpha_1}{\alpha_1}\left\{\frac{1 - \alpha_1}{1 + \alpha_1} - \left(\frac{1 - \alpha_1}{1 + \alpha_1}\right)^{1/2}\right\} \qquad (9.3.14e)$$

It is now possible to relate the conversions p_A and p_B in terms of α_1 and α_2^* as

$$p_A = \frac{[A_1^r](1 + \alpha_2^*) + A_2^r(1 + \alpha_1^*)}{2[A]_0} = \frac{1}{(1 + \alpha_1)}(1 + \alpha_2^*) \qquad (9.3.15a)$$

$$p_B = \frac{p_B\{f[B_f]_0 + 2[B_2]_0\} + [C]_0\{p_C + p_{C1}\}}{f[B_f]_0 + 2[B_2]_0 + 2[C]_0} \qquad (9.3.15b)$$

Then the stoichiometric relation and the gelation conditions [Eqs. (9.3.13) and (9.3.12)] reduce to

$$rp_A = p_B \qquad (9.3.16a)$$

$$(f - 1)\frac{r_{B_f}p_A p_B}{1 - (1 - r_{B_f})p_A p_B} \geq 1 \qquad (9.3.16b)$$

Substituting Eq. (9.3.16a) into (9.3.16b), the gelation criterion reduces to

$$r \geq \frac{1}{p_A^2\{(f - 1)r_{B_f} + (1 - r_{B_f})\}} \qquad (9.3.17)$$

If r is less than unity and all reactions are assumed irreversible, isocyanate groups would be completely consumed for large times of polymerization. It can thus be seen that the value of r calculated from the above equation would give the minimum value r_{min}, below which there is no gelation, i.e.,

$$r_{min} = \frac{1}{(f-1)r_{B_f} + (1 - r_{B_f})} \qquad (9.3.18)$$

Similarly if $r > 1$, ultimately p_B should be equal to unity. To obtain r_{max} for this case, it is necessary to rewrite the gelation criteria in terms of p_B, i.e.,

$$r_{max} = (f-1)r_{B_f} - (1 - r_{B_f}) \qquad (9.3.19)$$

Equations (9.3.18) and (9.3.19) for r_{min} and r_{max} define the gelation envelope, and this has been plotted in Fig. 9.2.

To assess the effect of R_3 and R_4 on r_{min} and r_{max} it is necessary to solve the general equation (9.3.10) for the equal reactivity case, and results have been included in Fig. 9.2 for comparison. As R_3 and R_4 are increased, the value of r_{min} for a given r_{B_f} falls and that of r_{max} increases, in this way increasing the area of the gelation envelope. Results are, however, found to be relatively insensitive to the variation in R_4, and at $r_{B_f} = 1$, r_{max} and r_{min} are independent of R_4. The latter occurs because at $r_{B_f} = 1$, there are

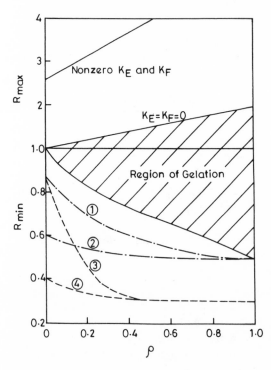

Figure 9.2. Gelation envelope shown by solid line for the case of equal reactivity ($R_1 = 2$, $R_2 = \frac{1}{2}$, $R_5 = R_6 = R_7 = 1$, $R_3 = R_4 = 0$). r_{max} shifts upwards and r_{min} downwards for nonzero R_3 and R_4. r_{min} lines have been computed for the following parameters: (1) $R_1 = R_2 = 1.0$, $R_3 = R_4 = 0.1$, (2) $R_1 = R_2 = 1.0$, $R_3 = 0.1$, $R_4 = 5$. (Reprinted from Ref. 3 with permission from Professor K. S. Gandhi.)

no urea linkages and therefore biurets cannot be formed. It is thus found that the presence of secondary reactions predicts the gelation of reactant mixtures which would otherwise not be predicted to gel. This is expected because the secondary reactions lead to formation of more cross-links.

Another interesting situation arises when diisocyanates are polymerized with polyols and blowing agents only. Presently, the secondary reactions are ignored and it is assumed that $r < 1$, in which case $\alpha_1 = \alpha_2 = \alpha_1^* = \alpha_2^* = 1$, $r_{B_2} = 0$, and Eqs. (9.3.12) and (9.3.13) reduce to

$$\frac{2(f-1)r_{B_f}p_B^2}{2r_{B_f}p_f + (1 - r_{B_f})(p_C - p_{C1})} \geq 1 \qquad (9.3.20a)$$

$$2r = 2r_{B_f}p_B + (1 - r_{B_f})(p_C - p_{C1}) \qquad (9.3.20b)$$

Equation (9.3.20a) can be rearranged as

$$2r_{B_f}\{(f-1)p_B^2 - p_B\} \geq (1 - r_{B_f})(p_C - p_{C1}) \qquad (9.3.21)$$

The way p_C and p_{C1} have been defined in Table 9.3 ($p_C - p_{C1}$) is always positive. Since $0 \leq r_{B_f} \leq 1$, Eq. (9.3.20b) implies that $\{(f-1)p_B^2 - p_B\}$ must always be positive, which means that for gelation to occur, it is necessary that

$$p_B \geq \frac{1}{(f-1)} \qquad (9.3.22)$$

Thus, if the polyol is trifunctional, the critical value of p_B is 0.5, if it is tetrafunctional $p_{B,\text{crit}}$ is 0.33, and so on.

The reactivity of the blowing agent plays a very important role, as can be seen from the following discussion. R_1 is the parameter which decides the relative reactivity of CD. When $R_1 = 0$, the blowing agent is noninteractive and simply acts as a diluent. Both p_C and p_{C1} are zero and for gelation to occur, Eq. (9.3.20) once again predicts that Eq. (9.3.22) must be satisfied and $r_{\min}$ is obtained by substituting this in Eq. (9.3.20), i.e.,

$$r_{\min} = \frac{p_B}{(f-1)} \qquad (9.3.23)$$

If the blowing reacts instantaneously and completely, $R_1 \to \infty$, $p_C = p_{C1} = 1$ and the critical value p_B is once again given by Eq. (9.3.21) and $r_{\min}$ can be calculated using

$$r_{\min} = \frac{r_{B_f}}{(f-1)} + (1 - r_{B_f}) \qquad (9.3.24)$$

This indicates that CD required for gelation for $R_1 = 0$ is less than that for $R_1 \to \infty$, which is evidently because in the latter case some of the CD monomer is used in reacting with monomer A_1A_2. Secondly, Eq. (9.3.24) predicts that r_{min} decreases as r_{B_f} is increased and A_1A_2 monomer required for gelation is decreased. This is because in the beginning of the reaction, CD is consumed to form $-AACD-$ type prepolymer but not resulting in cross-linking. As r_{B_f} is increased, the fraction of CD monomer decreases and the requirement of A_1A_2 monomer is reduced.

It is thus seen that for finite R_1, CD acts as nonreactive diluent as well as the chain extender. As r_{B_f} is increased, the former effect tends to increase r_{min} whereas the latter effect tends to reduce r_{min}. The effects of R_1 and r_{B_f} have been examined in Fig. 9.3, which shows that for $R_1 > 2$, the diluent effect is almost absent and r_{min} decreases monotonically with increasing r_{B_f}. However, for small values of R_1 (such as 0.1), a balance of both these effects is observed. For small R_1, CD reacts slowly for small values of r_{B_f}, a large amount of the blowing agent is present, and the lower reactivity is compensated by its higher concentration. Thus at low values of r_{B_f}, the blowing agent behaves like a chain extender and higher values of r_{B_f} lead to a decrease in r_{min}.

The upper boundary of the gelation envelope is obtained when $r > 1$, in which case A groups are in stoichiometric excess and $p_B = p_C = p_{C1} = 1$.

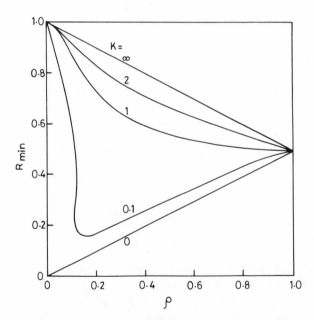

Figure 9.3. Effect of R_1 and r_{B_f} on r_{min} for trifunctional polyol. $R_5 = R_6 = R_7 = 1.0$, $R_3 = R_4 = 0$. (Reprinted from Ref. 3 with permission from Professor K. S. Gandhi.)

Substituting these in Eqs. (9.3.12) and (9.3.13) gives

$$\frac{4(f - 1)(\alpha_2^* + R_5\alpha_1^*)(1 - r_{B2})r_{B_f}}{2(1 - r_{B_2})\{1 + \alpha_2^* + R_5(1 + \alpha_1^*)\} - 4(\alpha_2^* + R_5\alpha_1^*)(1 - r_{B_2})(1 - r_{B_f})} \geq 1$$

(9.3.25a)

$$\frac{\alpha_1}{(1 + R_5\alpha_1)}\{1 + \alpha_2^* + R_5(1 + \alpha_1^*)\} = \frac{2}{r}\frac{(1 - r_{B_2})}{(1 - r_{B_2})} = \frac{2}{r} \quad (9.3.25b)$$

It is thus seen that r_{max} is independent of r_{B_2}. Equation (9.3.25) can be further simplified to

$$\frac{(1 + R_5)(\alpha_2^* + R_5\alpha_1^*) - 2(\alpha_2^* + R_5\alpha_1^*) + 2(\alpha_2^* + R_5\alpha_1^*)r_{B_f}}{2(f - 1) * (\alpha_2^* + R_5\alpha_1^*)r_{B_f}} \leq 1 \quad (9.3.26)$$

Since r_{B_f} is a positive number, multiplying both the sides of the inequality above by $2(f - 1)r_{B_f}$, Eq. (9.3.26) simplifies to

$$\left\{\frac{1 + R_5}{\alpha_2^* + R_5\alpha_1^*} - 1 + 2r_{B_f}\right\} \leq (2f - 1)r_{B_f} \quad (9.3.27)$$

The parameters affecting r_{max} are R_5, R_6, and R_7, and their effect upon it is shown in Fig. 9.4. If $R_5 > 1$ and monomer A_1A_2 is present in large quantity, all B groups are consumed by A_2, leaving unreacted A_1 groups in the reaction mass. To promote cross-linking, A_1 groups have to be forced to react, and this can be achieved by decreasing r at constant r_{B_f} as seen in Fig. 9.4. Similarly when R_5 or R_6 is greater than unity, polymeric A groups react faster, which implies that processes leading to formation of branch points and chain extension are promoted. Hence larger amount of AA monomer can be utilized or r_{max} is increased at a given r_{B_f}.

One now considers the gelation in presence of bifunctional alcohol B_2. The various results discussed above remain exactly the same and the exact numerical values can be calculated using Eqs. (9.3.12) and (9.3.13). In presence of B_2, $r_{B_2} \neq 0$ and has a finite value. Since r_{max} is independent of r_{B_2} [Eqs. (9.3.26) and (9.3.27)], the upper part of the gelation envelope (i.e., r_{max}) is not affected by the presence of B_2. It may be seen from Eqs. (9.3.23) and (9.3.24) that r_{min} is dependent only on r_{B_f}. Therefore, the effect of B_2 is only to change r_{B_f} and the maximum permissible value r_{B_2} is $(1 - r_{B_2})$ when $r_c = 0$, and its effect on r_{min} is shown in Fig. 9.5.

The gel point for incomplete conversion lies within the gelation envelope. For a given value of α_1, the kinetic model discussed in Section 9.2 yields p_B, p_C, p_{C1}, α_1^*, α_2, and α_2^*. From these probabilities, the set of values of r, r_{B_f}, and r_{B_2} satisfying the gelation condition and the

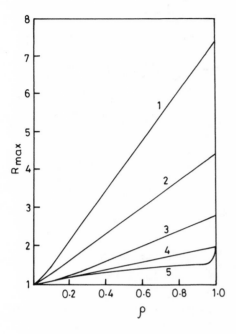

Figure 9.4. Effect of R_5, R_6, and R_7 on r_{max} versus r_{B_f}. $R_3 = R_4 = 0$, $f = 3$, R_5, R_6, and R_7 values for (1) 2.5, 10, 10; (2) 1, 4.4; (3) 10, 10, 10; (4) 1, 1, 1; and (5) 5, 1, 1. (Reprinted from Ref. 3 with permission from Professor K. S. Gandhi.)

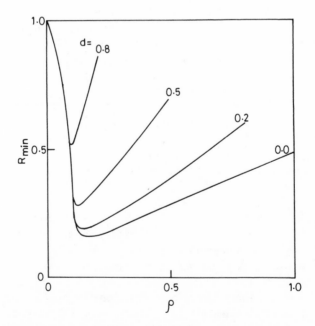

Figure 9.5. Effect of addition of B_2 on r_{min}. $R_1 = 0.1$, $R_2 = 1.0$, $R_3 = R_4 = 0$, $f = 3$. (Reprinted from Ref. 3 with permission from Professor K. S. Gandhi.)

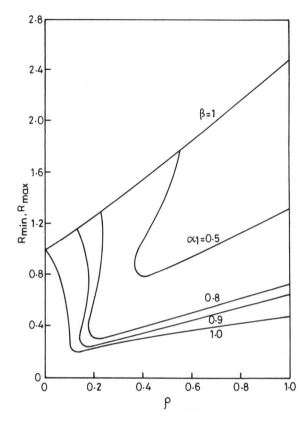

Figure 9.6. Gelation envelopes for incomplete conversions of monomers. $R_3 = R_4 = 0$, $R_5 = 2.0$, $R_2 = 2.0$, $R_6 = 1.5$, $R_7 = 3.0$, $R_1 = 0.1$, $f = 3$. (Reprinted from Ref. 3 with permission from Professor K. S. Gandhi.)

stoichiometric relation can be found as has been done in Fig. 9.6. Such parametric curves give an indication of the extents of conversion attained when gelation occurs for any given initial compositions specified by r, r_{B_f}, and r_{B_2}.

9.4. APPROXIMATE RATE EXPRESSION FOR URETHANE FORMATION

In the studies of reaction injection molding, because of the various physical processes, it becomes necessary to solve the rate equations along with transport equations. This makes the mathematical modeling of the process extremely difficult, particularly if one were to use the relatively detailed reaction scheme presented in Section 9.3. In order to simplify the analysis, an extremely simplified kinetic scheme of urethane polymerization must first be assumed, which can explain most of the gross features. In the literature, a simple Michaelis–Menten type kinetic expression has been used to approximate the more exact kinetics described in Sections 9.2 and 9.3. Macosko et al.[8,9] have measured the kinetic data for the polymerization of a thermoplastic polyurethane under conditions similar to those existing in commercial reaction injection molding. They have ignored the effect of substituents on the reactivity of an isocyanate group and have treated all isocyanate groups as kinetically equivalent. The following simplified two-step mechanism[10] has been used to fit the experimental data:

$$\text{\small$\sim$NCO} + \text{Cat} \xrightarrow{k_{13}} (\text{NCO}) - \text{Cat} \qquad (9.4.1a)$$

$$\text{\small$\sim$NCO} - \text{Cat} + \text{OH} \xrightarrow{k_{15}} \text{\small$\sim$NHCO\small$\sim$} + \text{Cat} \qquad (9.4.1b)$$

where $-$Cat denotes a molecule of catalyst and $\sim$NCO$)-$Cat is an intermediate formed through the reversible combination of an isocyanate group and the catalyst. Since the net rate of formation of an intermediate is zero (from the steady state approximation), therefore

$$\frac{d[\sim\text{NCO}) - \text{Cat}]}{dt}$$

$$= k_{13}[\text{NCO}][\text{Cat}] - [\sim\text{NCO}) - \text{Cat}]\{k_{14} + k_{15}[\text{OH}]\} = 0 \quad (9.4.2)$$

where $\sim$NCO], [$\sim$OH], and [Cat] represent the total concentrations of these groups in the reaction mass. One can solve for [$\sim$NCO$)-$Cat], and the rate of consumption of $\sim$OH] and $\sim$NCO] functional groups can be found as

$$\frac{d\sim\text{NCO}]}{dt} = \frac{d\sim\text{OH}]}{dt} = k_{15}[\sim\text{NCO}) - \text{Cat}][\sim\text{OH}]$$

$$= \frac{k_{13}k_{15}[\sim\text{NCO}][\sim\text{OH}][\text{Cat}]}{k_{14} + k_{15}\sim\text{OH}]} \qquad (9.4.3)$$

If an Arrhenius-type temperature dependence of various rate constants is assumed, when

$$\frac{k_{13}k_{15}}{k_{14}} = \frac{k_{13}^* k_{15}^*}{k_{14}^*} \exp\left\{ -\frac{E_{13} + E_{15} - E_{14}}{R}\left(\frac{1}{T} - \frac{1}{T_r} \right) \right\} \qquad (9.4.4a)$$

$$\frac{k_{15}}{k_{14}} = \frac{k_{15}^*}{k_{14}^*} \exp\left\{ -\frac{E_{15} - E_{14}}{R}\left(\frac{1}{T} - \frac{1}{T_r} \right) \right\} \qquad (9.4.4b)$$

where T_r is a reference temperature (373 K has been used in this study) for the kinetic rate and k_{13}^*, k_{14}^*, and k_{15}^* are the rate constants evaluated at T_r. If the extent of reaction, p, and the initial molar ratio r are defined as

$$p = \frac{\text{↓∿OH]}_0 - \text{↓∿OH]}}{\text{↓∿OH]}_0} \qquad (9.4.5a)$$

$$r = \frac{\text{∿NCO]}_0}{\text{↓∿OH]}_0} \qquad (9.4.5b)$$

Eqn. (9.4.3) can be rewritten in terms of these to give the conversion of OH group as

$$\frac{dp}{dt} = \frac{K_{c1} \exp\left\{ -\frac{E_{13}}{R}\left(\frac{1}{T} - \frac{1}{T_r} \right) \right\}[\text{Cat}][\text{∿∿OH}]_0(1 - p)(r - p)}{1 + K_{c2}[\text{∿∿OH}]_0(1 - p)} \qquad (9.4.6)$$

where

$$K_{c1} = \frac{k_{13}^* k_{15}^*}{k_{14}^*} \qquad (9.4.7a)$$

$$K_{c2} = \frac{k_{15}^*}{k_{14}^*} \qquad (9.4.7b)$$

In deriving Eq. (9.4.6), it has been assumed that[9] $E_{14} \simeq E_{15}$. The suitable monomers have been assumed to be charged into a suitable mold, and under no flow conditions it is assumed that there is no temperature gradient in the reaction mass. An energy balance on the mold gives

$$Mc_{pm}\frac{dT}{dt} = (\Delta H) V[\text{∿∿OH}]_0 \frac{dp}{dt} - hA(T - T_a) \qquad (9.4.8)$$

where M is the mass of the mold, h is the mean heat transfer coefficient, and A is the surface area available for the heat transfer. Above, V is the volume of the mold and T_a is the ambient temperature. The h and specific heat $c_{p,m}$ for the reacting system were estimated,[9] and using Eqs. (9.4.6), K_{c1} and K_{c2} were found. For various catalyst systems, these are determined and the results are summarized in Table 9.5. The degree of fit with experimental rate data is shown in Fig. 9.7.

It may be emphasized that in writing Eq. (9.4.1), it has been assumed that all functional groups have the same reactivity. Polymerization in molds, where there is no fluid motion after it is filled, can be well described by the kinetics in batch reactors, and in view of the experimental results of Table 9.5, Eq. (9.4.3) can be regarded as a two-parameter rate expression to fit

TABLE 9.5.[a] Estimated Values of Parameters in Eq. (9.4.6)

Catalyst	K_{c1} (g/sec mole)	K_{c2} (g/sec mole)	E_{13} (kcal/mole)	ΔH (kcal/mole)
Noncatalyzed	19.1	4.5	5.8	22.1
Dibutyl tin dilaurate (DBTL)	26.70	0.58	3.3	19.7
Triethylene diamine (Dabco)	303	0.02	10.1	19.8
Phenylmercuric propanate (PHgP)	27	−0.43	15.6	22.1

[a] References 9, 10. Reprinted from Ref. 9 with permission of John Wiley and Sons, New York.

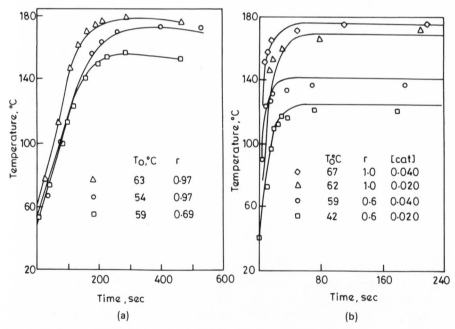

Figure 9.7. Temperature rise data in urethane polymerization, (a) noncatalyzed, (b) catalyzed by dibutyl tin dilaurate (DBTDL) catalyst. (Reprinted from Ref. 9 with permission of John Wiley & Sons, New York.)

the experimental data. This empiricism may be the reason why K_{c2} for phenylmercuric propanate catalyst (PHgP) was found to be negative.[9]

With the empirical expression for the rate of urethane formation, one is now in a position to analyze a far more complex process of reaction injection molding (RIM). In the following section, it is shown how momentum and energy transfer play important roles in determining the final property of the polymer.

9.5. REACTION INJECTION MOLDING

Reaction injection molding (RIM) is a method by which two highly reacting monomers or prepolymers are mixed in a known proportion and polymerized *in situ* in the molds, to form described articles directly from the liquid. The overall process for urethane formation can be represented as

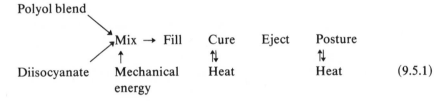

$$(9.5.1)$$

The schematic diagram of the RIM machine is shown in Fig. 9.8, in which a diisocyanate and a polyol blend (a mixture of a polyol, diol, a catalyst, and a blowing agent) are separately stored in chambers A and B. These are mixed in chamber G through impingement mixing, and the resultant liquid

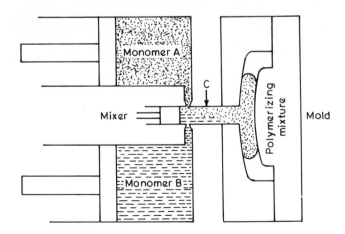

Figure 9.8. Schematic diagram of reaction injection molding (RIM).

mixture is pumped rapidly into the mold, where it is cured. After the reaction mass has reached a high conversion and the molded article is strong enough to maintain its shape, it is ejected from the mold and if necessary it is subjected to postcure treatment as shown in Eq. (9.5.1).

Evidently, reaction injection molding is a periodic process like any other molding operation, with the difference that the feed to RIM is monomer and one operation cycle takes no more than a few minutes. For this reason, the design of the nozzle and the value of the Reynolds number, Re, of the fluid at its exit are extremely important. Equation (9.4.8) shows that the rate of formation of urethane polymer in the mold can be followed by the measurement of the adiabatic rise in temperature, ΔT_{ad}, in it. In order to study the effect of Reynolds number (Re) of the polyol at the jet on the polymerization, this was systematically varied and ΔT_{ad} in the mold was measured and plotted in Fig. 9.9.[10] The curves are found to shift upwards as Re is increased, and for values beyond 200, results are found to become independent of the flow conditions in the mixing chamber.

The rate of reaction of the monomeric fluids in urethane formation can be varied by the use of catalyst. As has been discussed in the previous section (as well as Chapter 4) gelation of urethane polymer is a very complex process, but the gelation time can be very roughly used to characterize the rate of polymer formation. In Fig. 9.10, various reacting fluids differing in

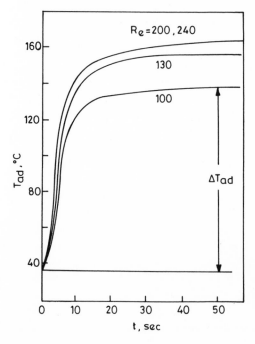

Figure 9.9. Degree of mixing in the mixing chamber measured by adiabatic temperature rise in the mold of RIM at different triol (B_3) Reynold numbers. (Reprinted from Ref. 13 with permission of SPE.)

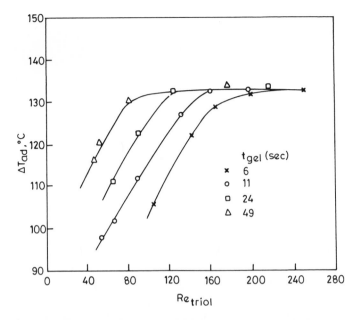

Figure 9.10. Degree of mixing in the mixing chamber measured by the adiabatic temperature rise at various reaction rates. The latter is characterized by the gelation time, t_g. (Reprinted from Ref. 13 with permission of SPE.)

the gelation time, t_g, have been used. Evidently, the monomeric compositions having lower t_g values are those that react faster. For these fluids, as the nozzle Reynolds number of the polyol is increased, ΔT_{ad} is found to increase almost linearly, ultimately leveling off to a value of 130°C beyond a Re value of 200. This indicates that for Re > 200 the mixing of the reacting fluids in the mixing chamber is complete and the overall polymerization in the mold can be assumed to be reaction controlled.

9.6. IMPINGEMENT MIXING IN RIM[11-17]

As shown in Eq. (9.5.1), polyols and diisocyanates must first be mixed in RIM, and the effect of mixing on polymer formation has been clearly demonstrated in Fig. 9.9. In the case of RIM, mixing becomes important because the time constants for mixing and polymerization are comparable. Since it is difficult to study the combination of both these phenomena, in the following a model experiment is described to explain mixing alone.

It is assumed that a very viscous fluid is placed between two concentric cylinders as in Fig. 9.11 and the outer cylinder rotated. If the fluid flow is strictly laminar and *nondiffusive*, a blank line of dye (as in Fig. 9.11a) would

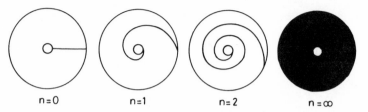

Figure 9.11. Laminar, nondiffusive mixing of a dye in polymeric fluid by the rotation of concentric cylinders.

deform into shapes shown in Figs. 9.11b, 9.11c and 9.11d after one, two, and infinite revolutions, respectively. If L represents the annular spacing and S_t the distance between two adjacent rings, then

$$S_t = \frac{L}{(n+1)} \qquad (9.6.1)$$

where n is the number of revolutions. The various rings are called striations and S_t, the striation thickness. As can be seen in the two-dimensional example of Fig. 9.11a, the decrease in striation thickness is associated with an increased mixing of the dye with the fluid and is used as a means of quantifying mixing of two fluids in more complex situations. Ideally, one would like to have $S_t = 0$ for the well mixed condition.

The experiment described above is highly idealized in the sense that *all* fluids mix on the molecular level. This can occur either through molecular diffusion or through turbulence. In the idealized experiment of mixing of dye with a viscous liquid, the presence of molecular diffusion would reduce the striation thickness and one cannot use Eq. (9.6.1) to find it. To predict S_t on a molecular level, in presence of diffusion and (or) turbulence, is difficult because of the lack of definite geometry as in Fig. 9.11. One still retains the concept of correlating mixing with striation thickness, and two theoretical approaches exist in the literature to predict the latter. One of them assumes that the striation thickness, in presence of turbulence, is determined by the "smallest eddy," which in turn is dependent on the Reynolds number. The development in this line has led to

$$S_t = c_1(\text{Re})^{-3/4} \qquad (9.6.2)$$

where c_1 is a constant determined by the nozzle geometry used in RIM and Re is the Reynolds number of the fluid at the nozzle given by

$$\text{Re} = \frac{4\mathring{Q}\rho_m}{D\eta_m} \qquad (9.6.3)$$

Above $\overset{\circ}{Q}$ is the total volumetric flow rate, ρ_m is the mean density, and η_m is the viscosity. The important assumption made in deriving Eq. (9.6.3) is that the fluid streams do not detach from the side wall of the nozzle and this would be strictly true at low Reynolds number (less than 100). At moderate Reynolds numbers (above 100), fluid streams are found to detach and a large vortex develops on impingement. As can be intuitively argued, this would lead to reduced striation thickness, hence increased degree of mixing, and the theoretical development in this line leads to the following relation:

$$S_t = c_2(\text{Re})^{-1/2} \qquad (9.6.4)$$

Baldyga and Bourne,[15] Lee et al.[12] and Kolodziej et al.[14] in recent publications point out that Eqs. (9.6.2) and (9.6.4) do not predict a correct Reynolds number dependence of the striation thickness. This is contrary to the experimental findings as seen in Figs. 9.9 and 9.10 for the following reason. If the polymerization was dependent upon the state of mixedness of the fluid, the adiabatic rise in temperature would change with Re, but it does not for Re > 200.

Baldyga and Bourne[15] have used the statistical theory of turbulent diffusion, and results derived based on this theory show that the striation thickness is independent of the Reynolds number but is strongly dependent upon (a) the ratio of reagent flow rates, (b) mixing head geometry, and (c) the residence time distribution of the flow in the mixing head. There is a need for considerable research in this area, as there are still several fundamental questions unanswered.

9.7. FILLING AND CURING IN MOLDS IN RIM[16-22]

In the mixing chamber, the polyol blend and the diisocyanate are brought into intimate contact, and since these are very reactive fluids, polymerization starts immediately after mixing. As the RIM molds become larger in size and more complex in geometry, it becomes important to understand the process of mold filling since polymerization is taking place simultaneously. For example, if there is a premature gelling during filling, the mold would never be completely filled with the material. In the following discussion, a simple mold is developed to explain the filling of a model rectangular mold and the interplay of various parameters is demonstrated. These studies can be extended to predict the performance of more complex geometries.

It may be recognized that as the polymerization progresses, the viscosity of the reaction mass increases. Lipshitz et al.[18] have developed the kinetic and viscosity relations for urethane polymerization using 1,6-hexamethylene diisocyanate (HDI) and tripropylene glycol (TPG). The reaction was carried out in the presence as well as the absence of catalyst. Viscosity and the conversion as a function of time were determined at various temperatures, and from this, the viscosity as a function of the molecular weight of the polymer and temperature was determined empirically to be

$$\eta_m = 3.983 \times 10^{-8} \bar{M}_w^{2.678} \exp\{(0.1259 - 0.02532 \ln \bar{M}_w)(T - 45)\} \quad (9.7.1)$$

where η_m is the viscosity of the reaction medium in poise and T is the temperature in °C. The effect of shear rate on the viscosity of the polymer was determined at three different times, and the results are shown[18] in Fig. 9.12. Evidently, the reaction mixture can be taken to be Newtonian.

The filling of rectangular molds has been studied by Macosko *et al.*[19-22] and the flow fronts have been photographed using a high-speed camera and the temperature and pressure recorded. Typical results are shown in Fig. 9.13 and the pressure is found to increase due to foaming. By varying the catalyst concentration, the gelation time, t_g, can be changed, whereas by varying the total flow rate of the fluid, the mold filling time, t_f, can be altered. A systematic variation of these operating variables reveals that it is not these individually but their ratio, G, defined as

$$G = \frac{t_f}{t_g} \quad (9.7.2)$$

which affects the filling of the mold. In Fig. 9.14, experimental results for various values of G have been plotted, and from the pressure rise data, it is evident that there has been premature gelling within the mold for $G = 0.19$. These experiments suggest that it may be advantageous to have very short gelling times, but that too runs into difficulty if the average velocity of filling is increased without limit, as seen in Fig. 9.15. A flow instability is found to develop for average filling velocities of about 0.5 m/sec.

The rectangular mold length L, width W, and thickness H is shown in Fig. 9.16 in which the x axis is placed along the direction of flow and the y axis at the center of the mold. If a drop of dye is introduced at a height, $-y$, it is found to drift towards the bottom plate (i.e., towards $y = -H/2$) and if it is placed at $+y$, it moves to the upper plate (i.e., $y = H/2$). If the entrance effect is neglected, the stream lines are fairly straight except near the front where they have a spreading nature as shown in Fig. 9.16. The flow near the front is at times called *fountain flow* and this is two dimensional in nature for the following reason. In the main flow, the velocity profile is expected to be parabolic, whereas it has been experimentally demonstrated that the flow front is flat[19] and the pressure drop across it is usually small.[20]

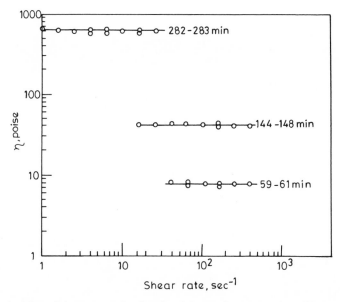

Figure 9.12. Effect of shear rate on the viscosity of the reaction mass at three different reaction times during the 45°C cure. (Reprinted from Ref. 16 with permission of SPE.)

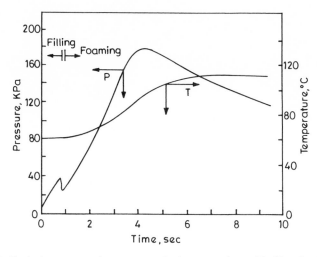

Figure 9.13. Typical pressure and temperature rise in rectangular molds. (Reprinted from Ref. 18 with permission of SPE.)

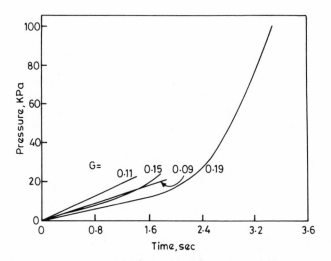

Figure 9.14. Pressure rise for various gelation parameter (G) values in the filling of rectangular mold. There has been gelling in the mold for $G = 0.19$.

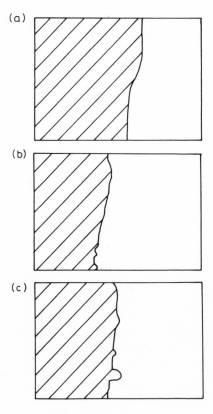

Figure 9.15. Flow front instability: (a) stable at $\bar{v}_x = 0.4$ m/sec; (b) semistable at $\bar{v}_x = 0.5$ m/sec; (c) unstable at $\bar{v}_x = 0.8$ m/sec.

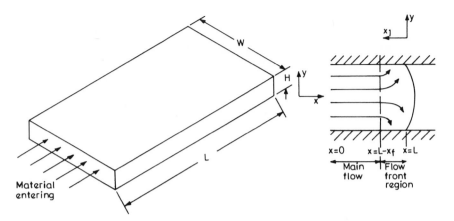

Figure 9.16. Schematic diagram of the rectangular mold. The flow can be divided into main flow and flow front regions.

For this transition in velocity profiles to occur, the fluid elements at the center (i.e., $y = 0$) should move towards the walls whereas fluid elements near the wall should move towards the center to give a flat front. It may be pointed out that the domain of the flow-front region lies in a very short length ($x_f \ll H$) but its analysis is important to predict the instability of the front as seen in Fig. 9.15. Thus in any analysis, the flow can be modeled in terms of these two regions which are characterized by different equations. In the following discussion, the main flow is analyzed first. The analysis presented here is in line with works of Macosco in which it is assumed that premixing of monomers has been done completely. Recent studies of Ottino have shown that this may not in general be true. If there is any segregation, then it should be properly accounted for, but the analysis of such a case is extremely complex. The segregation can also occur when monomers and polymers are thermodynamically incompatible; the discussion of such systems has begun to fascinate researchers only recently.

9.7a. Main Flow Region

In the main flow, it can be assumed that only v_x exists and that v_x is a function of y only. In this case, the linear momentum equation in dimensionless form in the x direction is given by

$$\frac{H \, \text{Re}}{2L} \frac{\partial v_x^*}{\partial t^*} = -\left(\frac{\partial P_{T1}}{\partial x}\right)^* + \frac{\partial}{\partial y^*}\left(\eta_m^* \frac{\partial v^*}{\partial y^*}\right) \qquad (9.7.3)$$

where P_{T1} is the pressure. The dimensionless variables are

$$\eta_m^* = \eta_m / \eta_{m,0} \tag{9.7.4a}$$

$$v_x^* = v_x / \bar{v}_x$$

$$x^* = x / L \tag{9.7.4b}$$

$$y^* = y / H \tag{9.7.4c}$$

$$t^* = t / t_f \tag{9.7.4d}$$

$$\mathrm{Re} = \frac{\rho_m \bar{v}_x (H/2)}{\eta_{m,0}}$$

$$\left(\frac{\partial P_{T1}}{\partial x}\right)^* = \frac{(\partial P_{T1}/\partial x)}{\dot{Q}\eta_{m,0}/H^3 W} \tag{9.7.4e}$$

where $\bar{v}_x$ is the average velocity, t_f is the filling time, $\eta_{m,0}$ is the viscosity at the entrance of the mold, and Q is the total volumetric flow rate. The filling time t_f is related to $\bar{v}_x$ by

$$t_f = \frac{L}{\bar{v}_x} \tag{9.7.5}$$

In general, for typical RIM flows

$$\frac{H \, \mathrm{Re}}{2L} \ll 1 \tag{9.7.6}$$

in which case, the linear momentum equation simplifies to

$$\left(\frac{\partial P_{T1}}{\partial x}\right)^* = \frac{\partial}{\partial y^*}\left(\eta_m^* \frac{\partial v_x^*}{\partial y^*}\right) \tag{9.7.7}$$

Since the flow is assumed to be fully developed, $(\partial P_{T1}/\partial x)^*$ can be assumed to be independent of y^* and Eq. (9.7.7) can be integrated once to give

$$\eta_m^* \frac{\partial v_x^*}{\partial y^*} = \left(\frac{\partial P_{T1}}{\partial x}\right)^* y^* + \mathscr{C}_1^0 \tag{9.7.8}$$

where C_1 is a constant of integration which would be zero since the flow is assumed to be symmetrical around the central plane $y = 0$. In addition to this, at any x, one has

$$\int_0^1 v_x^* \, dy^* = 1 \qquad (9.7.9)$$

On integrating this equation by parts and using the no-slip boundary condition (i.e., $v_x^* = 0$ at $y^* = 1$), one gets

$$-\int_0^1 y^* \frac{\partial v_x^*}{\partial y^*} \, dy^* = 1 \qquad (9.7.10)$$

On eliminating $(\partial v_x^*/\partial y^*)$ with the help of Eq. (9.7.8),

$$\left(\frac{\partial P_{T1}}{\partial x}\right)^* \int_0^1 \frac{y^{*2}}{\eta_m^*} \, dy^* = -1 \qquad (9.7.11)$$

Or, the pressure drop per unit axial length is given by

$$\left(\frac{\partial P_{T1}}{\partial x}\right)^* = -\frac{1}{\int_0^1 (y^{*2}/\eta_m^{*\prime}) \, dy^*} \qquad (9.7.12)$$

It may be observed that as the conversion increases along the length of the mold, η_m^* is going to increase, which implies that $\int_0^1 y^{*2}/\eta_m^* \, dy^*$ is a function of x^*. The total pressure drop ΔP_{T1} along the length of the reactor can be easily obtained by integrating Eq. (9.7.12),

$$\Delta P_{T1} = -\frac{4\eta_{m,0}QL}{H^3 W} \int_0^{(L-x_f)/L} \frac{dx^*}{\int_0^1 \dfrac{y^{*2} \, dy^*}{\eta_m^*}} \qquad (9.7.13)$$

In order to make a molar balance, the molecular diffusion of the species is neglected because the reaction is known to be very fast. In RIM operations, foaming, and overall density changes are assumed to be small. The rate of polymerization, $\mathscr{R}$, is assumed to be given approximately by

$$\mathscr{R} = k \, e^{-E/RTc^2} \qquad (9.7.14)$$

after neglecting K_{c2} in Eq. (9.4.7) and assuming $r = 1$. In the above equation, k is equal to $K_{c1}[\text{\small$\sim\!\!\!\sim$}OH]_0[\text{Cat}]$ and c is the concentration of functional groups. The mole balance of the function groups is given by

$$\frac{\partial p}{\partial t^*} + v_x^* \frac{\partial p}{\partial x^*} = k^* G(1-p)^2 \qquad (9.7.15)$$

where

$$p = \frac{c_0 - c}{c_0} \tag{9.7.16a}$$

$$k^* = \frac{k}{k_T} \frac{p_c}{1 - p_c} \tag{9.7.16b}$$

$$t_g = \frac{p_c^*}{(1 - p_c^*)(k_{T_0} c_0)} \tag{9.7.16c}$$

$$G = \frac{t_f}{t_g} \tag{9.7.16d}$$

In the above equation t_g has the dimension of time and is equal to the time taken for the fluid to reach the conversion p_c at the gel point at temperature T_0. The variable G is a dimensionless number which is the same as that in Eq. (9.7.2) and is sometimes known as the gelation number.

In writing the energy balance, it is assumed that there is negligible heat conduction in the axial direction. The resulting equation can be made dimensionless using

$$Gz = \frac{\text{heat transport by convection}}{\text{heat transport by conduction}} = \frac{\rho_m c_{pm} \bar{v}_x / L}{\dfrac{k_{km}}{H/2}} \tag{9.7.17a}$$

$$Br = \frac{\text{heat production by viscous dissipation}}{\text{heat transport by conduction}} = \frac{\eta_{m,0} \dfrac{\bar{v}_x}{(H/2)}}{\dfrac{k_m}{(H/2)^2} \Delta T_{ad}} \tag{9.7.17b}$$

$$T^* = \frac{T - T_0}{\Delta T_{ad}} \tag{9.7.17c}$$

where k_m and c_{pm} are thermal conductivity and heat capacity of the reaction medium. ΔT_{ad} is the adiabatic rise in temperature and can be calculated from Eq. (9.4.9) by substituting the heat transfer coefficient, h, in it to be zero. On doing this and integrating Eq. (9.4.9) once, one obtains the following equation for batch reactors:

$$\Delta T_{ad,b} = \left(\frac{-\Delta H_r}{\rho_m c_{pm}} \right) c_0 p \tag{9.7.18}$$

The gel number, G, defined in Eq. (9.7.16b) can be rewritten with the help of this relation as

$$G = \frac{t_f}{t_g}$$

$$= \frac{(L)/t_c}{\bar{v}_x} \frac{(-\Delta H_r)c_0 p}{\rho_m c_{pm} \Delta T_{ad}}$$

$$= \frac{\text{heat production by chemical reaction}}{\text{heat transported by convection}} \qquad (9.7.19)$$

In terms of these variables, the dimensionless form of the energy equation can be derived as

$$\frac{\partial T^*}{\partial t^*} + v_x^* \frac{\partial T^*}{\partial x^*} = \frac{1}{\text{Gz}} \frac{\partial^2 T^*}{\partial y^{*2}} + \frac{\text{Br}}{\text{Gz}} \eta_m^* \left(\frac{\partial v_x^*}{\partial y^*} \right)^2 + Gk^*(1-p)^2 \qquad (9.7.20)$$

Br and Gz are the Brinkman and Graetz numbers, and for typical RIM operations, the Brinkman number is much smaller than the Graetz number. In view of this, the viscous dissipation term $(\text{Br}/\text{Gz})\eta_m^*(\partial v_x^*/\partial y^*)^2$ can be neglected and the energy balance equation reduces to

$$\frac{\partial T^*}{\partial t^*} + v_x^* \frac{\partial T^*}{\partial x^*} = \frac{1}{\text{Gz}} \frac{\partial^2 T}{\partial y^{*2}} + Gk^*(1-p)^2 \qquad (9.7.21)$$

The various initial and boundary conditions are

$$t^* = 0, \quad T^* = 0 \qquad (9.7.22a)$$

$$x^* = 0, \quad T^* = 0 \qquad (9.7.22b)$$

$$y^* = 0, \quad \frac{\partial T^*}{\partial y^*} = 0 \qquad (9.7.22c)$$

$$t^* = 0, \quad p = 0 \qquad (9.7.22d)$$

$$x^* = 0, \quad p = 0 \qquad (9.7.22e)$$

Equations (9.7.7), (9.7.15), and (9.7.22) must be solved along with transport equations for the flow front region as developed below, in order to predict the mold filling phenomena.

9.7b. Flow Front Region

It has already been argued that in the flow front region the velocity distribution is two dimensional in nature due to the "fountain" effect. There are two components, v'_x and v'_y, of the velocity. The pressure in this region is not only a function of x but is a function of y also. If the flow front is moving at some average velocity $\bar{v}_x$ in the forward direction, the velocities v_x and v_y in the flow front region *with respect to the moving front* are given by

$$v_x = v'_x - \bar{v}_x \qquad (9.7.23a)$$

$$v_y = v'_y \qquad (9.7.23b)$$

The frame of reference has been assumed to be fixed with the moving front. The Navier–Stokes equation in the flow front region can be simplified after neglecting inertial and gravity forces to give

$$-\frac{\partial P_{T1}}{\partial x_1} + \eta_m \left(\frac{\partial^2 v_x}{\partial x_1^2} + \frac{\partial^2 v_x}{\partial y^2} \right) = 0 \qquad (9.7.24a)$$

$$-\frac{\partial P_{T1}}{\partial y} + \eta_m \left(\frac{\partial^2 v_y}{\partial x_1^2} + \frac{\partial^2 v_y}{\partial y^2} \right) = 0 \qquad (9.7.24b)$$

In writing these equations, the flow front region has been assumed to be small such that the viscosity change in this region could be neglected. The continuity equation is given by

$$\frac{\partial v_x}{\partial x_1} + \frac{\partial v_y}{\partial y} = 0 \qquad (9.7.25)$$

It is assumed that there is no slip at the wall except at the interface between the wall and flow front, which means that in the frame of reference (x_1, y),

$$\text{at } y = H/2, \qquad x_1 > 0, \qquad v_x = \bar{v}_x \qquad (9.7.26a)$$

$$\text{at } x_1 = 0, \qquad y = 0 \quad \text{and} \quad H/2, v_x = 0 \qquad (9.7.26b)$$

Since at the free surface, the shear stress tangential to the interface is zero, therefore

$$(\tau_{xy}/\eta_m) = \frac{\partial v_x}{\partial y} + \frac{\partial v_y}{\partial x_1} = 0 \qquad (9.7.27)$$

However, at $x_1 = 0$ (i.e., at the interface), $v_x = 0$, independent of y and so the above relation reduces to

$$\frac{\partial v_y}{\partial x_1} = 0 \qquad\qquad (9.7.28a)$$

$$\text{at } y = 0 \quad \text{and} \quad \pm H/2, \qquad v_y = 0 \qquad\qquad (9.7.28b)$$

Using a series expansion containing terms of up to the fourth order, Eqs. (9.7.24) and (9.7.25) can be solved, and the velocity distribution in the flow front region has been found to be[24]

$$\frac{v_x}{\bar{v}_x} = -\left(\frac{1}{2} - 6\frac{y^2}{H^2}\right)\{1 - 1.45\, e^{-5x_1/H}\sin(0.76 + 2x_1/H)\}$$

$$-0.53(1 - 80x_1^4 H^4)\, e^{-5x_1/H}\sin(2x_1/H) \qquad\qquad (9.7.29a)$$

$$\frac{v_y}{\bar{v}_x} = (y/H)(1 - 4y^2/H^2)\{3.63 e^{-5x_1/H}\sin(0.76 + 2x_1/H)$$

$$- 1.45 e^{-5x_1/H}\cos(0.76 + 2x_1/H)\}$$

$$-\frac{2y}{H}\left(1 - 16\frac{y^4}{H^4}\right)\{1.32 e^{-5x_1/H}\sin(2x_1/H)$$

$$-0.53 e^{-5x_1/H}\cos(2x_1/H)\} \qquad\qquad (9.7.29b)$$

It has been shown[21,24] that the flow front region extends from the interface up to a distance of about H upstream, i.e., x_f in Fig. 9.16 is equal to H. The coordinate x_1 can be eliminated by writing it in terms of x easily. One defines the following dimensionless lengths:

$$x_{ff} = \frac{L - x}{H} = \frac{H - x_1}{H} = \left(1 - \frac{x_1}{H}\right) \qquad\qquad (9.7.30)$$

Thus Eq. (9.7.29) can be rewritten in terms of the frame of reference fixed at the *entrance* of the mold very easily.

It is now assumed that the velocity distribution in the flow front region is negligibly affected by the chemical reaction, in which case the mole and energy balance equations can be easily written as

$$\frac{\partial p}{\partial t^*} + v_x^*\frac{\partial p}{\partial x_{ff}} + v_y^*\frac{\partial p}{\partial y^*} = Gk^*\frac{H}{L}(1 - p)^2 \qquad\qquad (9.7.31a)$$

$$\frac{\partial T^*}{\partial t^*} + v_x^* \frac{\partial T^*}{\partial x_{ff}} + v_y^* \frac{\partial T^*}{\partial y^*} = \frac{x_0}{LGz}\frac{\partial^2 T^*}{\partial y^{*2}} + \frac{1}{Gz}\frac{H}{L}\frac{\partial^2 T^*}{\partial x_{ff}^2} + G_L^H k^*(1-p)^2$$

$$(9.7.31b)$$

Neglecting terms containing Gz^{-1} in the energy balance equation in the flow front region, Eq. (9.7.31b) becomes

$$\frac{\partial T^*}{\partial t^*} + v_x^* \frac{\partial T^*}{\partial x_{ff}} + v_y^* \frac{\partial T^*}{\partial y^*} = G\frac{H}{L}k^*(1-p)^2 \qquad (9.7.32)$$

The boundary and initial conditions for these equations are

$$\text{at } t^* = 0; \qquad p = 0 \quad \text{and} \quad T^* = 0 \qquad (9.7.33a)$$

$$\text{at } x_{ff}^* = 0; \qquad p = p \text{ from the main flow} \qquad (9.7.33b)$$

$$T^* = T^* \text{ from the main flow} \qquad (9.7.33c)$$

9.7c. Curing

When the mold is full, there is no flow within it and only molar and energy balance equations need be solved. These are as follows:

$$\frac{\partial c}{\partial t_1} = \mathcal{R} \qquad (9.7.34a)$$

$$\rho_m c_{vm}\frac{\partial T}{\partial t_1} = \frac{\partial}{\partial y}\left(k_m\frac{\partial T}{\partial y}\right) + (-\Delta H_r)\mathcal{R} \qquad (9.7.34b)$$

The following dimensionless variables are defined:

$$\alpha_m = \frac{k_m}{\rho_m c_{vm}} \qquad (9.7.35a)$$

$$k_c^* = k/k_{T_0} \qquad (9.7.35b)$$

$$t_1^* = \frac{\alpha_m}{H^2}t \qquad (9.7.35c)$$

$$\text{Da} = \frac{\text{heat production by chemical reaction}}{\text{heat transport by conduction}}$$

$$= \frac{k_{T_0} c_0^2 p(-\Delta H_r)}{\dfrac{k_m \Delta T_{\text{ad}}}{(H/2)^2}} \tag{9.7.35d}$$

where Da is the Damkoehlor number. In terms of these, the mole and energy balance equations are

$$\frac{\partial p}{\partial t_1^*} = \text{Da} \, k_c^*(1 - p)^2 \tag{9.7.36a}$$

$$\frac{\partial T^*}{\partial t^*} = \frac{\partial^2 T^*}{\partial y^{*2}} + \text{Da} \, k_c^*(1 - p)^2 \tag{9.7.36b}$$

It is now an easy matter to solve for the performance of the RIM. During mold filling, for a specified viscosity of the reaction mass [for example, Eq. (9.7.1)], momentum, molar, and energy balance equations in the main flow region [Eqs. (9.7.7), (9.7.15), and (9.7.21)] are solved numerically. At $x = (L - H)$, the region of the flow-front starts and one switches over to the balance equations for this region [Eqs. (9.7.29), (9.7.31a), and (9.7.32)]. After the mold is filled, the equations to be solved are (9.7.36).

It is found that the gelling number is one of the most important parameters which determines whether premature gelling occurs in the mold. If this number is high, premature gelling could occur and the pressure required would begin to shoot up as shown in Fig. 9.17. If the flow front effect is ignored, for the same mold length, it amounts to extending the main flow region in length and in the same figure results computed after doing this (marked as "lengthening the reactor") have been shown. It is found that for small gelling numbers the fountain effect has negligible effect; however, for large G, the flow front region cannot be ignored.

The effect of the gelling number and the Graetz number has been examined qualitatively in Fig. 9.18. When G is small, there is little polymerization during the filling of the mold and the computed results are found to be close to those calculated assuming constant viscosity (in subsequent figures such curves are marked as $\Delta P \eta_{T_0}$). For small Graetz numbers and wall temperature T_w less than the center line temperature T_0, results predicted by the model are larger than those for large Gz. However for $T_0 > T_w$, the reverse effect is observed as shown qualitatively in Fig. 9.18. In Fig. 9.19, experimental results for the variation of the Graetz number for low values of G have been shown and the effect of Gz in predicting the pressure

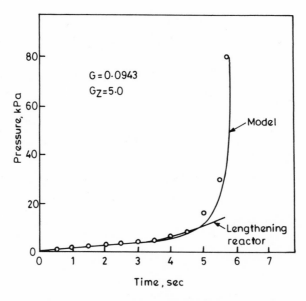

Figure 9.17. Comparison of the pressure rise predicted by the flow model (consisting of main flow and flow from region) with experimental data (shown by ○). Line marked "lengthening reactor" gives the computation results when the fountain effect is neglected. (Reprinted from Ref. 22 with permission of American Institute of Chemical Engineers.)

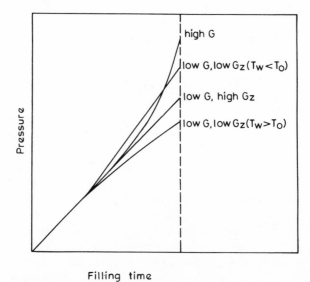

Figure 9.18. Influence of gelling number (G) and Graetz number (Gz) on the pressure rise in the rectangular mold. (Reprinted from Ref. 22 with permission of American Institute of Chemical Engineers.)

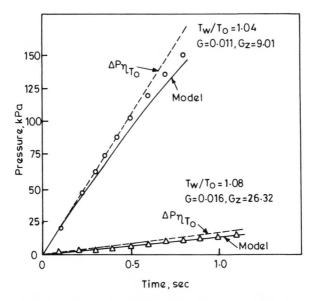

Figure 9.19. Experimental pressure rise (shown by ○) for low G and high Gz and low G and moderate Gz. Dashed line gives the prediction after assuming constant viscosity (also indicated by ΔP_{η_T}) and solid lines give the results when the fountain effect is included. (Reprinted from Ref. 22 with permission of American Institute of Chemical Engineers.)

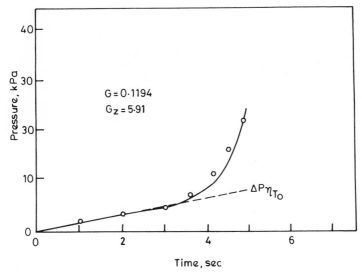

Figure 9.20. Experimental pressure rise (shown by ○) for high gelling number (G). Dashed line gives the prediction after assuming constant viscosity and solid line gives the prediction from the model. (Reprinted from Ref. 22 with permission of American Institute of Chemical Engineers.)

in the mold can be seen to be considerable. On the same figure, results assuming constant viscosity, marked as $\Delta P \eta_{T_0}$, have also been shown, and as pointed out earlier, results are not too different from the model results for small values of gelling time. However, there is an entirely different situation for large G as in Fig. 9.20, where $\Delta P \eta_{T_0}$ is found to give a very poor fit for the experimental data and the theoretical model presented in this section describes the physical situation very well.

The temperature rise during filling and curing has been shown in Fig. 9.21. During filling, the temperature is found to increase first and then decrease slightly. Near the center line, it decreases because the residence time near the center decreases when the front moves past a given point. During the curing, the temperature rises to a maximum value and then decreases due to the heat conduction to the mold wall. The thicker the mold, the closer is the center line temperature to the adiabatic rise in temperature.

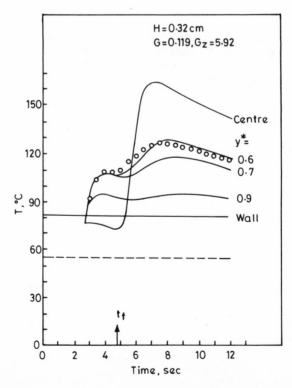

Figure 9.21. Temperature rise at various thicknesses during filling and curing of 0.32-cm-thick mold. Experimental points have been shown by (○). (Reprinted from Ref. 22 with permission of American Institute of Chemical Engineers.)

9.8. CONCLUSIONS

In this chapter, a rigorous kinetic model accounting for the unequal reactivity of functional groups and the various side reactions has been presented. From the kinetic model, expressions for the conversions of various functional groups have been written, and using Macosko's expectation theory, the criterion for gelation has been derived. It has been shown that there exists a gelation envelope of concentrations of reacting species in the feed, and conditions that lead to gelation have been established.

It is now a current practice to prepare the molded objects from urethane polymer using reaction injection molding in which a diisocyanate is mixed with a polyol blend and filled in the mold. As the shape of the mold becomes bigger and more complex, its filling becomes important. In this chapter, it has been shown that for a simple mold geometry, near the moving interface, the overall flow is two dimensional, whereas away from it, the flow is essentially one dimensional. The conditions existing near the flow front are sometimes known as the fountain effect and can never be ignored if the gelling number is high. As long as the gelling number, G, is small, polymerization during the filling can be neglected, and the fluid can be assumed to have a constant viscosity. For small G, as the Graetz number is increased, the pressure drop in the mold is found to reduce considerably. During the curing step, there is no flow within the mold and the necessary moles and energy balances can be written and solved numerically. The temperature at any position along the thickness is found to increase first and then decrease slightly due to the conduction of heat to the walls. The thicker the mold, the closer is the center line temperature to the adiabatic rise temperature.

The model presented in this chapter has been tested experimentally using a simple mold geometry. There is a definite need to examine the effect of unequal reactivity of functional groups on the mold design. It is possible to solve transport equations using a more realistic kinetic scheme described in Section 9.2. As the mold geometry becomes more complex (as it is usually employed commercially), the solution of transport equations even numerically is expected to present an intractable problem, and an effort should be directed in developing scale-up rules. Lastly the flow front instability even in rectangular molds remains an unsolved problem.

REFERENCES

1. J. K. Backus, in *Polymerization Processes* (C. E. Schildknecht and I. Skeist, Eds.), 1st ed., Interscience, New York (1977).
2. T. H. Saunders and K. C. Frisch, *Polyurethanes, Chemistry and Technology*, Interscience, New York (1962).

3. S. Mall, M. Tech. thesis, Modelling of crosslinked polyurethane systems, Department of Chemical Engineering, Indian Institute of Technology, Kanpur, India (1982).
4. P. Wright and A. P. C. Cummings, *Solid Polyurethanes Elastomers*, 1st ed., Macaren & Sons, New York (1962).
5. R. A. Martin, K. L. Hoy, and R. H. Peterson, Computer simulation of the tolylene diisocyanate-polyol reaction, *Ind. Eng. Chem. Prod. R&D* **6**, 218-222 (1967).
6. D. R. Miller and C. W. Macosko, Average property relations for nonlinear polymerization with unequal reactivity, *Macromolecules* **11**, 656-662 (1978).
7. D. R. Miller and C. W. Macosko, Substitution effects in property relations for stepwise polyfunctional polymerization, *Macromolecules* **13**, 1063-1069 (1980).
8. E. B. Richter and C. W. Macosko, Viscosity changes during isothermal and adiabatic urethane network polyering, *Polym. Eng. Sci.* **20**, 921-924 (1980).
9. E. C. Steinle, F. C. Critchfield, J. M. Castro, and C. W. Macosko, Kinetics and conversion monitoring in a RIM thermoplastic polyurethane system, *J. Appl. Polym. Sci.* **25**, 2317-2329 (1980).
10. L. J. Lee, Polyurethane reaction injection molding: Process materials and properties, *Rubber Chem. Tech.* **53**, 542-599 (1980).
11. J. M. Ottino and R. Chella, Laminar mixing of polymeric liquids: A brief review and recent theoretical developments, *Polym. Eng. Sci.* **23**, 357-379 (1983).
12. L. Erwin and F. Mokhtarian, Analysis of mixing in modified single screw extruders, *Polym. Eng. Sci.* **23**, 49-56 (1983).
13. W. D. Mohr, Mixing and dispersing, in *Processing of Thermoplastic Materials* (E. C. Bernherdt, Ed.), Reinhold, New York (1959).
14. J. M. McKelvey, *Polymer Processing*, Wiley, New York (1962).
15. L. J. Lee, J. M. Ottino, W. E. Ranz, and C. W. Macosko, Impingement mixing in reaction injection molding, *Polym. Eng. Sci.* **20**, 868-874 (1980).
16. P. Kollodzeij, C. W. Macosko, and W. E. Ranz, The influence of impingement mixing on striation thickness distribution and properties in fast polyurethane polymerization, *Polym. Eng. Sci.* **22**, 388-395 (1982).
17. J. Baldyga and J. R. Bourne, Distribution of striation thickness from impingement mixers in RIM, *Polym. Eng. Sci.* **23**, 556-559 (1983).
18. S. D. Lipshitz, F. G. Mussati, and C. W. Macosko, Kinetic and viscosity relations for urethane, network polymerizations, *SPE Technical Papers* **21**, 239-241 (1975).
19. J. H. Castro, C. W. Macosko, F. E. Critchfield, E. C. Steinle, and L. P. Tackett, Reaction injection molding: filling of rectangular mold, *J. Elastomers & Plastics* **12**, 3-17 (1980).
20. J. M. Cartro, C. W. Macosko, L. P. Tackett, E. C. Steinle and F. E. Critchfield, Premature gelling in RIM, *SPE Tech. Papers* **26**, 423-427 (1980).
21. E. Broyer, C. W. Macosko, F. E. Critchfield, and L. R. Lawler, Curing and heat transfer in polyurethane reaction molding, *Polym. Eng. Sci.* **18**, 382-387 (1978).
22. E. Broyer, C. Gutfinger, and Z. Tadmore, A theoretical model for the cavity filing process in injection molding, *Trans. Soc. Rheol.* **19**, 423-444 (1975).
23. J. P. Domine and C. G. Gogos, Simulation of reactive injection molding, *Polym. Eng. Sci.* **20**, 847-858 (1980).
24. J. M. Castro and C. W. Macosko, Studies of mold filling and curing in the reaction injection molding process, *AIChE J.* **28**, 250-260 (1982).

EXERCISES

1. Verify the mole balance relations of Table 9.3 for batch reactors and write the corresponding relations for HCSTRs. Derive Eq. (9.2.2).

2. Integrate proper equations to derive Eqs. (9.2.3)–(9.2.5). Verify Eq. (9.2.6).

3. Derive Eqs. (9.3.5) and (9.3.7), from which verify Eq. (9.3.10).

4. Assuming that the rate can be approximated as $k^*G(1-p)$, solve Eq. (9.7.15) analytically. Then approximating the mold as adiabatic and $\bar{M}_w \simeq \bar{M}_n$, find μ_w as a function of the mold length.

5. Discuss qualitatively how the flow conditions discussed in Section 9.7 would affect the gelation envelope given in Section 9.3.

6. If it is decided to have a circular mold with feed at the center of the mold, modify the analysis presented in Section 9.7.

EPOXY POLYMERS

10.1. INTRODUCTION

Epoxy polymers are formed by the reaction of an epoxy group,

$$\underset{O}{\overset{\displaystyle\diagup\!\!\!\diagdown}{-CH-CH_2}}$$

with an —OH group. The most important epoxy[1] is manufactured by reacting bisphenol A,

$$HOH_5C_6-\underset{\underset{CH_3}{|}}{\overset{\overset{CH_3}{|}}{C}}-C_6H_5OH \quad (\text{or } HO-R-OH)$$

with epichlorohydrin,

$$\underset{O}{\overset{\displaystyle\diagup\!\!\!\diagdown}{H_2C-CH}}-CH_2Cl$$

Two common industrial processes are used for manufacturing this polymer, the taffy process and the advancement process. In the taffy process,[1-3] bisphenol A is reacted at 90–95°C with a controlled excess of epichlorohydrin (to give polymer molecules with glycidyl ether groups,

$$-O-CH_2\underset{O}{\overset{\displaystyle\diagup\!\!\!\diagdown}{CH}}-CH_2$$

on both ends) in the presence of NaOH and an inert solvent. The following reactions take place in a heterogeneous medium:

$$\sim R-OH + H_2C\overset{O}{\underset{}{\diagup\!\diagdown}}CH-CH_2Cl \xrightarrow{k_1} \sim R-O-CH_2-\overset{OH}{\underset{|}{C}}H-CH_2Cl$$

$$HCl + \sim R-O-CH_2-CH\overset{O}{\underset{}{\diagup\!\diagdown}}CH_2 \xleftarrow{k_d \text{ (fast)}} \qquad\qquad (10.1.1a)$$

$$\sim R-OH + H_2C\overset{O}{\underset{}{\diagup\!\diagdown}}CH-CH_2-O-R\sim \longrightarrow$$

$$\xrightarrow{k_2} \sim R-O-CH_2-\overset{OH}{\underset{|}{C}}H-CH_2-OR\sim$$

$$(10.1.1b)$$

The HCl formed reacts with NaOH to give NaCl. The final product formed by reaction (10.1.1), called taffy, is an emulsion of molten polymer in brine of about 30% concentration. The average molecular weight of the polymer ranges from about 900 to 3000 (2-4 repeat units per molecule), depending on the initial concentrations of the reactants. This product is first washed with water and then stripped of solvent and water under vacuum at about 150°C. Still higher molecular weight resins can be made by treating this polymer in a second reactor with additional bisphenol A in the presence of suitable catalysts (e.g., lithium salts).

In reaction (10.1.1), k_d is usually very large and is assumed to be infinity, while the reactivity ratio, k_2/k_1, of the glycidyl ether and the epoxy groups, has been estimated from experiments[4,5] as 0.5. A common side reaction is that occurring between the secondary hydroxyls and the glycidyl ether groups, leading to branching, as follows:

$$\sim R-O-CH_2-CH\overset{O}{\underset{}{\diagup\!\diagdown}}CH_2 + HO-\Big\{ \xrightarrow{k_3} \sim R-O-CH_2-\overset{OH}{\underset{|}{C}}H-CH_2-O-\Big\}$$

$$(10.1.2)$$

Using NMR spectroscopy, Batzer and Zahir[6] have found that about 0.09-0.6 branches are formed for every polymer molecule of molecular weight 1500-4000. In view of this low value, the branching reaction is usually neglected in the modeling of reactors.

In the advancement process on the other hand, bisphenol A is reacted with a prepolymer, usually the diglycidyl ether of bisphenol A (DGEBA) along with some higher oligomers, in the presence of suitable catalysts. The reaction occurs in the homogeneous phase and is represented schematically as

$$HO-R-OH + H_2\overset{O}{\overset{/\backslash}{C}}-CH-CH_2-O-R-O-CH_2-\overset{O}{\overset{/\backslash}{CH}}-CH_2 \xrightarrow{NaOH}$$

DGEBA

$$\underset{OH}{\underset{|}{\left[R-O-CH_2-CH-CH_2-O\right]_n}}$$

$$(10.1.3)$$

This is a typical AA + BB type step growth polymerization and the equations characterizing this system are a special case of AA + BC polymerizations discussed in Section 3.4. Branching reactions of the type given in Eq. (10.1.2) are also found to occur.

The curing of epoxy resins to give a hard, cross-linked product can be carried out with several compounds, e.g., organic acids, anhydrides, amines, etc., of which diamines (e.g., hexamethylene diamine) are most common. The reactions taking place between the glycidyl ether and diamine groups are

$$\sim\!\!\sim R-O-CH_2-\overset{O}{\overset{/\backslash}{CH}}-CH_2$$

$$+ \underset{H}{\overset{H}{\underset{|}{H-N-\sim\!\!\sim}}} \xrightarrow{2k_4} \sim\!\!\sim ROCH_2-\underset{OH}{\overset{|}{CH}}-CH_2-\underset{H}{\overset{|}{N}}-\sim\!\!\sim$$

$$(10.1.4a)$$

$$\sim\!\!\sim ROCH_2-\overset{O}{\overset{/\backslash}{CH}}-CH_2$$

$$+ \sim\!\!\sim ROCH_2-\underset{OH}{\overset{|}{CH}}-CH_2-\underset{H}{\overset{|}{N}}-\sim\!\!\sim \xrightarrow{k_5}$$

$$\sim\!\!\sim ROCH_2-\underset{OH}{\overset{|}{CH}}-CH_2-\overset{\xi}{N}-CH_2\underset{OH}{\overset{|}{CH}}-CH_2O-R-\sim\!\!\sim$$

$$(10.1.4b)$$

where the rate constants k_4 and k_5 differ due to induction and steric effects.[7] A cross-linked structure is formed because both the amine hydrogens are reactive. In addition to these reactions, the hydroxyl groups can also react with the glycidyl ether groups according to reaction (10.1.2). Since the advancement process can be modeled as a special case of AA + BC polymerization, attention is focused in this chapter on the modeling of the taffy process and the curing stage only.

10.2. TAFFY PROCESS

The kinetic scheme for epoxy formation in this process can be represented by Eq. (10.1.1), with $k_d \to \infty$. This means that there are no chlorohydrin groups present at any time. The progress of the reaction can be represented schematically[8,9] by Eqs. (a)–(f) of Table 10.1, with the various molecular species defined in Eqs. (g)–(j) of the table. It is observed that A represents the hydroxyl end group, E represents the glycidyl ether end group, and AA, AE, and EE represent the fact that at the two ends of the molecules lie A and A, A and E, and E and E groups, respectively. According to this

TABLE 10.1. Reactions in the Taffy Process

$$EP + AA_n + NaOH \xrightarrow{k_1} AE_n + NaCl + H_2O \qquad (a)$$

$$EP + AE_n + NaOH \xrightarrow{k_1} EE_n + NaCl + H_2O \qquad (b)$$

$$EE_n + AE_m \xrightarrow{k_2} EE_{m+n+1} \qquad (c)$$

$$EE_n + AA_m \xrightarrow{k_2} AE_{m+n+1} \qquad (d)$$

$$AE_n + AE_m \xrightarrow{k_2} AE_{m+n+1} \qquad (e)$$

$$AE_n + AA_m \xrightarrow{k_2} AA_{m+n+1} \qquad (f)$$

where

$$AA_n: \quad H-(OROCH_2\underset{|}{C}HCH_2)_{\overline{n}}-OROH \qquad (g)$$
$$OH$$

$$AE_n: \quad H(-OROCH_2CH\ CH_2)_{\overline{n}}-OROCH_2CH-CH_2 \qquad (h)$$
$$OH \phantom{CH_2)_{\overline{n}}-OROCH_2}O$$

$$EE_n: \quad CH_2-CHCH_2-(OROCH_2CHCH_2)_{\overline{n}}-OROCH_2CH-CH_2 \qquad (i)$$
$$O OH \phantom{CH_2)_{\overline{n}}-OROCH_2CH}O$$

$$EP: \quad H_2C-CH-CH_2Cl \qquad (j)$$
$$O$$

representation, bisphenol A is AA_0 and DGEBA is EE_0. The reactions in Table 10.1 are of the same form as those encountered in the reaction between an anhydride and a diol, referred to in Chapter 3 as an example of unequal reactivity in step growth polymerization. The technique of Case[10] as extended by Gandhi and Babu,[11] discussed in Section 3.4, can be used to describe this system.

Mole balance equations can now be written for an isothermal batch reactor for the various species $\{EP, AA_n, AE_n, \text{ and } EE_n\}$ present and summed up appropriately to give the equations for $[EP]$, $[A]$ $\{=\sum_{m=0}^{\infty}(2[AA_m] + [AE_m])\}$ and E $\{=\sum_{m=0}^{\infty}(2[EE_m] + [AE_m])\}$ as

$$\frac{d[EP]}{dt} = -k_1[EP][A] \qquad (10.2.1a)$$

$$\frac{d[A]}{dt} = -k_1[EP][A] - k_2[E][A] \qquad (10.2.1b)$$

$$\frac{d[E]}{dt} = -k_2[E][A] + k_1[EP][A] \qquad (10.2.1c)$$

Equation (10.2.1) is consistent with the following schematic representation of the various reactions in terms of functional groups:

$$EP + A \xrightarrow{k_1} E + H_2O$$

$$A + E \xrightarrow{k_2} -AE- \qquad (10.2.2)$$

It is often more convenient to work in terms of the conversions of functional groups, since they can be conveniently related to the probabilities of reaction. These are defined as

$$p_A \equiv \text{fraction of hydroxyl groups reacted} = \frac{[A]_0 - [A]}{[A]_0} \qquad (10.2.3a)$$

$$p_{EP} \equiv \text{fraction of epichlorohydrin molecules reacted} = \frac{[EP]_0 - [EP]}{[EP]_0}$$
$$(10.2.3b)$$

$$p_E \equiv \frac{\text{moles/liter of reacted glycidyl ether } (-AE-) \text{ groups}}{\text{moles/liter of epichlorohydrin in feed}}$$

$$= \frac{\{[EP]_0 - [EP]\} - [E]}{[EP]_0} \qquad (10.2.3c)$$

where the subscript 0 represents the initial concentration. The expression for p_E follows from the stoichiometry of reaction (10.2.2). In Eq. (10.2.3c), $[EP]_0 - [EP]$ represents the epichlorohydrin molecules that are reacted to give E groups, some of which react further to give $-AE-$ groups. From Eq. (10.2.3c), one obtains

$$\frac{[E]}{[EP]_0} = p_{EP} - p_E \tag{10.2.4}$$

Equation (10.2.1) can be represented in terms of p_A, p_{EP}, and p_E as

$$\frac{dp_{EP}}{d\tau} = (1 - p_A)(1 - p_{EP}) \tag{10.2.5a}$$

$$\frac{dp_A}{d\tau} = \frac{r}{2}\{(1 - p_A)(1 - p_{EP}) + R(1 - p_A)(p_{EP} - p_E)\} \tag{10.2.5b}$$

$$\frac{dp_E}{d\tau} = R(p_{EP} - p_E)(1 - p_A) \tag{10.2.5c}$$

where

$$\tau = k_1[A]_0 t \tag{10.2.6a}$$

$$r = 2[EP]_0/[A]_0 \tag{10.2.6b}$$

$$R = k_2/k_1 \tag{10.2.6c}$$

Equation (10.2.5) can be solved numerically with appropriate initial conditions (e.g., for feed consisting of epichlorohydrin and bisphenol A in ratio r, at $\tau = 0$, p_A, p_{EP}, and p_E are all 0). It may be mentioned that p_A, p_{EP}, and p_E are not all independent but are related by the stoichiometry of the reaction (10.2.2) as

$$p_A = \frac{r}{2}(p_{EP} + p_E) \tag{10.2.7}$$

The molecular weight distribution of the polymer formed can be obtained in terms of the conversions using statistical arguments similar to those used in Section 3.4. The concentration of all the end groups, N_e, can be obtained by summing up the concentration of A groups [Eq. (10.2.3a)], E groups [Eq. (10.2.4)], and twice the concentration of unreacted epichlorohydrin molecules [Eq. (10.2.3b)] as

$$N_e = 2N_t = [A]_0(1 - p_A) + [EP]_0(p_{EP} - p_E) + 2[EP]_0(1 - p_{EP})$$

$$= [A]_0(1 - 2p_A + r) \equiv [A]_0\phi \tag{10.2.8}$$

The probability of the sequence —AEEA— in a polymer molecule can be written following arguments developed in Section 3.4 as

Prob(—AEEA—) = 2{Probability that A is reacted}

$$\times \{\text{fraction of total reacted A groups} \atop \text{which have reacted with EP}\}$$

$$\times \{\text{Probability that E group has reacted with A}\}$$

$$= 2\{p_A\}\left\{\frac{p_{EP}}{p_{EP} + p_E}\right\}\left\{\frac{-AE-\text{ groups present}}{\text{total E groups formed}}\right\}$$

$$= \frac{2p_A p_E}{p_{EP} + p_E} = rp_E \qquad (10.2.9)$$

The factor of 2 in the above equation is necessary because units —AAEE— and —EEAA— are indistinguishable in a polymer molecule. The same expression is obtained if the product (probability that A has reacted times fraction of total reacted A which have reacted with E) is used instead of that used in Eq. (10.2.9). It may be noted that $p_A p_{EP}/(p_{EP} + p_E)$ is the probability that an A group has reacted with E.

The probability (or mole fraction) of finding a molecule of AA_n {$= A(AEEA)_n A$} is then given by

$$\text{Prob}(AA_n) = \left\{\frac{[A]_0(1 - p_A)}{[A]_0 \phi}\right\}\{rp_E\}^n(1 - p_A)$$

$$= \frac{(1 - p_A)^2 (rp_E)^n}{\phi} \qquad (10.2.10)$$

Similarly, the probabilities of finding molecules AE_n, EP, and EE_n are easily written as

$$\text{Prob}(AE_n) = \left\{\frac{[A]_0(1 - p_A)}{[A]_0 \phi}\right\}\{rp_E\}^n\left\{p_A \frac{p_{EP}}{p_{EP} + p_E}\right\}\left\{\frac{p_{EP} - p_E}{p_{EP}}\right\}\{2\}$$

$$= \frac{r(1 - p_A)(rp_E)^n(p_{EP} - p_E)}{\phi} \qquad (10.2.11a)$$

$$\text{Prob}(EP) = \frac{2[EP]}{N_e} = \frac{r(1 - p_{EP})}{\phi} \qquad (10.2.11b)$$

$$\text{Prob}(\text{EE}_n) = \left\{ \frac{[\text{EP}]_0 (p_{\text{EP}} - p_{\text{E}})}{[\text{A}]_0 \phi} \right\} \{ r p_{\text{E}} \}^n \left\{ p_{\text{A}} \frac{p_{\text{EP}}}{p_{\text{EP}} + p_{\text{E}}} \right\} \left\{ \frac{p_{\text{EP}} - p_{\text{E}}}{p_{\text{EP}}} \right\}$$

$$= r^2 (p_{\text{EP}} - p_{\text{E}})^2 (r p_{\text{E}})^n (p_{\text{E}} / p_{\text{EP}}) / (4\phi) \qquad (10.2.11\text{c})$$

In this equation $(p_{\text{EP}} - p_{\text{E}}) / p_{\text{EP}}$ is the probability of finding an unreacted E group and $[\text{EP}]_0 (p_{\text{EP}} - p_{\text{E}}) / ([\text{A}]_0 \phi)$ is that of finding an unreacted E end group. The various terms in parentheses can easily be recognized by referring to Section 3.4. Equation (10.2.11) can also be derived by starting from the E end of the molecule. The factor of 2 in Eq. (10.2.11a) can be justified using arguments of Section 3.4, wherein it was mentioned that in order to write the probability of A⌇B, one must sum up the two expressions obtained, one starting from the A end and the other starting from the B end. It may be observed that in Eq. (10.2.11c), once the term for the probability of finding an unreacted E end group is determined, there is no need to write another term for an E group *reacted* with an A because the latter condition is automatically satisfied once the former is taken care of.

With the various probabilities (equal to the number or mole fractions) of the molecules now known, expressions for $\bar{M}_n$ and $\bar{M}_w$ can easily be written. These are given in Table 10.2. Here, M_{AA_0}, M_{EP}, and M_{HCl} are the molecular weights of bisphenol A, epichlorohydrin, and HCl, respectively. In the equations for $\bar{M}_n$ and $\bar{M}_w$, EP has not been included in the summations since unreacted epichlorohydrin is usually distilled off before the polymer is used. Also, the removal of HCl by reaction has been accounted for.

The set of equations can be solved numerically to give p_{A}, p_{EP}, p_{E}, $\bar{M}_n$, $\bar{M}_w$, Q, and the mole fractions of the various molecules as a function of the dimensionless time τ, for a specified value of the initial reactant ratio, r, and the reactivity ratio, R. Detailed numerical results have been obtained by Ravindranath and Gandhi[8,9] and some presented in Fig. 10.1. These workers have also studied the case when k_d is not infinity and find that the

TABLE 10.2. Equations for $\bar{M}_n$ and $\bar{M}_w$ for the Taffy Process

$$\bar{M}_n = \frac{\sum_{n=0}^{\infty} \{ \text{Prob}(\text{AE}_n) M_{\text{AE}_n} + \text{Prob}(\text{EE}_n) M_{\text{EE}_n} + \text{Prob}(\text{AA}_n) M_{\text{AA}_n} \}}{\sum_{n=0}^{\infty} \{ \text{Prob}(\text{AE}_n) + \text{Prob}(\text{EE}_n) + \text{Prob}(\text{AA}_n) \}} \qquad (\text{a})$$

$$\bar{M}_w = \frac{\sum_{n=0}^{\infty} \{ \text{Prob}(\text{AE}_n) M_{\text{AE}_n}^2 + \text{Prob}(\text{EE}_n) M_{\text{EE}_n}^2 + \text{Prob}(\text{AA}_n) M_{\text{AA}_n}^2 \}}{\sum_{n=0}^{\infty} \{ \text{Prob}(\text{AE}_n) M_{\text{AE}_n} + \text{Prob}(\text{EE}_n) M_{\text{EE}_n} + \text{Prob}(\text{AA}_n) M_{\text{AA}_n} \}} \qquad (\text{b})$$

with

$$M_{\text{AE}_n} = (n+1) M_{\text{AA}_0} + (n+1)(M_{\text{EP}} - M_{\text{HCl}}) \qquad (\text{c})$$

$$M_{\text{EE}_n} = (n+1) M_{\text{AA}_0} + (n+2)(M_{\text{EP}} - M_{\text{HCl}}) \qquad (\text{d})$$

$$M_{\text{AA}_n} = (n+1) M_{\text{AA}_0} + n(M_{\text{EP}} - M_{\text{HCl}}) \qquad (\text{e})$$

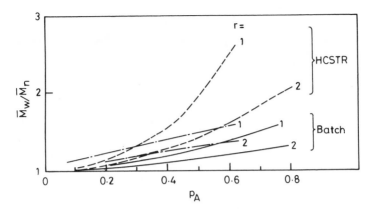

Figure 10.1. Polydispersity index vs. conversion of —OH groups for batch reactors (———) and HCSTRs (- - -) for the reaction between epichlorohydrin and bisphenol A (taffy process).[9] $k_d \rightarrow \infty$, $R = 0.6$. Results for batch advancement process with p_A = fractional conversion of —OH groups are shown by — - — ($r = [EP]_0/[AA_0]_0$ for taffy process and $[DGEBA]_0/[AA_0]_0$ for the advancement process). (Reprinted from Ref. 9 with permission of Professor K. S. Gandhi.)

values of the polydispersity index are slightly lower. Also, the results are relatively insensitive to changes in k_d. Thus, the assumption of infinitely fast dehydrohalogenation appears to be justified.

Ravindranath and Gandhi[8] have found that their theoretical results agree fairly well with some limited experimental GPC results of Batzer and Zahir[3,5] on the polydispersity index. However, more extensive experimental results are required, followed by theoretical curve-fitting to conclude if the model is really justified. In addition, it has been reported that the reaction mass in the taffy process[1,5] is heterogeneous, with the active reaction site being the interphase of the water and the organic media. It is possible that the more soluble chlorohydrin functional groups (since k_d is really not ∞, some chlorohydrin groups will be present; in fact these have been experimentally detected) have a higher concentration near this interphase. Thus, concentration gradients must be accounted for and mass transfer concepts may be applied in modeling the taffy process to improve agreement between theory and experiment.

Figure 10.1 also shows some results on taffy polymerization in HCSTRs for which Case's analysis is inapplicable. Mole balance equations for all the species have been written, assuming that the feed contains epichlorohydrin and bisphenol A in a molar ratio of r. These equations have been appropriately summed up to give $\bar{M}_n$ and $\bar{M}_w$. As found for ARB polymerization, higher values of Q are observed for the same conversion of hydroxyl groups in an HCSTR producing epoxy resins compared to that obtained in batch reactors.

10.3. ADVANCEMENT PROCESS

In the advancement process, commercial DGEBA (EE_0) is reacted with bisphenol A. The commercial DGEBA usually contains[3] (about 15%) diepoxides of higher molecular weights (EE_1, EE_2, etc.) as well as some[3] (3%–5% by weight) monofunctional epoxides represented by XE, where X is an unreactive group. As a first approximation, the presence of these compounds can be neglected and the polymerization can be modeled as the reaction between AA and EE. Results of Section 3.4 on AA + BC polymerization (for the special case of $k_1 = k_2$) can then be directly applied to this system. Identical results are obtained either using the technique of Grethlein[12] or by applying Macosko and Miller's[13] expectation theory (see Chapter 4) to the simplified situation of linear polymerization. Figure 10.1 also includes some results for the advancement process for batch reactors in the absence of monofunctional compounds or higher molecular weight diepoxides. These results compare favorably with some experimental results of Batzer and Zahir.[3] Incorporation of the effect of higher molecular weight diepoxide impurities using the method of Macosko and Miller[13] does not lead to any significant change in the results. In contrast, if one assumes that about 3% of the monofunctional epoxide,

$$\underset{\text{Cl}}{\overset{}{|}}\ \underset{\text{OH}}{\overset{}{|}}\qquad\qquad\qquad\qquad\qquad \text{O}$$
$$CH_2-CH-CH_2-O-R-O-CH_2-CH-CH_2,$$

is present in the commercial DGEBA used (the exact analysis of this compound has not been reported), the agreement between theory and experiment is considerably improved,[8] as shown in Table 10.3.

TABLE 10.3. Results for Some Epoxy Polymerizations in the Advancement Process[a]

	Experimental[b]			Theoretical[c]			
				Pure DGEBA		DGEBA + 3% monoepoxide	
	Epoxide value[d]	$\bar{M}_n$ VPO[e]	$\bar{M}_w$ GPC	$\bar{M}_n$	$\bar{M}_w$	$\bar{M}_n$	$\bar{M}_w$
1	1.9	1150	1838	1050	1920	1020	1834
2	1.1	1706	3648	1818	3481	1731	3285
3	0.65	2688	5609	3077	6015	2847	5529

[a] Reprinted from Ref. 8 with permission of John Wiley & Sons, New York.
[b] Reference 3.
[c] Reference 8.
[d] Epoxide value is defined as the fractional number of epoxy groups in 100 g of resin. It can usually be obtained by potentiometric titration with perchloric acid[14] or ir spectroscopy. Values of r used have not been given in Ref. 3.
[e] Vapor phase osmometry.

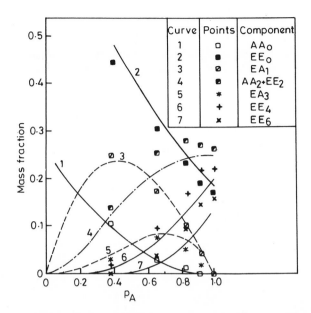

Figure 10.2. Mass fraction of various oligomers in the reaction of *pure* DGEBA (EE_0) with pure bisphenol A (AA_0) at 145°C (advancement process). $[DGEBA]_0/[AA_0]_0 = 2$. p_A is the conversion of OH groups. Solid and dashed lines are theoretical curves, while various points represent experimental values. (Reprinted from Ref. 14 with permission of John Wiley & Sons, New York.)

Recently, Antal *et al.*[14] have presented experimental GPC results on the concentrations of the various oligomers when pure DGEBA is reacted with pure bisphenol A in the molar ratio of 2 (i.e., $r \equiv [DGEBA]_0/[AA_0]_0 = 2$), at 145°C. The catalyst used was tetramethyl ammonium chloride.[15] Their experimental results are shown in Fig. 10.2 along with the theoretical predictions obtained using either Grethlein's[12] or Case's method.[10] The agreement with experimental results is fairly good, and the slightly lower concentrations of DGEBA compared to the theoretical values have been attributed to side reactions.[16] Similar *well documented* results on the variations of the species concentrations with time under different conditions are required for the taffy process also. It may be observed from Fig. 10.2 that as $p_A \to 1$, the concentrations of molecules of type AE_n and AA_n tend towards zero, and that the product contains mostly molecules with epoxy ends on both sides.

10.4. CURING OF EPOXIES

One of the most important properties of epoxy resins is their ability to transform from the thermoplastic "prepolymer" state to a hard thermoset form, after suitable chemical reaction. This is called the hardening or curing process. Two useful characteristics of epoxy hardening are that it can be carried out at room temperatures and that no volatiles are given off. Organic acids,[17] anhydrides and amines are commonly used as hardeners. Of these, curing with diamines is most commonly encountered and is discussed in this section. The analysis of curing with other compounds can similarly be performed.

Dušek et al.[18-21] have done extensive work on the curing of epoxies using the vector probability generating function theory.[22-24] More recently, however, identical results have been obtained by Bokare and Gandhi[25] using the far simpler expectation theory of Macosko and Miller[13] (presented earlier in Chapter 4). It is this latter development which is discussed here, since it can easily be extended to account for side reactions.

Equation (10.1.4) represents the main curing reaction with amines. It has been found that the rate constants for the primary and secondary amines differ,[7] a phenomenon referred to in Chapter 3 as induced asymmetry or first-shell substitution effect.[24] It has also been found that the hydroxyl groups catalyze both the reactions in Eq. (10.1.4). Since these OH groups are generated by the reaction, an autocatalytic effect is observed. To simplify the analysis, it is assumed that both k_4 and k_5 are equally influenced by the presence of OH groups, implying that the ratio k_4/k_5 is constant with time.[18,26] The hydroxyl groups formed can also react with the glycidyl ether groups[1,20,27-30] according to reaction (10.1.2). Since, on reaction, one OH group generates another OH group, this reaction has characteristics similar to those for chain growth polymerization (polyaddition). Thus, the curing of epoxies with diamines is a combination of step growth and chain growth polymerizations (as was nylon 6). The value of k_3 has been found[1,27] to be relatively small compared to k_4 or k_5. It is further assumed that k_3/k_4 or k_3/k_5 is constant in view of the low (but finite) value of k_3.

The analysis of the curing of a general epoxide

$$R_1 \!\!\left(\!\! \underset{\displaystyle CH}{\overset{\displaystyle \overset{\textstyle O}{\diagup\!\!\diagdown}}{}}\!\!-\!\! CH_2\right)_{\!\!g}$$

(R_1E_g, with g usually 2) with a general amine $R_2(NH_2)_{f/2}$ {$R_2A_{f/2}$, with $f = 4$ for hexamethylene diamine} can be performed using the expectation theory. The reaction mass at any time contains molecules having structures of the type S_n and T_n as shown in Fig. 10.3, in addition to unreacted compounds. S_n is a secondary amine group with n units on a nitrogen while T_n is a tertiary amine group with m units on one side of the nitrogen and $n - m$ units on the other side. Since, on reaction, each glycidyl ether group E on R_1E_g (it may be noted that R_1, as defined, includes the $-OCH_2-$

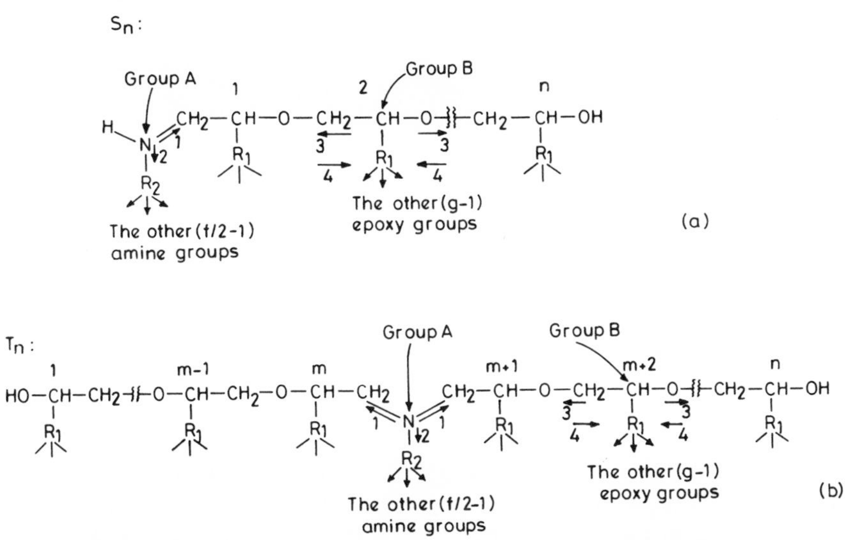

S_n:

Group A   1   Group B   n
H   CH$_2$—CH—O—CH$_2$—CH—O—CH$_2$—CH—OH
  N   R$_1$ R$_1$ R$_1$
  R$_2$

The other$(f/2-1)$
amine groups

The other$(g-1)$
epoxy groups   (a)

T_n:

              Group A   Group B
        1   m-1   m   m+1   m+2   n
HO—CH—CH$_2$—O—CH—CH$_2$—O—CH—CH$_2$ CH$_2$—CH—O—CH$_2$—CH—O—CH$_2$—CH—OH
   R$_1$ R$_1$ R$_1$ N   R$_1$ R$_1$ R$_1$
                                 R$_2$

The other$(f/2-1)$
amine groups

The other$(g-1)$
epoxy groups   (b)

Figure 10.3. Structures S_n and T_n present in various molecules in the reaction mass at time t.

parts of the glycidyl ether groups) must show up either in a T_n or an S_n, a stoichiometric balance gives

$$[E] = [E]_0 - \sum_{n=1}^{\infty} n[S_n] - \sum_{n=2}^{\infty} n[T_n] \qquad (10.4.1)$$

Similarly, a balance on the amino groups, A, gives

$$[A] = [A]_0 - \sum_{n=1}^{\infty} [S_n] - \sum_{n=2}^{\infty} [T_n] \qquad (10.4.2)$$

where the subscript 0 indicates initial values. The concentrations $[A]_0$ and $[E]_0$ are equal to $(f/2)[R_2(\mathrm{NH}_2)_{f/2}]_0$ and

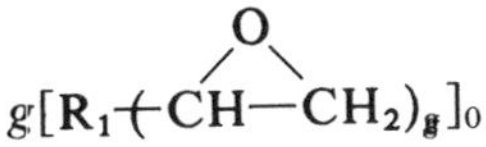

$$g[R_1(\overset{O}{\overset{\diagup\diagdown}{CH-CH_2}})_g]_0$$

respectively.

It is possible, now, to write an expression for the expected value $E(M_E^{out})$ of M_E^{out}, the mass attached to a randomly selected E group when looking *out* from the molecule of which it is a part. This will consist of several terms [see Eq. (4.2.2)], corresponding to the following conditions:

(a) The selected E group is unreacted;
(b) The selected E group is reacted and is part of one of the groups in S_n (with $n = 1, 2, \ldots, \infty$), e.g., B in Fig. 10.3a;
(c) The selected E group is reacted and is part of T_n ($n = 2, 3, \ldots$), e.g., B in Fig. 10.3b.

The probability that the selected E is reacted (and is part of *any* group) is $([E]_0 - [E])/[E]_0$. The probability that once reacted, this E is part of S_n is

$$\{([E]_0 - [E])/[E]_0\} \frac{n[S_n]}{\sum_{n=1}^{\infty} n[S_n] + \sum_{n=2}^{\infty} n[T_n]}$$

which, on using Eq. (10.4.1), gives this probability as $n[S_n]/[E]_0$. Similarly, the probability that E is reacted and is part of T_n is $n[T_n]/[E]_0$. The total expected mass looking out in the two directions 3 of Figs. 10.3a and 10.3b, is $(n - 1)E(M_E^{in}) + E(M_A^{in})$, where M_E^{in} and M_A^{in} are the masses attached to the epoxy and amino groups looking *into* the molecules (directions 4 and 2 of Figs. 10.3a and 10.3b) respectively. Thus, Eq. (a) in Table 10.4 can be obtained.

Further simplifications and developments can easily be followed from Table 10.4. Here, $M_{R_1E_g}$ is the molecular weight of the epoxide molecule

$$R_1-(\overset{\displaystyle O}{\overset{\displaystyle \diagup\!\!\diagdown}{CH-CH_2}})_g$$

and $M_{R_2A_{f/2}}$ is that of the amine. Equation (d) of Table 10.4 can easily be deduced by recognizing the following possibilities relevant to $E(M_A^{out})$:

(a) The selected A group is unreacted;
(b) The selected A group is reacted and is part of S_n ($n = 1, 2, \ldots$);
(c) The selected A group is reacted and is part of T_n ($n = 2, 3, \ldots$).

Once again, the probability that a selected A group is reacted and is part of S_n is

$$\{([A]_0 - [A])/[A]_0\} \frac{[S_n]}{\sum_{n=1}^{\infty} [S_n] + \sum_{n=2}^{\infty} [T_n]} = [S_n]/[A]_0$$

Similarly, the probability that A is on T_n is $[T_n]/[A]_0$. The expected mass looking out in direction 1 for an S_n type molecule is $nE(M_E^{in})$. The expected mass looking out in the two directions 1 in a T_n molecule is also $nE(M_E^{in})$.

The set of recursive relations so developed can be solved to give Eqs. (e) and (f) in Table 10.4. Here, $r = [A]_0/[E]_0$, $p_E = ([E]_0 - [E])/[E]_0$, the

TABLE 10.4. Expectation Theory for Curing of Epoxies

Development

$$E(M_E^{out}) = \frac{[E]}{[E]_0} \times O + \sum_{n=1}^{\infty} \frac{n[S_n]}{[E]_0} \{(n-1)E(M_E^{in}) + E(M_A^{in})\}$$

$$+ \sum_{n=2}^{\infty} \frac{n[T_n]}{[E]_0} \{(n-1)E(M_E^{in}) + E(M_A^{in})\} \qquad (a)$$

$$E(M_E^{in}) = M_{R_1E_g} + (g-1)E(M_E^{out}) \qquad (b)$$

$$E(M_A^{in}) = M_{R_2A_{f/2}} + (f/2-1)E(M_A^{out}) \qquad (c)$$

$$E(M_A^{out}) = \frac{[A]}{[A]_0} \times O + \sum_{n=1}^{\infty} \frac{[S_n]}{[A]_0}\{nE(M_E^{in})\} + \sum_{n=2}^{\infty} \frac{[T_n]}{[A]_0}\{nE(M_E^{in})\}$$

$$= \frac{[E]_0}{[A]_0}\left(1 - \frac{[E]}{[E]_0}\right)E(M_E^{in}) \qquad (d)$$

Final equations

$$E(M_E^{out}) = \frac{rM_{R_1E_g}\left\{\dfrac{f/2-1}{r^2}p_E^2 - \dfrac{p_E}{r} + \dfrac{\lambda_{S,2}+\lambda_{T,2}}{[E]_0 r}\right\} + M_{R_2A_{f/2}}p_E}{1 - r(g-1)\left\{\dfrac{f/2-1}{r^2}p_E^2 - \dfrac{p_E}{r} + \dfrac{\lambda_{S,2}+\lambda_{T,2}}{[E]_0 r}\right\}} \qquad (e)$$

$$E(M_A^{out}) = \frac{p_E}{r}\{M_{R_1E_g} + (g-1)E(M_E^{out})\} \qquad (f)$$

conversion of E groups,

$$\lambda_{S,2} = \sum_{n=1}^{\infty} n^2[S_n] \quad \text{and} \quad \lambda_{T,2} = \sum_{n=2}^{\infty} n^2[T_n]$$

The weight average molecular weight, $\bar{M}_w$, can be written using Eq. (4.2.11) as

$$\bar{M}_w = (\text{mass fraction of } R_1E_g \text{ molecular units}) \times E(M^*_{R_1E_g})$$

$$+ (\text{mass fraction of } R_1A_{f/2} \text{ units}) \times E(M^*_{R_2A_{f/2}})$$

$$= \frac{fM_{R_1E_g}E(M^*_{R_1E_g}) + 2rgM_{R_2A_{f/2}}E(M^*_{R_2A_{f/2}})}{fM_{R_1E_g} + 2rgM_{R_2A_{f/2}}} \qquad (10.4.3)$$

In Eq. (10.4.3), $E(M^*_{R_1E_g})$ and $E(M^*_{R_2A_{f/2}})$ are the expected masses attached to randomly selected epoxide and amine units and are given by

$$E(M^*_{R_1E_g}) = M_{R_1E_g} + gE(M_E^{out})$$

$$E(M^*_{R_2A_{f/2}}) = M_{R_2A_{f/2}} + (f/2)E(M_A^{out}) \qquad (10.4.4)$$

Equations (e) and (f) of Table 10.4 can be solved simultaneously with Eqs. (10.4.3) and (10.4.4) to compute $\bar{M}_w$ provided $\lambda_{S,2}$ and $\lambda_{T,2}$ are known (expressions for these are established later). It is observed that $\bar{M}_w$ is a function of r and p_E for a given system.

At the gel point (denoted by subscript c) $\bar{M}_w \to \infty$, which occurs when [see Eq. (e) of Table 10.4 and Eqs. (10.4.3) and (10.4.4)]

$$\left(\frac{f}{2} - 1\right)\left(\frac{p_{E,c}}{r}\right)^2 + \frac{\lambda_{S,2c} + \lambda_{T,2c}}{r[E]_0} - \frac{p_{E,c}}{r} = \frac{1}{r(g-1)} \qquad (10.4.5)$$

Appropriate expressions can be similarly derived for the probability of a finite chain, mass fraction sol (or gel), and the concentration of elastically effective network points or chains, etc., following the methodology suggested in Chapter 4. These are left as exercises. It may be added that in the literature, expressions for these molecular properties have been obtained by Dušek et al.[18] only for the case when $k_3 \to 0$.

In order to complete the analysis of the curing of epoxides, it is necessary to establish expressions for the variation of p_E, $\lambda_{S,2}$, and $\lambda_{T,2}$ with time. These can be obtained kinetically using the reaction scheme shown by Eqs. (a)-(d) of Table 10.5. The factor of 2 used in Eqs. (a) and (d) of this table occurs because of two primary hydrogens being present in the amino group A or the two OH groups present in T_n. The mole balance equations for the various species can easily be written and then summed up to give, finally, Eqs. (e)-(l) in Table 10.5. Equations (f), (h), and (k) of Table 10.5 are consistent with Eq. (10.4.1), and likewise, Eqs. (e), (g), and (j) of Table 10.5 are consistent with Eq. (10.4.2). The equations for [E], [A], $\sum_{n=1}^{\infty}[S_n]$, $\sum_{n=1}^{\infty} n[S_n]$, $\sum_{n=1}^{\infty} n^2[S_n]$, $\sum_{n=2}^{\infty}[T_n]$, $\sum_{n=2}^{\infty} n[T_n]$, and $\sum_{n=2}^{\infty} n^2[T_n]$ form a complete set and can be integrated numerically to give [E] (and so p_E), [A], $\lambda_{S,2}$, and $\lambda_{T,2}$ as functions of time. These can then be used in the earlier set of equations to give $\bar{M}_w$ as a function of time (before gelation), p_{Ec} [Eq. (10.4.5) must be solved by trial and error], the mass fraction of the gel, w_g, and the concentration of elastically effective network strands, μ_e.

The number average molecular weight, $\bar{M}_n$, of the polymer before gelation can easily be written by noting that for each E group reacted, either by the reaction in Eq. (10.1.2) or (10.1.4), the total number of molecules in the system reduces by one. Hence,

$$\bar{M}_n = \frac{[R_1E_g]_0 M_{R_1E_g} + [R_2A_{f/2}]_0 M_{R_2A_{f/2}}}{[R_1E_g]_0 + [R_2A_{f/2}]_0 - [E]_0 p_E} \qquad (10.4.6)$$

Bokare and Gandhi[25] find that for the reaction between a monoepoxide and a primary amine ($f = 2$, $g = 1$) with nonreactive secondary hydrogen

TABLE 10.5. Kinetic Scheme and Balance Equations for the Curing of Epoxies

$$A + E \xrightarrow{2k_4} S_1 \tag{a}$$

$$S_n + E \xrightarrow{k_5} T_{n+1}, \qquad n \geq 1 \tag{b}$$

$$S_n + E \xrightarrow{k_3} S_{n+1}, \qquad n \geq 1 \tag{c}$$

$$T_n + E \xrightarrow{2k_3} T_{n+1}, \qquad n \geq 2 \tag{d}$$

Balance equations

$$\frac{d[A]}{dt} = -2k_4[A][E] \tag{e}$$

$$\frac{d[E]}{dt} = -2k_4[A][E] - k_5[E] \sum_{n=1}^{\infty} [S_n] - k_3[E] \sum_{n=1}^{\infty} [S_n] - 2k_3[E] \sum_{n=2}^{\infty} [T_n] \tag{f}$$

$$\frac{d}{dt} \sum_{n=1}^{\infty} [S_n] = 2k_4[E][A] - k_5[E] \sum_{n=1}^{\infty} [S_n] \tag{g}$$

$$\frac{d}{dt} \sum_{n=1}^{\infty} n[S_n] = 2k_4[A][E] + k_3[E] \sum_{n=1}^{\infty} [S_n] - k_5[E] \sum_{n=1}^{\infty} n[S_n] \tag{h}$$

$$\frac{d}{dt} \sum_{n=1}^{\infty} n^2[S_n] = 2k_4[A][E] - 2k_3[E] \sum_{n=1}^{\infty} n[S_n] - k_3[E] \sum_{n=1}^{\infty} [S_n] - k_5[E] \sum_{n=1}^{\infty} n^2[S_n] \tag{i}$$

$$\frac{d}{dt} \sum_{n=2}^{\infty} [T_n] = k_5[E] \sum_{n=1}^{\infty} [S_n] \tag{j}$$

$$\frac{d}{dt} \sum_{n=1}^{\infty} n[T_n] = k_5[E] \sum_{n=1}^{\infty} n[S_n] + k_5[E] \sum_{n=1}^{\infty} [S_n] + 2k_3[E] \sum_{n=2}^{\infty} [T_n] \tag{k}$$

$$\frac{d}{dt} \sum_{n=2}^{\infty} n^2[T_n] = k_5[E] \left\{ \sum_{n=1}^{\infty} n^2[S_n] + 2 \sum_{n=1}^{\infty} n[S_n] + \sum_{n=1}^{\infty} [S_n] \right\}$$

$$+ 4k_3[E] \sum_{n=2}^{\infty} n[T_n] + 2k_3[E] \sum_{n=2}^{\infty} [T_n] \tag{l}$$

$(k_5 = 0)$ and with OH groups reacting as fast as the primary amine ($k_3 = 2k_4$), the above set of equations gives results that are identical with those obtained by the kinetic approach discussed in Chapter 2.[31,32] Also, for the reaction between a diamine and a diepoxide ($f = 4$, $g = 2$) with the OH group nonreactive, the results obtained from the equations derived above match those obtained by Dušek et al.[18] using the more difficult vector-probability generating functions. These results are shown in Fig. 10.4 for $k_5/k_4 = 0.5$. Experimental results[19] on the DGEBA–HMDA system at 50°C on $\bar{M}_n$ (vapor-phase osmometry) vs. conversion, p_{Ec} vs. r, w_g vs. p_E and r, μ_e (obtained from mechanical properties) vs. r at total conversion of the limiting functional group, and $\sum_{n=2}^{\infty} [T_n]$ vs. p_E, have also been found to be explained fairly well by using $k_5/k_4 = 0.6$–0.7.

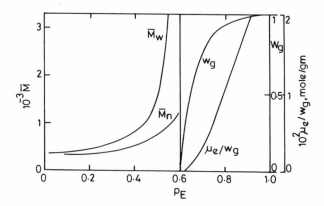

Figure 10.4. Various properties of interest in the reaction between hexamethylene diamine and DGEBA with stoichiometric amounts ($r = [A]_0/[E]_0 = 0.5$) of functional groups initially.[18] $k_5/k_4 = 0.6$, $k_3 = 0$. (Reprinted from Ref. 18 with permission of John Wiley & Sons, New York.)

Figure 10.5 shows[25,33] the effect of r and k_5/k_3 on the variation of $\bar{M}_w$ with p_H, the conversion of the hydrogen atoms on the amine given by

$$p_H = \frac{2[A]_0 - (2[A] + \sum_{n=1}^{\infty}[S_n])}{2[A]_0} \tag{10.4.7}$$

The effect of incorporating the reaction with OH groups is found to be significant for values of r below 0.5, though for $r > 0.5$, the effect is not much. Figure 10.6 shows[25,33] the effect of k_5/k_3 and the initial composition

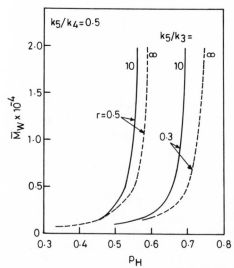

Figure 10.5. $\bar{M}_w$ vs. p_H for two different values of r for the DGEBA–HMDA system.[25,33] Solid line, $k_5/k_3 = 10$; dashed line, $k_5/k_3 = \infty$. (Reprinted from Ref. 25 with permission of John Wiley & Sons, New York.)

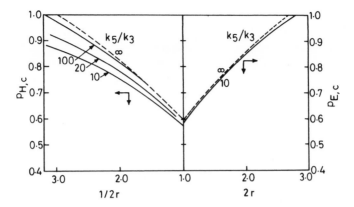

Figure 10.6. Effect of k_5/k_3 and r on the critical conversion for the DGEBA–HMDA system.[25,33] $k_5/k_4 = 0.5$. Dashed curves same as those of Dušek *et al.*[18] (Reprinted from Ref. 25 with permission of John Wiley & Sons, New York.)

on the critical conversion ($p_{H,c}$ or $p_{E,c}$) for the HMDA–DGEBA system. For mixtures containing an excess of amine hydrogens initially ($2r > 1$), the critical epoxide group conversion, $p_{E,c}$, increases as r increases, and is negligibly affected by k_5/k_3. At some value of r, denoted as r_{max}, the value of $p_{E,c}$ becomes unity. No gelation is possible for $r > r_{max}$, which implies that the roots of Eq. (10.4.5) do not lie in the physically meaningful range of $0 \leqslant p_{E,c} \leqslant 1$. When the epoxide groups are in excess initially (i.e., when $r < 0.5$), the limiting functional group is the amine hydrogen and its critical conversion, $p_{H,c}$, increases as r decreases. The critical point is found to be sensitive to the value of k_5/k_3 in this region. Once again, as r decreases, $p_{H,c}$ increases till it attains a value of 1 at $r = r_{min}$. For $r < r_{min}$, no physically meaningful solutions of Eq. (10.4.5) exist and gelation would not occur. The range $r_{min} \leqq r \leqq r_{max}$ thus represents a set of initial compositions for which gelation occurs. Figure 10.6 shows that this range is wider when the reactivity of OH groups is accounted for than when it is neglected. In fact, when the OH group reactivity is neglected, no gelation is predicted for $[E]_0/2[A]_0 > 3$ (even for other values of k_5/k_4 than 0.5)[18] and the experimental observation[19] of $p_{H,c} = 1$ at $[E]_0/2[A]_0 \simeq 3.2$ cannot be explained theoretically. This poses no problem, however, if the OH group reactivity is accounted for. Bokare and Gandhi[25,33] suggest a value of k_5/k_3 of 100 to explain some of the available experimental data.

10.5. CONCLUSIONS

The prepolymerization stages of epoxy polymer formation for both the industrially important taffy process and the advancement process have been modeled[8] using the method of Case.[10] Recent experimental data[14] on the bisphenol A-DGEBA system have been found to be fairly well explained by this model and one can easily extend the development to account for the presence of higher oligomers in the feed. *Detailed* experimental results on the taffy process are not available, but the kinetic model developed does improve the agreement with some semiquantitative experimental results of Batzer and Zahir.[3,5,6]

The expectation theory of Macosko and Miller[13] has been used[25,33] to describe the curing of epoxides with amines. It is found that a large amount of experimental data[19] on the curing of DGEBA with HMDA can be explained fairly well by this theory, even if the OH group reactivity is neglected. However, in order to explain some data on critical conversions, one must incorporate the reactivity of OH groups.

REFERENCES

1. H. Lee and K. Neville, *Handbook of Epoxy Resins*, 1st ed., McGraw-Hill, New York (1967).
2. C. E. Hutz, in *Manufacture of Plastics* (W. M. Smith, Ed.), 1st ed., pp. 492–511, Reinhold, New York (1964).
3. H. Batzer and S. A. Zahir, Studies in the molecular weight distribution of epoxide resins. I. Gel permeation chromatography of epoxide resins, *J. Appl. Polym. Sci.* **19**, 585–600 (1975).
4. W. Fisch, Über den verlauf der umsetzung von epichlorhydrin mit zweiwertigen phenolen (kinetik der epoxyharzherstellung), *Chimia* **16**, 66–71 (1962).
5. H. Batzer and S. A. Zahir, Studies in the molecular weight distribution of epoxide resins. IV. Molecular weight distributions of epoxide resins made from bisphenol A and epichlorohydrin, *J. Appl. Polym. Sci.* **21**, 1843–1857 (1977).
6. H. Batzer and S. A. Zahir, Studies in the molecular weight distributions of epoxide resins. II. Chain branching in epoxide resins, *J. Appl. Polym. Sci.* **19**, 601–607 (1975).
7. H. Wesslau, Strukturabhängige ringschlubreaktionen bei der vernetzenden copolymerization, *Makromol. Chem.* **93**, 55–68 (1966).
8. K. Ravindranath and K. S. Gandhi, Molecular weight distributions in epoxy resins, *J. Appl. Polym. Sci.* **24**, 1115–1123 (1979).
9. K. Ravindranath, M.Tech. dissertation, IIT, Kanpur, India (1978).
10. L. C. Case, Molecular distributions in polycondensations involving unlike reactants. II. Linear distributions, *J. Polym. Sci.* **29**, 455–495 (1958).
11. K. S. Gandhi and S. V. Babu, Kinetics of step polymerization with unequal reactivities, *AIChE J.* **25**, 266–272 (1979).
12. H. E. Grethlein, Exact weight fraction distribution in linear condensation polymerization, *Ind. Eng. Chem., Fundam.* **8**, 206–210 (1969).
13. C. W. Macosko and D. R. Miller, A new derivation of average molecular weights of nonlinear polymers, *Macromolecules* **9**, 199–206 (1976).
14. I. Antal, L. Füzes, G. Samay, and L. Csillag, Kinetics of epoxy resin synthesis on the basis of GPC measurements, *J. Appl. Polym. Sci.* **26**, 2783–2786 (1981).
15. German Patent No. 2263175 (1972).
16. M. Lidarik, High molecular epoxy resins by melting low-molecular bisphenol, *Kunstst. Rundsch.* **1**, 6–10 (1959).

17. L. Shechter and J. Wynstra, Glycidyl ether reactions with alcohols, phenols, carboxylic acids, and acid anhydrides, *Ind. Eng. Chem.* **48**, 86-93 (1956).

18. K. Dušek, M. Ilavský, and S. Luňák, Curing of epoxy resins. I. Statistics of curing of diepoxides with diamines, *J. Polym. Sci. Symp. No. 53*, 29-44 (1975).

19. S. Luňák and K. Dušek, Curing of epoxy resins. II. Curing of bisphenol A diglycidyl ether with diamines, *J. Polym. Sci. Sym. No. 53*, 45-55 (1975).

20. K. Dušek, M. Bleha and S. Luňák, Curing of epoxide resins: Model reactions of curing with amines, *J. Polym. Sci., Polym. Chem. Ed.* **15**, 2393-2400 (1977).

21. K. Dušek and W. Prins, Structure and elasticity of non-crystalline polymer networks, *Adv. Polym. Sci.* **6**, 1-102 (1969).

22. I. J. Good, Cascade theory and the molecular weight averages of the sol fraction, *Proc. R. Soc., London. A* **272**, 54-59 (1962).

23. D. S. Butler, G. N. Malcolm, and M. Gordon, Configurational statistics of copolymer systems, *Proc. R. Soc. London A* **295**, 29-54 (1966).

24. M. Gordon and G. R. Scantlebury, Statistical kinetics of polyesterification of adipic acid with pentaerythritol or trimethylol ethane, *J. Chem. Soc. London B*, 1-13 (1967).

25. U. M. Bokare and K. S. Gandhi, Effect of simultaneous polyaddition reaction on the curing of epoxies, *J. Polym. Sci., Polym. Chem. Ed.* **18**, 857-870 (1980).

26. K. Horie, H. Hiura, M. Sawada, I. Mita, and H. Kambe, Calorimetric investigation of polymerization reactions. III. Curing reaction of epoxides with amines, *J. Polym. Sci. A-1* **8**, 1357-1372 (1970).

27. D. Schechter, J. Wynstra, and R. P. Kurkjy, Glycidyl ether reactions with amines, *Ind. Eng. Chem.* **48**, 94-97 (1956).

28. H. C. Anderson, The effect of an amine cured epoxy resin on the stability of trinitrotoluene, *SPE J.* **16**, 1241-1245 (1960).

29. T. K. Kwei, Swelling of highly crosslinked network structure, *J. Polym. Sci. A-1* **1**, 2977-2988 (1963).

30. C. A. May and Y. Tanaka, *Epoxy Resins, Chemistry and Technology*, 1st ed., Marcel Dekker, New York (1973).

31. W. H. Abraham, Path-dependent distribution of molecular weight in linear polymers, *Ind. Eng. Chem., Fundam.* **2**, 221-224 (1963).

32. H. Kilkson, Effect of reaction path and initial distribution on molecular weight distribution of irreversible condensation polymers, *Ind. Eng. Chem. Fundam.* **3**, 281-293 (1964).

33. U. M. Bokare, M.Tech. dissertation, I.I.T. Kanpur, India (1978).

EXERCISES

1. Write mole balance equations for isothermal batch reactors for the species EP, AA_n, AE_n, and EE_n for the taffy process using the kinetic scheme of Table 10.1 Sum these up appropriately to obtain Eq. (10.2.1), and then represent these in the form given in Eq. (10.2.5).

2. Integrate the mole balance equations developed in Exercise 1 and obtain the MWD. Compare with results shown in Fig. 10.1. Also extend your equations for an HCSTR operating at steady state.

3. Provide the intermediate steps in the analysis of Section 10.4. In particular, derive Eqs. (e) and (f) of Table 10.4, Eq. (10.4.3) and Eqs. (e)-(l) of Table 10.5.

4. Derive expressions for the mass fraction of sol after gelation, as well as for the concentration of elastically effective network strands for the epoxy curing process discussed in Section 10.4.

POLYMERIZATION WITH FORMALDEHYDE

11.1. INTRODUCTION

Formaldehyde is a gas at room temperature and is commercially available as 37% solution in water, which is called formalin. As a gas, it polymerizes into a white powder called paraformaldehyde:[1,2]

$$n\,\text{HCHO} \rightleftharpoons (\text{CH}_2\text{O})_n \tag{11.1.1}$$

where n is usually about 10. Under alkaline conditions, *para*-formaldehyde can easily depolymerize into a gas. It also depolymerizes in presence of water and formalin is formed.

In formalin, formaldehyde molecules can either exist as methylene glycol or as low molecular weight polymeric molecules and the equilibrium between these molecular species can be represented by[2]

$$\text{HCHO} + \text{H}_2\text{O} \xrightleftharpoons{K_1} \text{HO}-\text{CH}_2-\text{OH} \tag{11.1.2a}$$
$$\text{(methylene glycol)}$$

$$n\,\text{HOCH}_2\text{OH} \xrightleftharpoons{K_2} \text{HO}\!\!-\!\!(\text{CH}_2\text{O})_n\!\!-\!\!\text{H} + (n-1)\text{H}_2\text{O} \tag{11.1.2b}$$

$$\text{OH}-\text{CH}_2\text{OH} + \text{OH}\!\!-\!\!(\text{CH}_2\text{O})_n\!\!-\!\!\text{H} \xrightleftharpoons{K_3} \text{OH}\!\!-\!\!(\text{CH}_2\text{O}))_{\overline{n+1}}\!\!-\!\!\text{H} + \text{H}_2\text{O} \tag{11.1.2c}$$

where K_1, K_2, and K_3 are the equilibrium constants. In aqueous solution, these constants are such that HCHO exists mostly as methylene glycol.

However, as the gas concentration in water is increased, larger concentrations of higher oligomers in the solution are formed.

Formaldehyde can polymerize with phenol, cresol, urea, and melamine, and depending upon the reaction conditions, it can form either a linear polymer or a network. For example, in phenol, there are two ortho and one para hydrogens which can participate in the polymerization. This would mean that the polymerization of trifunctional phenol with bifunctional methylene glycol is expected to lead to network formation. However, experiments have shown that under acidic conditions, the polymer formed is essentially linear with small branching, and under alkaline conditions, the resultant polymer is a network.

A urea molecule, on the other hand, has four hydrogens which can participate in polymerization. All these are known to exhibit varying reactivities and under usual conditions of polymerization, the fourth hydrogen rarely reacts with the OH of methylene glycol. Similar reaction anomalies are found in the polymerization of melamine. The polymerizations of formaldehyde with these compounds are found to be extremely complex and their kinetic descriptions have been summarized in Table 11.1.[3-24]

A careful examination of Table 11.1 reveals that one of the important features of the polymerization of formaldehyde with phenol, urea, and melamine is that the various reactive sites on these compounds have different reactivities. In this chapter, the unequal reactivities of the ortho and para hydrogens of phenol and phenyl rings on the polymer chain have been modeled mathematically so as to carry out a realistic reactor simulation. Similar mathematical developments have been carried out for the polymerization of urea and melamine but are not included here since the analysis is similar.

Commercially, two grades of prepolymers, novolacs and resoles, are made through the polymerization of phenol and formaldehyde. Novolacs are linear polymer chains with little branching and are formed when the pH of the reaction mass is low (~2 or 3). Resole prepolymers are manufactured at high pH (~9-11) and are highly branched. These prepolymers are then used to make molded products by reacting them with suitable cross-linking agents. In the following, the polymerization of these prepolymers is discussed. The analysis of Chapter 4 can be easily adapted to account for the cross-linking stage, and is not discussed in this chapter.

11.1. MODELING OF NOVOLAC REACTORS

As seen from Table 11.1, when phenol and formaldehyde are polymerized in presence of an acid catalyst, the polymer that results is called *novolac.* Usually sulfuric acid is used as the catalyst and the pH of the reaction

TABLE 11.1.[a] Some Reaction Characteristics of Polymers
with Formaldehyde

Novolacs

1. The para hydrogens of phenol are about 2.4 times more reactive than its ortho positions.
2. The growth of molecules occurs mainly by the reaction of end groups, thus indicating a lower reactivity of internal positions of polymer chains.
3. The ortho-to-ortho linkages can be promoted by using directive catalysts. The effectiveness of the latter have been found to be in the order Mn^{+2}, Zn^{+2}, Cd^{+2}, Mg^{+2}, Co^{+2}, and Ba^{+2}.
4. The use of paraformaldehyde improves the rate of polymerization, indicating that excess water serves as a retardant.
5. In an organic medium when phenol and formaldehyde combine, reaction at one ortho position of phenol occurs. Polymer formation occurs exclusively through the ether reaction.

Resoles

1. Step growth reaction between bound formaldehyde (CH_2OH groups on phenyl ring) occurs in addition to reactions at the ortho and para positions.
2. In the early stages, trimethylol phenol is formed in large excess compared to polynuclear species.
3. The rate constants of the different reaction steps are dependent upon the pH of the reaction mass.

Urea

1. Three of the hydrogens are highly reactive compared to the fourth one and the rate constants for the forward and reverse reactions of each of these are different.
2. Polymerization can be acid or base catalyzed. The equilibrium constants are independent of pH, but the rate constants are not.
3. Methanol serves as a retarder. Methylol derivatives can undergo intermolecular condensation.

Melamine

1. Melamine is hexafunctional, with three $-NH_2$ groups on melamine.
2. Polymerization proceeds through monomethylation with formaldehyde *reversibly* and the condensation reaction forms the methylene bridge. The primary H's in $-NH_2$ groups have a different reactivity from the secondary H's (induced asymmetry).
3. The rate constants depend upon the pH, temperature, and formaldehyde concentration in the reaction mass.[3]
4. Polymerization occurs over a wide range of pH. Basic conditions promote further addition of CH_2O until the functionality is satisfied. As opposed to this, acid conditions lead to rapid step growth polymerization. For pH above 9, there is a substantial loss of formaldehyde to form formic acid.

---

[a] References 3–24.

medium is kept below 5. Novolacs are essentially linear chain polymer with average molecular weights of 600, and usually very little branching is found to occur. In the following discussion, an attempt has been made to model the molecular weight distribution of novolac polymers.

It has already been stated in Section 11.1 that formaldehyde exists in the aqueous medium as methylene glycol, $OH-CH_2-OH$, and therefore it shows a functionality of 2. In phenol, the two ortho and one para hydrogens

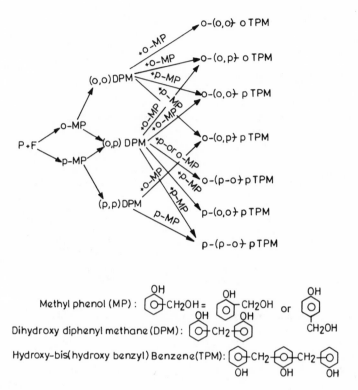

Figure 11.1. Various reaction routes for the formation of TPM in phenol formaldehyde polymerization.

can undergo reaction with the hydroxyl group of methylene glycol, and therefore, it exhibits a functionality of 3. The initial efforts in modeling the polymerization have been in the direction of enumerating the various molecular species in the reaction mass and finding out the rate constants of some of these steps. The formation of up to trinuclear species (i.e., molecules having three phenyl rings) has been shown in Fig. 11.1.

In Fig. 11.1, when phenol reacts with formaldehyde, the reaction can occur either at the ortho or the para position forming either ortho- or para-methylol phenol. These have been denoted by

these two being jointly indicated as

$$
\begin{array}{c}
OH \\
\bigcirc\!\!-\!\!CH_2OH
\end{array}
$$

where

$$
\begin{array}{c}
OH \\
\bigcirc\!\!-
\end{array}
$$

indicates that the $-CH_2OH$ group could occur either at the ortho or the para position.[3] Dinuclear species (DPM) are formed when an MP reacts with phenol and the various isomers that could be formed are (o, o) DPM, (o, p) DPM, and (p, p) DPM. Here (o, o), (o, p), and (p, p) specify the positions of the linkage on the two phenyl rings and these are the three isomers of DPM. There are seven isomers of trinuclear species (TPM), and in Fig. 11.1 these are assumed to be formed only by the reactions of DPMs with MPs.[25-27] Evidently, the scheme presented in Fig. 11.1 is an oversimplification because DPMs could also react with formaldehyde and the CH_2OH substituted product can very well react with phenol to form trinuclear species. This approach of the kinetic representation of novolac formation runs into difficulty because, for higher oligomers, the number of isomers for a given chain length increases very rapidly. As given in Table 11.2, for oligomers of a given chain length, the number of isomers increases very rapidly. For example, for oligomers of chain length 9, the total number of isomers is approximately 8500 and the detailed kinetic representation of the polymerization in terms of the molecular species can thus become an impossible task.

TABLE 11.2.[26] Structural Isomers in Novolacs

Chain length n	Number of unbranched isomers	Total including branched isomers
4	21	27
5	57	99
6	171	439
9	4,401	8500 (extrapolated)
10	13,203	—

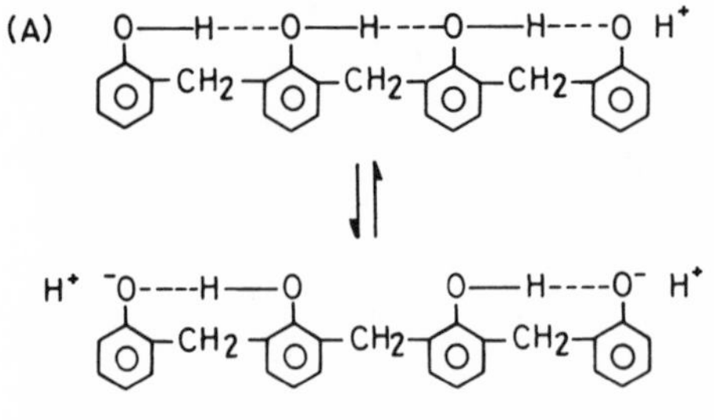

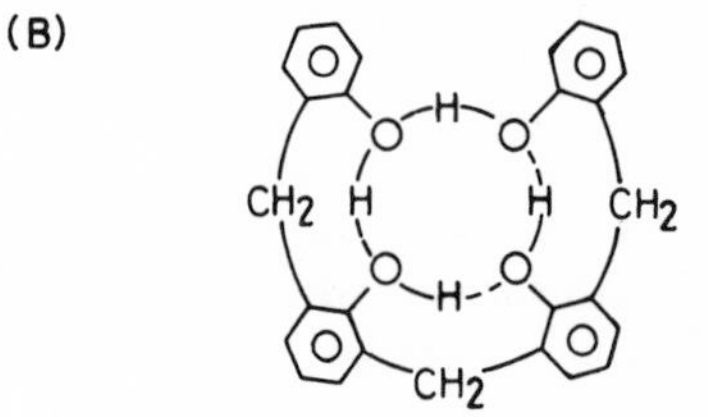

Figure 11.2. Stereoregular structures arising because of the intramolecular hydrogen bonding in novolac polymer chains.

In addition to the difficulty of accounting for various species in modeling the polymerization of phenol formaldehyde, an uncertainty exists about the rate constants for the reaction of higher homologs. This is because the polymerization is normally carried out in an aqueous medium and there is a prevalence of hydrogen bonding, which could either be intramolecular or intermolecular in nature.[28-31] In Fig. 11.2, some of the stereoregular structures arising because of the intramolecular hydrogen bonding are given and it can be visualized easily that the oligomers of novolac would be highly coiled in the reaction mass because of this effect. In polymer molecules with chain lengths of three and beyond, *internal* reactive hydrogen atoms are distinguished from the reactive hydrogens on the chain *ends* (called external sites). For example, in

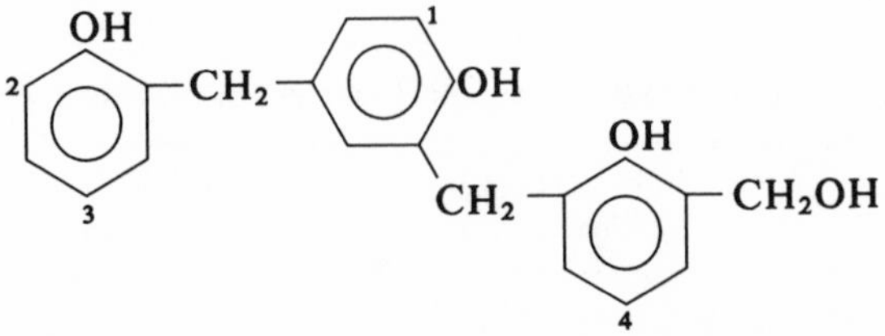

site 1 is an internal ortho site, site 2 is an external ortho site and sites 4 and 3 are external para positions. Because of the coiled nature of the molecules, the internal sites lying generally on the inside of the spherical molecules are expected to react at lower rates compared to the external hydrogens, and this phenomenon is ascribed to the effect of molecules shielding.[6] As a result, researchers involved with the study of novolac polymers sometimes claim that the growth of polymer chains occurs mainly through the reaction of the external hydrogens and that molecular shielding is responsible for the very small amount of branching usually present in novolacs.

In view of this discussion, any kinetic model for novolac formation should distinguish between the external and internal sites because of their different reactivities. Since the ortho and para positions have already been shown to have different reactivities (Table 11.1), one should, therefore, devise a mechanism to distinguish the following sites in the reaction mass:

(a) external ortho sites, o_{eT} (having reactivity k_1);
(b) internal ortho site, o_i (having reactivity k_2);
(c) internal para sites, p_i (having reactivity k_3);
(d) external para sites, p_e (having reactivity k_4).

In the following section, a kinetic model is proposed which does just this.

11.3. CONCENTRATIONS OF SITES o_{eT}, o_i, p_i AND p_e IN NOVOLAC FORMATION

It has been pointed out in Section 11.1 that formaldehyde exists as methylene glycol ($OH-CH_2-OH$) in aqueous solution and, therefore, exhibits a functionality of 2. On its reaction with any hydrogen, whether located on phenol or on novolac molecules, a reacted CH_2OH (as in methylol phenol) is formed. These CH_2OH groups can also react with hydrogen to give a reacted $-CH_2-$ bond between two phenyl rings.

To keep track of various reactive sites in the reaction mass, five species (A–E) are defined in Fig. 11.3. The species differ on the basis of their reacted sites and the polymerization of novolacs can be described in terms of these. No distinction is made as to whether the linkages at these reacted sites is a reacted $-CH_2-$ bond or a reacted $-CH_2OH$ group only. Thus, species A, denoted as

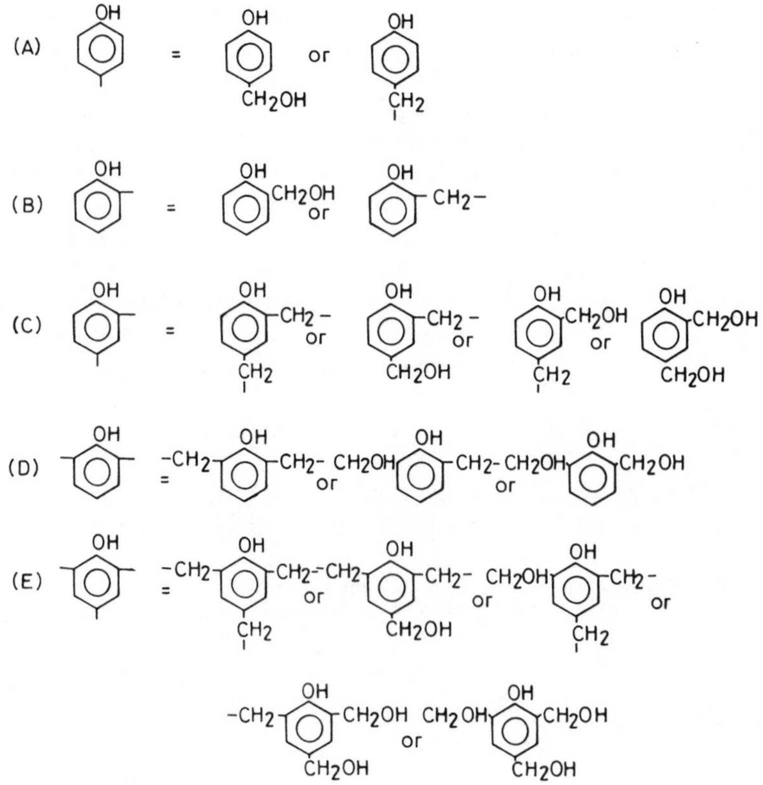

Figure 11.3. Various reactive species (A-E) in novolac formation. The $-CH_2OH$ and $-CH_2-$ have not been distinguished.

could either be part of a polymer chain

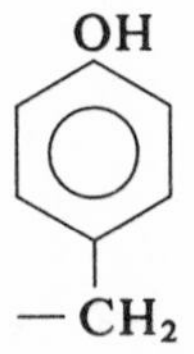

or could just be *p*-methylol phenol,

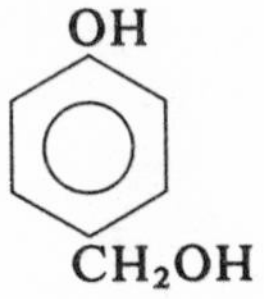

Similarly, species B, denoted as

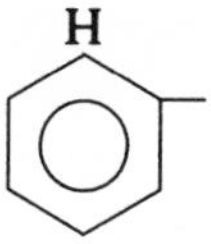

could be either

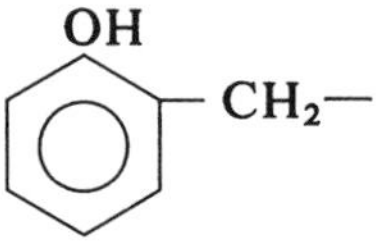

or

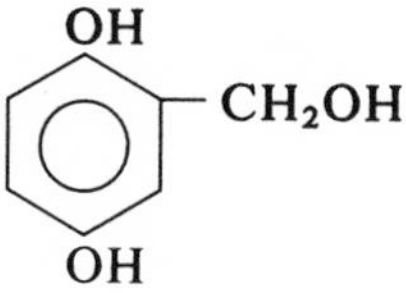

species C, denoted as

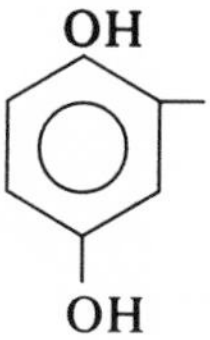

could be either

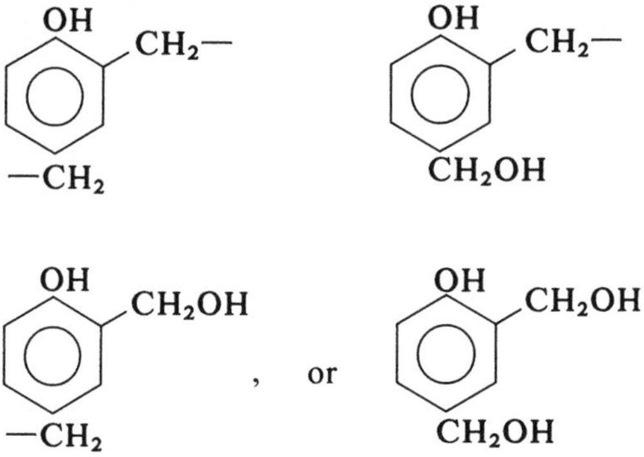

etc. The various equivalent structures for these species are shown in Fig. 11.3. Species A has two unreacted external ortho sites, B has an external ortho and an external para site, C has an internal ortho site, D has an internal para site, and E represents a branch point.

Species A-E can be used to represent any polymer molecules. For example, the following molecule in the reaction mass

(11.3.1)

can be represented by

$$D-C-E-D-A$$
$$|$$
$$C-B$$

(11.3.2)

Also, for a specified arrangement, as in Eq. (11.3.2), there is rarely any confusion in getting back the structure of the polymer molecule as in Eq. (11.3.1). Thus there is a one-to-one correspondence between the two representations. It has already been shown in Section 11.2, that there is a considerable increase in the number of isomers as the chain length of the polymer increases. Therefore, instead of attempting to find out the concentrations of these individual isomers, in the following analysis, an effort is made to find out the conversion of phenol and formaldehyde in the reaction mass as a function of time in terms of these five species.

When polymerization is carried out for some time starting with a feed consisting of phenol and formaldehyde, novolac molecules of various chain lengths and structures are formed. One plausible description of the progress of reaction could be to follow the concentrations of species to A-E in the reaction mass. This approach is similar to the analysis of A–R–B polymerization in which the extent of reaction is written in terms of the concentrations of functional groups as presented in Chapters 2 and 3.

Keeping the mechanism of novolac formation in mind the forward reactions leading to the formation of species A-E can be written. The various reactive sites can either react with formaldehyde (F) or with bound CH_2OH

(denoted as $-CH_2OH$) and A and B are formed whenever a molecule of phenol (P) reacts with F or $-CH_2OH$, i.e.,

$$P + F \xrightarrow{4k_1} B + -CH_2OH + H_2O \qquad (11.3.3)$$

$$P + -CH_2OH \xrightarrow{2k_1} B + H_2O \qquad (11.3.4)$$

$$P + F \xrightarrow{2k_4} A + -CH_2OH + H_2O \qquad (11.3.5)$$

$$P + -CH_2OH \xrightarrow{k_4} A + H_2O \qquad (11.3.6)$$

In writing these reactions, it has been assumed that the overall reactivity of a given reaction is completely governed by the site involved. Therefore, when the two ortho positions (which would naturally be regarded as external sites) of phenol react with the two $-OH$ groups of formaldehyde (or methylene glycol), the overall reactivity is $4k_1$ as shown in Eq. (11.3.3). As opposed to this, when P reacts with bound CH_2OH, the reactivity is only $2k_1$ as in Eq. (11.3.4). In addition to the formation of species B and A there is a simultaneous production of a bound CH_2OH group, as shown in Eqs. (11.3.3) and (11.3.5), whenever F reacts. This must be accounted for because the $-CH_2OH$ groups thus formed can further react.

Species C and D are formed by the forward reaction whenever A and B undergo reactions with F or with bound CH_2OH. Since A and B contain only external ortho and para reactive sites, the reactivities used in these reactions are k_1 and k_4 and the schematic representation of these reactions are

$$A + F \xrightarrow{4k_1} C + -CH_2OH + H_2O \qquad (11.3.7)$$

$$A + -CH_2OH \xrightarrow{2k_1} C + H_2O \qquad (11.3.8)$$

$$B + F \xrightarrow{2k_1} D + -CH_2OH + H_2O \qquad (11.3.9)$$

$$B + -CH_2OH \xrightarrow{k_1} D + H_2O \qquad (11.3.10)$$

$$B + F \xrightarrow{2k_4} C + -CH_2OH + H_2O \qquad (11.3.11)$$

$$B + -CH_2OH \xrightarrow{k_4} C + H_2O \qquad (11.3.12)$$

Species C and D have internal ortho and para sites, respectively, and when they react, branch points, E, are formed as follows:

$$C + F \xrightarrow{2k_2} E + -CH_2OH + H_2O \qquad (11.3.13)$$

$$C + -CH_2OH \xrightarrow{k_2} E + H_2O \qquad (11.3.14)$$

$$D + F \xrightarrow{2k_3} E + -CH_2OH + H_2O \qquad (11.3.15)$$

$$D + -CH_2OH \xrightarrow{k_3} E + H_2O \qquad (11.3.16)$$

With the kinetic scheme given in Eqs. (11.3.3)–(11.3.16), one can easily write the mole balances for species A–E, for batch reactors. These are given in Table 11.3, and can easily be derived.

After a certain time of polymerization, the total count of phenyl rings, C_1, in the reaction mass is given by

$$C_1 = [P] + [A] + [B] + [C] + [D] + [E] = [P]_0 \qquad (11.3.17)$$

TABLE 11.3. Molecular Weight Distribution of Novolac in Batch Reactors

---

a. Mole balance of species A–E

$$\frac{d[A]}{dt} = k_4[P](2[F] + [-CH_2OH]) - 2k_1[A](2[F] + [-CH_2OH]) \qquad (a)$$

$$\frac{d[B]}{dt} = 2k_1(2[F] + [-CH_2OH])[P] - k_1[B](2[F] + [-CH_2OH])$$
$$- k_4[B](2[F] + [-CH_2OH]) \qquad (b)$$

$$\frac{d[C]}{dt} = 2k_1[A](2[F] + [-CH_2OH]) + k_4[B](2[F] + [-CH_2OH])$$
$$- k_2[C](2[F] + [-CH_2OH]) \qquad (c)$$

$$\frac{d[D]}{dt} = k_1[B](2[F] + [-CH_2OH]) - k_3[D](2[F] + [-CH_2OH]) \qquad (d)$$

$$\frac{d[E]}{dt} = k_2[C](2[F] + [-CH_2OH]) + k_3[D](2[F] + [-CH_2OH]) \qquad (e)$$

$$\frac{d[H_2O]}{dt} = 2k_1[A](2[F] + [-CH_2OH]) + (k_1 + k_4)[B](2[F] + [-CH_2OH])$$
$$+ 2k_2[C](2[F] + [-CH_2OH])$$
$$+ k_3[D](2[F] + [-CH_2OH]) + (2k_1 + k_4)[P](2[F] + [-CH_2OH]) \qquad (f)$$

(*continued*)

TABLE 11.3. (*continued*)

$$\frac{d[P]}{dt} = -(2k_1 + k_4)[P](2[F] + [-CH_2OH]) \tag{g}$$

$$\frac{d[F]}{dt} = -2[F]\{(2k_1[A] + (k_1 + k_4)[B] + k_2[C] + k_3[D] + (2k_1 + k_4)[P]\} \tag{h}$$

$$\frac{d[-CH_2OH]}{dt} = 2[F]\{2k_1[A] + (k_1 + k_4)[B] + k_2[C] + k_3[D] + (2k_1 + k_4)[P]\} - [-CH_2OH]$$

$$\times \{2k_1[A] + (k_1 + k_4)[B] + k_2[C] + k_3[D] + (2k_1 + k_4)[P]\} \tag{i}$$

b. *Molecular weight distribution*

$$\frac{d[Q_1]}{dt} = 2(2k_1 + k_4)[P][F] - (2k_1 + k_4)[P][Q_1] - [Q_1] \sum_{i=2}^{\infty} H_{P_i}$$

$$- [Q_1] \sum_{i=1}^{\infty} H_{Q_i} - H_{Q_1} \sum_{i=1}^{\infty} [Q_i] \tag{j}$$

$$\frac{d[Q_n]}{dt} = 2[F]H_{P_n} + \sum_{i=1}^{n-1} [Q_i]H_{Q_{n-i}} - [Q_n] \sum_{i=2}^{\infty} H_{P_i}$$

$$- (2k_1 + k_4)[P][Q_n] - [Q_n] \sum_{i=1}^{\infty} H_{Q_i} - H_{Q_n} \sum_{i=1}^{\infty} [Q_i], \qquad n \geq 2 \tag{k}$$

$$\frac{d[P]}{dt} = - (2k_1 + k_4)[P] \sum_{i=1}^{\infty} [Q_i] - 2(2k_1 + k_4)[P][F] \tag{l}$$

$$\frac{d[P_2]}{dt} = (2k_1 + k_4)[P][Q_1] - 2H_{P_2}[F] - H_{P_2} \sum_{i=1}^{\infty} [Q_i] \tag{m}$$

$$\frac{d[P_n]}{dt} = \sum_{i=1}^{n-2} [Q_i]H_{P_{n-i}} + (2k_1 + k_4)[P][Q_{n-1}] - 2H_{P_n}[F] - H_{P_n} \sum_{i=1}^{\infty} [Q_i]; \qquad n \geq 3 \tag{n}$$

$$H_{P_2} = 2k_1[A_{P_2}] + (k_1 + k_4)[B_{P_2}] \tag{o}$$

$$H_{P_n} = 2k_1[A_{P_n}] + (k_1 + k_4)[B_{P_n}] + k_2[C_{Pn}] + k_3[D_{P_n}]; \qquad n \geq 3 \tag{p}$$

$$H_{Q_1} = 2k_1[A_{Q_1}] + (k_1 + k_4)[B_{Q_1}] \tag{q}$$

$$H_{Q_n} = 2k_1[A_{Q_n}] + (k_1 + k_4)[B_{Q_n}] + k_2[C_{Q_n}] + k_2[C_{Q_n}] + k_3[D_{Q_n}], \qquad n \geq 2 \tag{r}$$

$$R_1 = (k_2/k_1)$$

$$R_2 = k_3/k_1$$

$$R_3 = k_4/k_1$$

$$\tau = k_1[F]_0 t$$

$$y_1 = [A]/[F]_0, \qquad y_2 = [B]/[F]_0$$

$$y_3 = [C]/[F]_0, \qquad y_4 = [D]/[F]_0$$

$$y_5 = [E]/[F]_0$$

where $[P]_0$ is the initial moles of phenol charged into the reactor. For C_1 to be independent of time, $dC_1/dt = 0$ and indeed, this is found to be so. Secondly, stoichiometry demands that at any time, the following relation be satisfied:

$$2[F] + [-CH_2OH] + [H_2O] = 2[F]_0 \qquad (11.3.18)$$

where $[F]_0$ is the initial concentration of formaldehyde in the reactor. This is also satisfied by the balance equations of Table 11.3.

Differential equations of Table 11.3 can be integrated numerically for isothermal batch reactors for given rate constants k_1–k_4 and phenol and formaldehyde concentrations in the feed. It is found (through a detailed sensitivity analysis) that the kinetic models are such that the phenomenon of molecular shielding in the forward step (i.e., internal sites having different

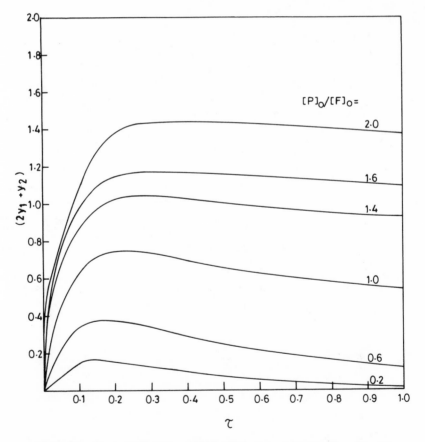

Figure 11.4. Effect of $[P]_0/[F]_0$ on $(2[A]+[B])$ versus τ, $R_1 = 1$, $R_2 = R_3 = 2.4$.

reactivities compared to the corresponding external sites) has negligible effect on the final results. Therefore, the following could be assumed without considerable effect on the numerical solution:

$$k_2 = k_1 \qquad (11.3.19a)$$

$$k_4 = k_3 \qquad (11.3.19b)$$

which means that $R_1 = 1$ and $R_2 = R_3$ in Table 11.3.

In generating the results in Figs. 11.4–11.9, R_3 has been taken as equal to 2.4, which is the same as the reactivity ratio of the ortho and para sites of phenol.[3] To find out the effect of phenol to formaldehyde ratio $[P]_0/[F]_0$ in the feed, results have been computed and plotted in Figs. 11.4–11.7. For a given $[P]_0/[F]_0$, the concentration of external ortho sites in the reaction mass quickly increases to a large value and then very slowly decreases for

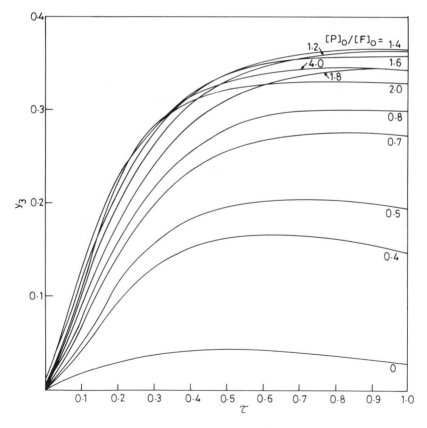

Figure 11.5. Effect of $[P]_0/[F]_0$ on [C] versus τ.[32] $R_1 = 1$, $R_2 = R_3 = 2.4$.

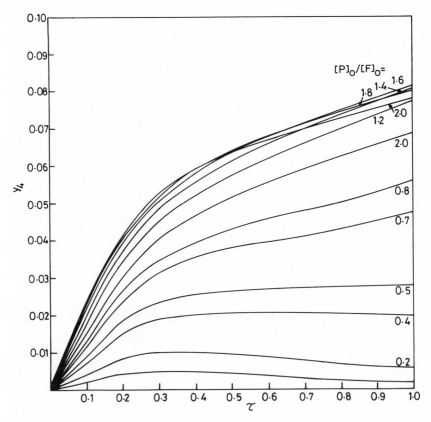

Figure 11.6. Effect of $[P]_0/[F]_0$ on $[D]$ versus τ.[32] $R_1 = 1$, $R_2 = R_3 = 2.4$.

large reaction times as seen in Fig. 11.4. The curve for external ortho sites versus τ shifts upwards as $[P]_0/[F]_0$ is increased to larger values. In Figs. 11.5 and 11.6, $[C]$ and $[D]$ versus τ, respectively, have been plotted. The concentrations of species C and D give a measure of the internal ortho and para sites in the batch reactors and indicate the formation of large linear chains. As $[P]_0/[F]_0$ is increased from a very small value of 0.1, $[C]$ and $[D]$ both are found to increase first but the asymptotic value of $[C]$ begins to fall for the $[P]_0/[F]_0$ beyond 1.4 and the asymptotic value of D begins to fall beyond 1.6. This suggests that at this $[P]_0/[F]_0$ ratio, linear novolac chains are maximized. This result is surprising if it is noted that the functionality of phenol is 3 whereas that of formaldehyde is 2. This would mean that a network structure (i.e., preponderance of species E) is expected to be favored at $[P]_0/[F]_0$ of 2/3, whereas the formation of linear polymer should be maximized at $[P]_0 = [F]_0 = 1$. The anomaly existing between the observed results in Figs. 11.5 and 11.6 and those expected from the analysis

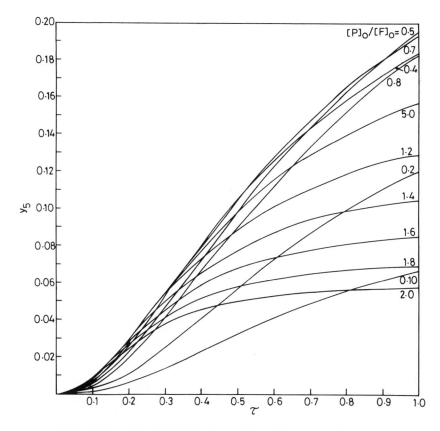

Figure 11.7. Effect of $[P]_0/[F]_0$ on $[E]$ versus τ.[32] $R_1 = 1$, $R_2 = R_3 = 2.4$.

based on stoichiometry is attributed to the unequal reactivity of various sites.

In Fig. 11.6, the concentration of species E versus the dimensionless time τ has been plotted with $[P]_0/[F]_0$ as the parameter. For a given $[P]_0/[F]_0$, $[E]$ starts from zero and is found to increase continuously for the values of τ studied. As the $[P]_0/[F]_0$ ratio is increased from a very low value of 0.1, the curves are found to shift upwards, but for ratios beyond 0.5, the curves begin to fall as shown in the figure. This would mean that the maximum branching occurs at $[P]_0/[F]_0 = 0.50$, whereas as expected from the stoiochiometry, it should occur at $[P]_0/[F]_0 = 2/3$. This anomaly is once again attributed to the unequal reactivity of the various sites of phenol.

It is possible to determine the molecular weight distribution of the novolac polymer as a function of reaction time with a knowledge of the concentrations of species A–E. This is discussed in the next section.

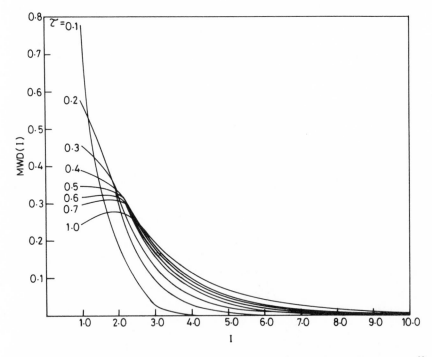

Figure 11.8. Weight fraction distribution of novolac as a function of time of polymerization.[32] $R_1 = 1$, $R_2 = 3 = 2.4$, $[P]_0/[F]_0 = 1.67$.

11.4. MOLECULAR WEIGHT DISTRIBUTIONS IN NOVOLAC POLYMER FORMATION

It has already been shown in Table 11.2 that the number of structural isomers of the novolac polymer molecules increases considerably with the chain length. Therefore, any approach based on an exhaustive enumeration of the various isomers is bound to become intractable. In view of this, in the analysis presented in this section, an average molecular structure is assumed.

Molecules of novolac polymer are first grouped into those (P_n) which do not have any bound $-CH_2OH$ group on their chain and those (Q_n) which have one or more bound $-CH_2OH$ groups. The analysis of the previous section shows that $[-CH_2OH]$ is always small in the reaction mass, and as a first approximation, it could be assumed that a given Q_n molecule can have only one bound CH_2OH group. Thus P_n and Q_n

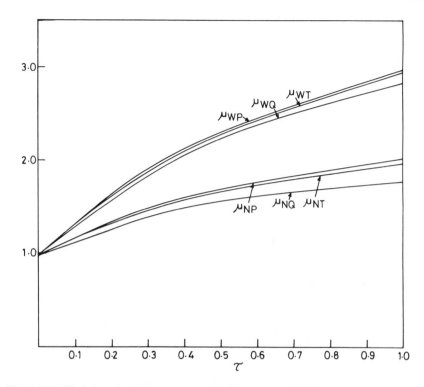

Figure 11.9. Chain length average μ_{NP}, μ_{NQ}, and μ_{NT} and weight average μ_{WP}, μ_{WQ}, and μ_{WT} molecular weight of P_n, Q_n, and polymer.[32]

molecules can approximately be represented by

P_n:

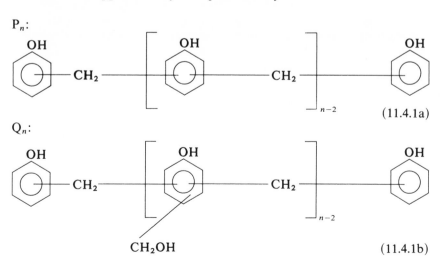

(11.4.1a)

Q_n:

(11.4.1b)

where the CH_2OH group on Q_n could be situated anywhere on the chain. The various reactions can be written equivalently in terms of these molecules as

$$P_n + F \rightarrow Q_n + H_2O \tag{11.4.2a}$$

$$P_n + Q_m \rightarrow P_{m+n} + H_2O \tag{11.4.2b}$$

$$Q_n + Q_m \rightarrow Q_{m+n} + H_2O \tag{11.4.2c}$$

Since Q_n can have only one CH_2OH group per chain, in the kinetic representation above, Q_n is assumed not to react with F.

In order to ascertain the reactivity of the molecular species P_n and Q_n in Eq. (11.4.2), it is necessary to have a knowledge of the number of species A–E on them. Since the specific number of species A to E is expected to depend upon the isomer, the overall reactivity is expected to vary from one isomer to the other. To overcome this difficulty, the number of these species on *all* different isomers of P_n and Q_n are added up and the total concentration of these species is determined. Let $[A_{P_n}]$, $[B_{P_n}]$, $[C_{P_n}]$, $[D_{P_n}]$, $[E_{P_n}]$, and $[A_{Q_n}]$, $[B_{Q_n}]$, $[C_{Q_n}]$, $[D_{Q_n}]$, $[E_{Q_n}]$ be the concentrations of these species. Evidently,

$$[C_{Q_1}] = [D_{Q_1}] = [E_{Q_1}] = 0 \tag{11.4.3a}$$

$$[E_{Q_2}] = [C_{P_2}] = [D_{P_2}] = [E_{P_2}] = 0 \tag{11.4.3b}$$

$$[E_{P_3}] = 0 \tag{11.4.3c}$$

When the CH_2OH group of molecule Q_m reacts with the sites of P_n through Eq. (11.4.2b), the rate of reaction, $\mathscr{R}_{Q_mP_n}$, is given by

$$\mathscr{R}_{Q_mP_n} = [Q_m]\{2k_1[A_{P_n}] + (k_1 + k_4)[B_{P_n}] + k_2[C_{P_n}] + k_3[D_{P_n}]\},$$
$$m \geq 1, \qquad n \geq 3 \tag{11.4.4}$$

Since P_2 does not have any internal sites, $\mathscr{R}_{Q_mP_2}$ is given by

$$\mathscr{R}_{Q_mP_2} = [Q_m]\{2k_1[A_{P_2}] + (k_1 + k_4)[B_{P_2}]\}, \qquad m \geq 1 \tag{11.4.5}$$

Similarly when the CH_2OH group of Q_m reacts with the reactive hydrogen of Q_n, the rate of reaction $\mathscr{R}_{Q_mQ_n}$ is given by

$$\mathscr{R}_{Q_mQ_n} = [Q_m]\{2k_1[A_{Q_n}] + (k_1 + k_4)[B_{Q_n}] + k_2[C_{Q_n}] + k_3[D_{Q_n}]\} \tag{11.4.6}$$

with

$$\mathscr{R}_{Q_m Q_1} = [Q_m] 2 k_1 \{ [A_{Q_1}] + (k_1 + k_4)[B_{Q_1}] \} \qquad (11.4.7)$$

In writing these, it has been assumed that the reactivity of Q_n is governed by the site where the reaction is occurring and not by the position of the CH_2OH group. In order to write mole balance equations for Q_n and P_n in batch reactors, the appropriate terms must be collected. Q_n is formed when

(a) P_n reacts with formaldehyde, and
(b) The CH_2OH group of Q_i $(i < n - 1)$ reacts with the various sites of the species of Q_{n-i}.

On the other hand, Q_n is depleted when

(c) The CH_2OH group of Q_n reacts with any of the active species of P_i $(i = 1, 2, \ldots, \infty)$;
(d) The CH_2OH group of Q_n reacts with any of the active sites of Q_i $(i = 1, 2, \ldots, \infty)$;
(e) The various sites of Q_n react with CH_2OH groups of Q_i $(i = 1, 2, \ldots, \infty)$.

Q_1, however, does not form through step (b). The mole balance equations for batch reactors are also given in Table 11.3.

To obtain $[A_{P_i}]$ to $[E_{P_i}]$ and $[A_{Q_i}]$ to $[E_{Q_i}]$, it is assumed that the fractions of these species on the molecules of P_i and Q_i are equal to *overall* fractions of these species in the reaction mass. This means that

$$[A_{Q_i}] = \frac{[A]}{\sum_{i=1}^{\infty} [Q_i] + \sum_{i=2}^{\infty} [P_i]} [Q_i] \qquad (11.4.8)$$

Similar relations are assumed to hold for $[B_{Q_i}]$ to $[E_{Q_i}]$ and $[A_{P_i}]$ to $[E_{P_i}]$.

It may be noted that the molar balance for phenol in Eqs. (g) and (l) of Table 11.3 are different. This difference arises because of the assumption that there is only one CH_2OH group on a given Q_n molecule, whereas no such assumption has been made in deriving Eq. (g). To show that the mole balance equations in Table 11.3 are consistent, it must be shown that the ring count must be constant for all time. It can analytically be shown that the rate of production of RC_2 is indeed identically zero.

The equations of Table 11.3 have been solved numerically using the Runge–Kutta algorithm of the fourth-order along with the mole balance equations for species A–E. Since P_n and Q_n cannot be separated physically the molecular weight distributions have been defined by the following relation:

$$w_i = \frac{i[P_i] + i[Q_i]}{\sum_{i=1}^{\infty} \{i[P_i] + i[Q_i]\}} \qquad (11.4.9)$$

where P_1 represents phenol. In Fig. 11.8, the weight fraction distribution for various dimensionless reaction times, τ, has been given. As the time of polymerization increases, the MWD of novolac becomes broader and broader. After $\tau = 0.5$, the curves begin to have maxima at $n = 2$ and concentrations of 8- and 9-mers are found to be very small as observed experimentally in typical novolac polymers.

In Fig. 11.9 the number average and weight average chain lengths of the P_i and Q_i distributions (μ_{nP}, μ_{wP} and μ_{nQ}, μ_{wQ}) and those of their combined distribution (μ_{nT} and μ_{wT}) have been plotted as a function of the dimensionless time τ. It is observed that the curves for μ_{nT} and μ_{wT} are close to those for μ_{nP} and μ_{wP}. This would imply that $\sum_{n=1}^{\infty} [Q_n]$ in the reaction mass must be small, thus further justifying the assumption of permitting only one CH_2OH group on a Q_n molecule. In Fig. 11.10, the

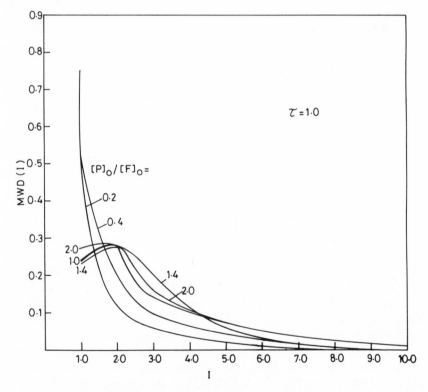

Figure 11.10. Effect of $[P]_0/[F]_0$ on the MWD of the novolac polymer.[32] $R_1 = 1$, $R_2 = R_3 = 2.4$.

effect of $[P]_0/[F]_0$ in the feed on the molecular weight distribution has been examined. For $[P]_0/[F]_0 = 1.4$, the MWD is found to be the broadest. In Fig. 11.11, the number average chain length calculated from the MWDs has been plotted as a function of τ and is found to be the largest for the $[P]_0/[F]_0$ ratio of 1.4. The analysis of the previous section showed that the concentration of internal sites C and D are maximized at $[P]_0/[F]_0 = 1.60$. [C] and [D] can be large only when there is a preponderance of linear chains in the reaction mass. The discrepancy between these two results can be attributed to the approximations made in the calculations of the MWD of novolacs.

An earlier model of novolac formation was first presented by Drumm et al.,[6] who assumed that the reaction mass consists of P_n and Q_n molecules

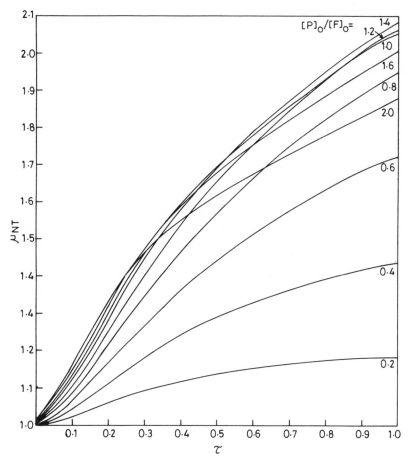

Figure 11.11. Effect of $[P]_0/[F]_0$ on the average chain length of novolac polymer.[32] $R_1 = R_2 = R_3 = 2.4$.

and the reactions are given by

$$P_n + F \xrightarrow{k_{a,n}} Q_n \tag{11.4.10a}$$

and

$$P_n + Q_m \xrightarrow{k_{c,n}} P_{m+n} \tag{11.4.10b}$$

where $k_{a,n}$ and $k_{c,n}$ are the reactivities of the molecules of P_n which are written in terms of the reactivities, $k'_{a,n}$ and $k'_{c,n}$ of formaldehyde and bound CH_2OH groups on Q_n, respectively, as

$$k_{a,n} = k'_a S_n \tag{11.4.11a}$$

$$k_{c,n} = k'_c S_n \tag{11.4.11b}$$

In this equation, S_n is the total number of reaction sites. If the polymer chains are essentially linear, it is given by

$$S_n = \{4 + \theta(n - 2)\} \tag{11.4.12}$$

where the factor of 4 is due to the external reaction sites and $(n - 2)$ are the number of internal sites. Since the internal ortho and para sites have lower reactivities due to molecular shielding, θ is a factor less than unity in Eq. (11.4.12). With the mechanism of Eq. (11.4.17) and rate constants given in Eqs. (11.4.18) and (11.4.19), the mole balance equations for various species for batch reactors and HCSTRs have been written and solved numerically.[6,35] This analysis is less accurate because it assumes polymer chains to be strictly linear and ignores the reaction between two polymer molecules of the Q_m type.

In defining species A–E in this section, a reacted bond and $-CH_2OH$ have not been distinguished. If one distinguishes between them, it is not possible to write the reverse reactions. For irreversible novolac formation,[28,36-39] one defines species given in Table 11.4, and writing various reactions leading to site formation, one can easily derive MWD of the polymer formed. It is observed that any reaction on internal ortho and para sites leads to branching of the novolac polymer. The model presented in this section as well that given in Table 11.4 both predict that branching rate is negligible.

Flory[40] has suggested that in novolac formation, phenol can be assumed to be trifunctional in the step growth polymerization with formaldehyde serving as a mere linkage. One can then employ the usual MWD relations (Chapter 4) for multifunctional polymerization. Borrajo et al.[41] observe that

TABLE 11.4. Model for Irreversible Novolac Formation

1. *Reactive sites and molecules in the reaction mass*

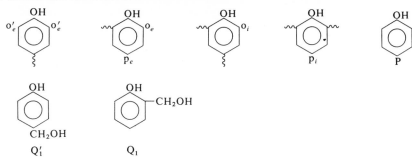

2. *Reaction of sites*

$$o'_e + F \xrightarrow{2k_1} (-CH_2OH) + o_i - o'_e \qquad o'_e + (-CH_2OH) \xrightarrow{k_1} o_i - o_e$$

$$o_e + F \xrightarrow{2k_1} (-CH_2OH) + p_i - p_e \qquad o_e + (-CH_2OH) \xrightarrow{k_1} p_i - p_e$$

$$o_i + F \xrightarrow{2k_2} (-CH_2OH) \qquad o_i + (-CH_2OH) \xrightarrow{k_2} \text{consumption}$$

$$p_i + F \xrightarrow{2k_3} (-CH_2OH) \qquad p_i + (-CH_2OH) \xrightarrow{k_3} \text{consumption}$$

$$p_e + F \xrightarrow{2k_4} (-CH_2OH) + o_i - o_e \qquad p_e + (-CH_2OH) \xrightarrow{k_4} o_i - o_e$$

3. *Reactions of P, Q_1, and Q'_1*

$$P + F \xrightarrow{4k_1} Q_1 \qquad\qquad P + (-CH_2OH) \text{ L} \xrightarrow{2k_1} o_e + p_e$$

$$P + F \xrightarrow{k_4} Q'_1 \qquad\qquad P + (-CH_2OH) \xrightarrow{k_4} 2o'_e$$

$$P + Q_1 \xrightarrow{2k_1} 2o_e + 2p_e \qquad\qquad P + Q'_1 \xrightarrow{2k_1} o_e + p_e + 2o'_e$$

$$P + Q_1 \xrightarrow{k_4} o_e + p_e + 2o'_e \qquad\qquad P + Q'_1 \xrightarrow{k_4} 4o'_e$$

$$o'_e + Q'_1 \xrightarrow{k_1} o_i + 2o'_e - o'_e \qquad\qquad o_e + Q_1 \xrightarrow{k_1} o_e + p_e + p_i - p_e$$

$$o'_e + Q_1 \xrightarrow{k_1} o_e + p_e + o_i - o'_e \qquad\qquad o_e + Q'_1 \xrightarrow{k_1} 2o'_e + p_i - p_e$$

$$o_i + Q'_i \xrightarrow{k_2} 2o'_e \qquad\qquad p_i + Q'_1 \xrightarrow{k_1} 2o'_e$$

$$o_i + Q_1 \xrightarrow{k_2} o_e + p_e \qquad\qquad p_i + Q_1 \xrightarrow{k_3} o_e + p_e$$

$$p_e + Q'_1 \xrightarrow{k_4} 2o'_e + o_i - o_e$$

$$p_e + Q_1 \xrightarrow{k_4} o_e + p_e + o_i - o_e$$

these relations would not hold because all the reactive sites are not kinetically equivalent and the various reactive sites and functional groups do not react independently. They used Stockmayer's equation to predict the MWD in which the functionality of phenol has been treated as a parameter. It was found that experimental data are well described by the Stockmayer equation if the average functionality of phenol is assumed to be 2.31.

Recently, there has been an effort to simulate the formation of novolac polymers on a computer using the Monte Carlo technique.[42] The ortho and para hydrogens on phenol (1OH and 1PH) and on the molecules of novolac polymer (2OH and 2PH) were distinguished as follows:

$$(11.4.13)$$

In this equation, the internal and external sites on the polymer were treated as equivalent. Once again, the reactive CH_2OH group could exist on phenol as well as on polymer chains as

$$(11.4.14)$$

and the CH_2OH groups on the ortho and para positions were represented by OM and PM, respectively. The various reactions occurring in novolac formation were written in terms of these as

$$F \begin{cases} \nearrow 1OH \\ \rightarrow 2OH \\ \rightarrow 1PH \\ \searrow 2PH \end{cases}$$

(11.4.15a)

$$OM \begin{cases} \nearrow 1OH \\ \rightarrow 2OH \\ \rightarrow 1PH \\ \searrow 2PH \end{cases}$$

(11.4.15b)

$$PM \begin{cases} \nearrow 1OH \\ \rightarrow 2OH \\ \rightarrow 1PH \\ \searrow 2PH \end{cases}$$

(11.4.15c)

Using Monte Carlo simulation, Ishida *et al.*[42] have calculated the relative ratios of o–o, o–p, and p–p linkages on a novolac polymer of a given chain length. They have claimed that the MWD of the novolac polymer can be calculated but they have not reported any experimental results or computations of the model.

11.5. MODELING OF RESOLE REACTORS[42–44]

Commercially resole polymer is formed by the step growth polymerization of phenol and formaldehyde in alkaline reaction medium (pH ~ 9). The monomers are normally present in the feed in a molar ratio $[P]_0/[F]_0$, less than unity. The resole molecules so formed are relatively more branched compared to novolacs and serve as a prepolymer which are further polymerized in suitable molds to produce a highly cross-linked thermoset polymer called Bakelite. In this section attention is focused on the modeling of reactors forming the prepolymer only.

Ideally a prepolymer having the molecular structure

would give the maximum branching in the thermoset. In practice, however, various oligomers of different chain lengths are formed on which the $-CH_2OH$ groups are randomly distributed. These oligomers need not necessarily be linear chains and are indeed expected to be sufficiently branched. Since formaldehyde is used in excess, at any time there is a large concentration of $-CH_2OH$ groups in the reactor. It may be recalled that while modeling novolac reactors in Sections 11.3 and 11.4, $[-CH_2OH]$ was assumed to be small enough so that there is no more than one CH_2OH group per polymer molecule, irrespective of its chain length. This would mean that for resole formation, the chain growth reactions cannot be represented by Eq. (11.4.2) as done for novolac formation. In addition to this difficulty, at high pH of the reaction medium, two CH_2OH groups can also undergo a condensation reaction as follows:

$$(11.5.1)$$

Nineteen species have been defined in Table 11.5 to be able to model the formation of resole molecules. The large number of species is essential to explain the branching characteristics of resole formation. Species A and B are phenol and formaldehyde, respectively, and species C–H are the various $-CH_2OH$ substituted phenols. Species I–T, on the other hand, represent a phenyl ring on resole *polymer* molecules, and M represents a branch point. On comparison of Table 11.5 with Fig. 11.3, it may be found that the former is a more detailed version of the latter.

The formation of resole can be represented in terms of several elementary reactions between species A–T of Table 11.5. Assuming irreversible reactions, there are, in all, 360 elementary reactions and the reactions involving A are listed in Table 11.6 in terms of the following five rate constants for illustration:

(a) k_1, the rate constant for the reaction between an external ortho site on a phenyl ring and a $-CH_2OH$ or an OH group on a formaldehyde molecule;

TABLE 11.5. Various Species Used in Modeling Resole Formation

(A) (B) (C) (D)

(E) (G) (H)

(I) (J) (K) (L) (M)

(N) (O) (P) (Q)

(R) (S) (T)

TABLE 11.6. Various Elementary Reactions between
Basic Entities of Table 11.5 along with Rate Constants
Involving Species A

1. $A + B = C \, (4k_1)$	17. $A + N = I + K \, (2k_1)$
2. $A + B = D \, (2k_4)$	18. $A + N = J + K \, (k_4)$
3. $A + C = 2I \, (2k_1)$	19. $A + O = I + L \, (2k_1)$
4. $A + C = I + J \, (k_4)$	20. $A + O = J + L \, (k_4)$
5. $A + D = I + J \, (2k_1)$	21. $A + P = I + L \, (2k_4)$
6. $A + D = 2J \, (k_4)$	22. $A + P = J + L \, (k_4)$
7. $A + E = I + N \, (4k_1)$	23. $A + Q = I + M \, (2k_1)$
8. $A + E = J + N \, (2k_4)$	24. $A + Q = J + M \, (k_4)$
9. $A + G = J + G \, (k_4)$	25. $A + R = I + M \, (2k_1)$
10. $A + G = J + P \, (k_4)$	26. $A + R = J + M \, (k_4)$
11. $A + G = I + O \, (2k_1)$	27. $A + S = I + Q \, (4k_1)$
12. $A + G = J + O \, (k_4)$	28. $A + S = J + Q \, (2k_4)$
13. $A + H = I + T \, (4k_1)$	29. $A + T = I + R \, (2k_1)$
14. $A + H = J + T \, (2k_4)$	30. $A + T = J + R \, (k_4)$
15. $A + H = I + S \, (k_1)$	31. $A + T = I + Q(2k_1)$
16. $A + H = J + S \, (k_4)$	32. $A + T = J + Q \, (k_4)$

(b) k_2, the rate constant for the reaction between an internal ortho (o_i) site and a CH_2OH group or the OH of formaldehyde;

(c) k_3, the rate constant for the reaction between an internal para site p_i and a CH_2OH group or the OH of formaldehyde;

(d) k_4, the rate constant between an external para site p_e and a CH_2OH group or the OH of formaldehyde; and

(e) k_5, the rate constant for the reaction between two CH_2OH groups giving a reacted $-CH_2-$ bond and formaldehyde; even though this step is postulated to occur in several steps, here it has been modeled as a single step to keep the analysis simple.

The complete list of reactions of these species can be found in Ref. 45. The mole balance equations for isothermal batch reactors and HCSTRs have been written and solved numerically and the concentrations of each of the species A–R obtained. The numerical technique used for batch reactors is the Runge–Kutta algorithm of the fourth order, whereas that for HCSTRs is the Gauss–Jordon method.

The number average chain length μ_n, of the resole formed can also be calculated as follows. A typical branched molecule is shown in Fig. 11.12 in which blb is a loop and bb is a branch connecting chain e_1e_1 and e_2e_2. A study of its structure reveals that the presence of every branch point not associated with a loop increases the number of end phenyl rings by unity. The presence of a loop, however, wastes two branch points. The concentration of polymer molecules containing two and more phenyl rings, $[X_p]$, can

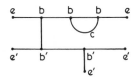

Figure 11.12. A typical branched molecule with a loop.

be written as

$$[X_p] = \frac{[X_e] - [X_{bl}]}{2} \tag{11.5.2}$$

where $[X_e]$ is the concentration of end rings and $[X_{bl}]$ is the concentration of those branch points which do not lead to a loop. An examination of Table 11.5 gives $[X_e]$ to be

$$[X_e] = [I] + [J] + [N] + [O] + [P] + [S] + [T] \tag{11.5.3}$$

The total concentration of branch points, $[X_b]$ is

$$[X_b] = [M] \tag{11.5.4}$$

and $[X_{bl}]$ is obtained by subtracting those branch points which lead to loop formation, i.e.,

$$[X_{bl}] = [X_b] - 2[X_l] \tag{11.5.5}$$

where $[X_l]$ is the concentration of cyclic loops in the reaction mass. The number average chain length μ_n (this is really the number average number of phenyl rings per molecule) is given by the ratio of the original number of phenolic rings, $[A]_0$, to be the total number of polymer molecules (including single ring species), i.e.,

$$\mu_n = \frac{[A]_0}{[X_p] + [A] + [C] + [D] + [E] + [G] + [H]} \tag{11.5.6}$$

μ_n can now be determined provided it is possible to determine the concentration of loops, $[X_l]$, in the reaction mass. Loops are formed by intramolecular reactions and can be estimated as follows. If the reactions of species I and N are taken as an example, the rate of loop formation due to this {abbreviated as $(d[X_1]/dt)_{I,N}$} can be approximated as

$$\left(\frac{d[X_l]}{dt}\right)_{I,N} = k_l[I][N]$$

$$\times \left\{ \frac{\text{Number of linkages formed that are intramolecular and are I–N type}}{\text{Total number of linkages formed of I–N type}} \right\}$$

$$\tag{11.5.7}$$

The term in the parentheses is the fraction of intramolecular links of the I–N type, and on an average, it is given by

Fraction of intramolecular links of I–N type

$$= \frac{\left\{\begin{matrix} \text{Average number} \\ \text{of I in a} \\ \text{polymer molecule} \end{matrix}\right\}\left\{\begin{matrix} \text{Average number} \\ \text{of N in a} \\ \text{polymer molecule} \end{matrix}\right\}\left\{\begin{matrix} \text{Number of} \\ \text{polymer molecules} \\ \text{per unit volume} \end{matrix}\right\}}{\left\{\begin{matrix} \text{Total number of I} \\ \text{per unit volume} \end{matrix}\right\}\left\{\begin{matrix} \text{Total number of N} \\ \text{per unit volume} \end{matrix}\right\}}$$

$$= \frac{\dfrac{[I]}{[X_p]}\dfrac{[N]}{[X_p]}\{[X_p]N_{Av}\}}{\{[I]N_{Av}\}\{[N]N_{Av}\}}$$

$$= \frac{1}{[X_p]N_{Av}} \tag{11.5.8}$$

where N_{Av} is the Avogadro number. Thus, the total rate of loop formation can be calculated by summing over all possible reactions as

$$N_{Av}[X_p]\frac{d[X_1]}{dt} = \{(k_1 + k_4)[I] + 2k_1[J] + k_3[K] + k_2[L]\}$$

$$\times \{[N] + [O] + [P] + [Q] + [R] + 2[S] + 2[T]\}$$

$$+ [N]\{k_1([O] + [P]) + k_4([N] + [O] + [P]$$

$$+ [Q] + [R] + 2[S] + 2[T])\}$$

$$+ k_1[O]\{[O] + 2[P] + [Q] + [R] + 2[S] + 2[T]\}$$

$$+ k_1[P]\{[Q] + [R] + 2[S] + 2[T] + [P]\}$$

$$+ k_5\{[N][[N]/2 + [O] + [P] + [Q] + [R] + 2[S] + 2[T]]$$

$$+ [O][[O]/2 + [P] + [Q] + [R] + 2[S] + 2[T]]$$

$$+ [P][[P]/2 + [Q] + [R] + 2[S]$$

$$+ 2[T]] + [Q][[Q]/2 + [R] + 2[S] + 2[T]]$$

$$+ [R][[R]/2 + 2[S] + 2[T]]$$

$$+ 2[S][[S] + 2[T]] + 2[T][T]\} \tag{11.5.9}$$

The various mole balance equations presented in this section can be written in terms of the following dimensionless parameters:

$$R_1 = k_2/k_1 \qquad (11.5.10a)$$

$$R_2 = k_3/k_1 \qquad (11.5.10b)$$

$$R_3 = k_4/k_1 \qquad (11.5.10c)$$

$$R_4 = k_5/k_1 \qquad (11.5.10d)$$

and

$$\tau = k_1[B]_0 t \qquad (11.5.10e)$$

where t is the time of polymerization and $[B]_0$ is the initial concentration of formaldehyde in batch reactors.

The values of parameters R_1-R_3 have been chosen to be the same as those for novolac polymer formation, viz., $R_1 = 0.125$, $R_2 = 0.3$, and $R_3 = 2.4$. This has been done because experimental data on resole formation are not available. It has been mentioned by Drumm and LeBlanc[6] that the reaction between two CH_2OH groups is fairly rapid, as a result of which, whenever not specified, R_4 has arbitrarily been chosen as 3. In the following analysis, the relative importance of the rate parameters R_1-R_4 is first established by systematically varying these.

Among the rate parameters, R_1 and R_4 are found to influence the branching considerably as seen in Fig. 11.13. The concentration of branch points rises very slowly initially, but after a sufficient number of polynuclear species are formed, branching increases very rapidly. Numerical computation has shown that R_2 has considerably less effect on branching. This probably occurs for the following reason. Since R_3 ($=2.4$) is greater than unity, it is expected that there is a relative abundance of internal ortho sites compared to internal para sites in the reaction mass and a lower sensitivity of $[M]$ towards the variation of R_2 can be attributed to the smaller concentration of p_i. For a similar reason, the effect of R_3 on branching is small. In Fig. 11.14, the number of average chain length, μ_n, has been plotted for variations of R_1 and R_4. Initially, μ_n increases very slowly but after some time, the curves shoot upwards giving a large increase in the number average chain length in a short time. This indicates the occurrence of the phenomena of gelation and is found to be delayed by the reduction of R_1 or R_4.

The ratio of phenol to formaldehyde in the feed of the batch reactor is found to be an extremely important parameter and determines the nature of the resole polymer formed. In Figs. 11.15 and 11.16, $[A]_0/[B]_0$ has been

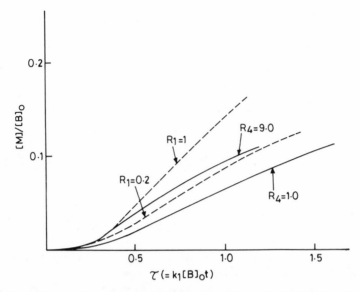

Figure 11.13. Concentration of branch points for various values of R_1 (dashed lines, $R_2 = 0.30$, $R_3 = 2.4$, $R_4 = 3.0$) and R_4 (solid lines $R_1 = 0.125$, $R_2 = 0.30$, $R_3 = 2.4$).[42]

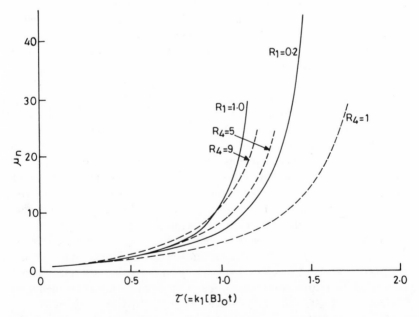

Figure 11.14. Average chain length as a function of dimensionless time τ. Parameters changed are R_1 (solid lines, $R_2 = 0.30$, $R_3 = 2.4$, $R_4 = 3.0$) and R_4 (dashed lines, $R_1 = 0.125$, $R_2 = 0.30$, $R_3 = 2.4$).[42]

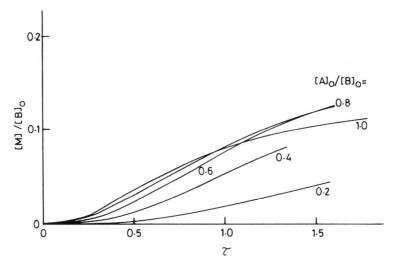

Figure 11.15. Dimensionless concentration of branch points for various ratios of phenol and formaldehyde in feed.[42]

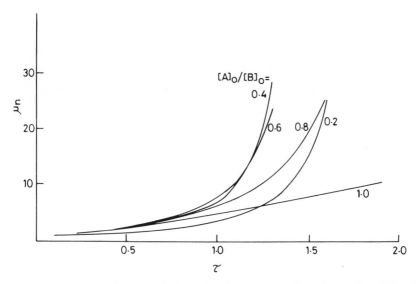

Figure 11.16. Average chain length of the resole polymer versus τ for various ratios of phenol and formaldehyde in feed.[42]

varied and the concentration of branch points and the number average chain length versus the time of polymerization have been plotted. It is observed that $[M]/[B]_0$ as well as μ_n at a given time first increase as $[A]_0/[B]_0$ increases from a low value of 0.2 to the value of 0.6 and thereafter begin to decrease. This means that $[A]_0/[B]_0 \simeq 0.6$ produces the maximum branching, and interestingly enough, commercial reactors are operated at about this phenol to formaldehyde ratio in the feed.

There have been relatively few attempts in the literature to model resole formation. Zavitsas *et al.*[15,16] have carried out a computer simulation for resole formation with the following kinetic scheme:

$$(11.5.11)$$

They, however, exclude the possibility of the formation of polynuclear species, which could be justified only when the reaction is carried out in large excess of formaldehyde (and is useful to study the rate constants for these initial reactions).

An effort to model the MWD of resoles runs into difficulty largely owing to the fact that polymer chains, on an average, have more than one CH_2OH group. As a result, it is almost impossible to generalize the chain growth reactions, as could be done in Eq. (11.4.2) for novolac formation. If it is assumed that all resole molecules either have no or one CH_2OH groups, and there is no condensation reaction between CH_2OH groups (i.e.,

$k_5 = 0$), results derived from this kinetic model match those for irreversible novolac formation exactly.

11.6. CONCLUSIONS

In this chapter, kinetic models for novolac and resole formation have been discussed. Novolacs are essentially linear chains and are formed in acidic reaction medium. As a result of excessive coiling of polymer molecules due to hydrogen bonding, it is expected that the internal reactive sites have lower reactivities compared to those for external sites, a phenomenon termed molecular shielding. In addition to this, experiments carried out in dilute solutions have clearly demonstrated that the ortho and para sites, whether external or internal, have different reactivities.

Keeping the basic reaction characteristics of novolac formation, a kinetic model has been proposed which involves five species A to E and four rate constants k_1-k_4. To examine the effect of molecular shielding, the rate constants k_2 and k_3 for internal ortho and para sites have been systematically varied and results have been found to be negligibly affected. A careful examination of mole balance equations of o_i and p_i reveal that this is largely due to the fact that terms involving k_2 and k_3 remain small for a wide variation of these rate constants. This seems to imply that novolac chains grow in size largely due to reactions involving external sites and the phenomena of molecular shielding could be easily neglected. Results for batch reactors and HCSTRs have been obtained, and it is found that the latter give shorter but more branched polymer chains compared to those from the former.

Resole formation has also been modeled through the use of 19 species and the conversion and the number average chain length of the polymer formed calculated. Industrially used values of the phenol to formaldehyde ratio in the feed are about 0.67, and this model shows that at this ratio, the branching is maximized.

A similar analysis has been attempted for the prepolymerization stage of melamine formaldehyde,[46] where the reversibility of the reaction adds an added dimension to the complexity of the problem.

REFERENCES

1. J. F. Walker, *Formaldehyde*, ACS Monograph No. 159, 3rd ed., Reinhold, New York (1964).
2. F. Auerbach and H. Barschall, *Studien über Formaldehyde-Die Festern Polymeren der Formaldehyde*, Springer, Berlin (1907).
3. S. Ishida, Polymerization of formaldehyde and physical properties of polymerization products, *J. Appl. Polym. Sci.* **26**, 2743-2750 (1981).

4. H. C. Malhotra and V. P. Tyagi, Kinetics and mechanism of acid catalyzed 2,5-xylenol-formaldehyde reaction, *J. Macromol Sci. Chem.* **A16**, 1183–1192 (1983).

5. R. T. Jones, The condensation of trimethylolphenol, *J. Polym. Sci. Polym. Chem.* **21**, 1801–1817 (1983).

6. M. F. Drumm and J. R. LeBlanc, in *Step Growth Polymerization* (D. H. Solomon, Ed.), 1st ed., Dekker, New York (1972).

7. D. A. Fraser, R. W. Hall, and A. L. J. Raum, Preparation of "high ortho" novolak resins. I. Metal ion catalysis and orientation effect, *J. Appl. Chem.* (*London*) **7**, 676–700 (1957).

8. H. E. Adabbo and R. J. J. Williams, The curing of novolacs with paraformaldehyde, *J. Appl. Polym. Sci.* **27**, 893–901 (1982).

9. A. J. Rojas and R. J. J. Williams, Novolacs from paraformaldehyde, *J. Appl. Polym. Sci.* **23**, 2083–2088 (1979).

10. M. I. Aranguren, J. Barrajo, and R. J. J. Williams, Some aspects of curing novolacs with hexamethylene tetramine, *J. Polym. Chem. Polym. Chem.* **20**, 311–318 (1982).

11. H. C. Malhotra and V. P. Tyagi, Kinetics of alkali-catalyzed 2,5-dimethylphenol-formaldehyde reaction, *J. Macrom. Sci. Chem.* **A14**, 675–686 (1980).

12. A. Knop and W. Shieb, *Chemistry and Application of Phenolic Resins*, Springer, Berlin (1979).

13. D. J. Francis and L. M. Yeddanapalli, Kinetics and mechanism of the alkali-catalysed condensations of di- and tri-methylol phenols by themselves and with phenol, *Makromol. Chem.* **125**, 119–125 (1969).

14. J. H. Freeman and C. W. Lewis, Alkali-catalyzed reaction of formaldehyde and the methylols of phenol: A kinetic study, *J. Am. Chem. Soc.* **76**, 2080–2087 (1954).

15. A. A. Zavitsas, Formaldehyde equilibria: Their effects on the kinetics of the reaction with phenol, *J. Polym. Sci. A-1* **6**, 2533–2540 (1968).

16. A. A. Zavitsas, R. D. Beaulieu, and J. R. LeBlanc, Base-catalyzed hydroxymethylation of phenol by aqueous formaldehyde, *J. Polym. Sci. A-1* **6**, 2541–2559 (1968).

17. A. A. Zavitsas, Acid ionization constants of phenol and some hydroxymethylphenols between 20 and 60°C, *J. Chem. Eng. Data* **12**(1), 94–97 (1967).

18. I. H. Updegraff and T. J. Suen, in *Polymerization Processes* (C. E. Schildknecht and I. Skeist, Eds.), 1st ed., Wiley Interscience, New York (1977).

19. N. J. L. Mazson, *Phenolic Resin Chemistry*, Academic, New York (1958).

20. T. S. Carswell, *Phenoplasts, Their Structure, Properties and Chemical Technology*, Wiley Interscience, New York (1947).

21. B. Meyer, *Urea Formaldehyde Resins*, Addison-Wesley, Reading, Massachusetts (1979).

22. S. Katuscak, M. Thomas, and O. Schiessel, Kinetics of polycondensation of urea with formaldehyde, M.W.D., average M.W. and polydispersity parameters, *J. Appl. Polym. Sci.* **26**, 381–394 (1981).

23. N. Landquist, On the reaction between urea and formaldehyde in neutral and alkaline solutions, *Acta Chem. Scand.* **11**, 786–791 (1957).

24. M. Okano and Y. Ogata, Kinetics of the condensation of melamine with formaldehyde, *J. Am. Chem. Soc.* **74**, 5728–5731 (1952).

25. P. F. Frontini, J. Barrajo, and R. J. J. Williams, Manufacture of urea-formaldehyde concentrates using gas-liquid technology, *Polym. Eng. Sci.* **24**, 1245–1248 (1984).

26. S. H. Hollingdale and N. J. L. Megson, Formaldehyde condensation with phenol and its homologues. XVII. The chemical complexity of resins, *J. Appl. Chem.* (*London*) **5**, 616–624 (1955).

27. S. R. Finn and J. W. James, Formaldehyde condensations with phenol and its homologues, XVIII Chromatographic analysis, *J. Appl. Chem.* (*London*) **6**, 466–476 (1956).

28. U. K. Phukan, Modelling and simulation of novolac type phenol formaldehyde polymerization, M.Tech. dissertation, I.I.T. Kanpur (1980).

29. S. Tyagi, Modelling of reversible urea formaldehyde polymerization, M.Tech. thesis, I.I.T. Kanpur (1986).

30. B. Tomita, Melamine-formaldehyde resins: molecular species distribution of methylol-melamine and some kinetics of methylolation, *J. Polym. Sci., Polym. Chem.* **15**, 2347-2363 (1977).

31. G. R. Sprengling and C. W. Lewis, Dissociation constants of some phenols and methylol phenols, *J. Am. Chem. Soc.* **75**, 5709-5711 (1953).

32. B. Kumar, Modelling of reversible novolac type phenol formaldehyde polymerization, M.Tech. dissertation, I.I.T. Kanpur (1982).

33. A. Kumar, S. K. Gupta, and B. Kumar, Modeling of reversible novolac type phenol formaldehyde polymerization, *Polymer* **23**, 1929-1936 (1982).

34. A. Kumar, S. K. Gupta, N. Somu, and B. Kumar, M.W.D. in novolac type phenol-formaldehyde polymerization, *Polymer* **24**, 1180-1188 (1983).

35. P. M. Frontini, T. R. Cuadrado, and R. J. J. Williams, Batch and continuous reactors for production of novolacs, *Polymer* **23**, 267-270 (1982).

36. A. Kumar, A. K. Kulshreshtha, and S. K. Gupta, Modeling of phenol formaldehyde polymerization, *Polymer* **21**, 317-324 (1980).

37. A. Kumar, A. K. Kulshreshtha, S. K. Gupta, and U. K. Phukan, M.W.D. in novolac type phenol formaldehyde polymerization, *Polymer* **23**, 215-221 (1982).

38. A. Kumar, S. K. Gupta, and U. K. Phukan, Modelling of condensation polymerization of novolac type polymerization in HCSTRs, *J. Appl. Polym. Sci.* **27**, 3393-3405 (1982).

39. A. Kumar, S. K. Gupta, and U. K. Phukan, M.W.D. in novolac type, polymerization in HCSTRs, *Polym. Eng. Sci.* **21**, 1218-1227 (1981).

40. P. J. Flory, *Frontiers in Chemistry*, Vol. VI, Wiley Interscience, New York (1949).

41. J. Berrajo, M. I. Aranguren, and R. J. J. Williams, Statistical aspects for the production of novolacs, *Polymer* **23**, 263-266 (1982).

42. S. Ishida, Y. Tsutsumi, and K. Kaneko, Studies of the formation of thermosetting resins XII, Computer simulation of the reactions of phenol with formaldehyde, *J. Polym. Sci. Polym. Chem.* **19**, 1609-1620 (1981).

43. P. K. Pal, Modelling of resole type phenol formaldehyde polymerization, M.Tech. dissertation, Indian Institute of Technology, Kanpur (1980).

44. S. K. Gupta, A. Kumar, and P. K. Pal, Modeling of resol type phenol formaldehyde polymerization, *Polymer* **22**, 1699-1704 (1981).

45. P. K. Pal, A. Kumar, and S. K. Gupta, Modeling of resol type phenol formaldehyde polymerisation in HCSTRs, *Brit. Polym. J.* **12**, 121-129 (1980).

46. S. K. Gupta, A. K. Gupta, and A. Kumar, in *Frontiers of Chemical Reaction Engineering* (L. K. Doraiswamy and R. A. Mashelkar, Eds.), Wiley Eastern, New Delhi (1984).

EXERCISES

1. From the mechanism given in Table 11.4, derive the mole balance relations for various species in batch reactors and HCSTRs. From the knowledge of concentration of these species, you can write MWD equations following the logic given in Section 11.4.

2. Assume that Novolac formation is a reversible reaction. Assume that $[CH_2OH]$ is negligibly small and water reacts with a bond by the reverse reaction. Using species A-E, rewrite mole balance for these species. After doing this, establish the MWD equations.

3. Please define suitable species in light of Section 11.3 for resole polymerization. In terms of these rewrite the mole balance of these species.

4. Complete Table 11.6. Complete list of reactions given in Ref. 45.

5. Melamine polymerization has been shown to be reversible in nature. Follow the logic given in Exercise 11.2 and write the MWD equations.

INDEX